About the Series

The Lecture Notes in Geoinformation and Cartography series provides a contemporary view of current research and development in Geoinformation and Cartography, including GIS and Geographic Information Science. Publications with associated electronic media examine areas of development and current technology. Editors from multiple continents, in association with national and international organizations and societies bring together the most comprehensive forum for Geoinformation and Cartography.

The scope of Lecture Notes in Geoinformation and Cartography spans the range of interdisciplinary topics in a variety of research and application fields. The type of material published traditionally includes:

- proceedings that are peer-reviewed and published in association with a conference;
- post-proceedings consisting of thoroughly revised final papers; and
- research monographs that may be based on individual research projects.

The Lecture Notes in Geoinformation and Cartography series also includes various other publications, including:

- tutorials or collections of lectures for advanced courses;
- contemporary surveys that offer an objective summary of a current topic of interest; and
- emerging areas of research directed at a broad community of practitioners.

More information about this series at http://www.springer.com/series/7418

Jamal Jokar Arsanjani
Alexander Zipf
Peter Mooney
Marco Helbich
Editors

OpenStreetMap in GIScience

Experiences, Research, and Applications

Springer

Editors
Jamal Jokar Arsanjani
GIScience Research Group
Institute of Geography
Heidelberg University
Heidelberg
Germany

Alexander Zipf
GIScience Research Group
Institute of Geography
Heidelberg University
Heidelberg
Germany

Peter Mooney
Department of Computer Science
Maynooth University
Maynooth, Co. Kildare
Ireland

Marco Helbich
Department of Human Geography and Spatial Planning
Utrecht University
Utrecht
The Netherlands

ISSN 1863-2246 ISSN 1863-2351 (electronic)
Lecture Notes in Geoinformation and Cartography
ISBN 978-3-319-14279-1 ISBN 978-3-319-14280-7 (eBook)
DOI 10.1007/978-3-319-14280-7

Library of Congress Control Number: 2014958977

Springer Cham Heidelberg New York Dordrecht London

Printed on acid-free paper

Springer International Publishing AG Switzerland is part of Springer Science+Business Media (www.springer.com)

Foreword

OpenStreetMap Studies and Volunteered Geographical Information

This book comes at an apt time to reflect on the growing role of OpenStreetMap (OSM) in Geographical Information Science. This summer, the OpenStreetMap project celebrated ten years of operation, which began on the date of the domain name registration. I first heard about the project when it was in its very early stages and, with the support of the Royal Geographical Society, carried out the first research project that focused on OpenStreetMap, with an attempt to develop a mobile data collection tool on an early GPS-enabled phone. As a result, I found myself writing, together with Patrick Weber, what is now the most cited paper on the project (Haklay and Weber 2008). This early exposure to the project provided me with opportunities to watch, with astonishment, how it has become an important source of geographical information, as well as the explosive growth in academic research with and about it.

Of course, in the early years the project was small, with an unclear future and too localised to have a wider impact. It is, therefore, unsurprising that, so far as academic publications indexing reveals, Nelson et al. (2006) 'Towards development of a high quality public domain global roads database' and Taylor and Caquard (2006) 'Cybercartography: Maps and Mapping in the Information Era' are the first peer-reviewed papers that mention OpenStreetMap. Yet, it is interesting that, within two years of establishment, researchers in Canada and the United States heard about it and realised its potential. Moreover, many chapters in the current volume attest to the foresight that these two papers demonstrated.

Since 2006, OpenStreetMap has received plenty of academic attention. As of August 2014, more 'conservative' academic search engines such as ScienceDirect or Scopus find 286 and 236 peer-reviewed papers (respectively) that mention the project. The ACM digital library finds 461 papers in the areas that are relevant to computing and electronics, while Microsoft Academic Research finds only 112. Google Scholar, probably the most expansive of the search engines, lists over

9000 (!). Even with the most conservative version from Microsoft, we can see an impact on fields ranging from social science to engineering and physics. In short, OpenStreetMap has facilitated major contributions to knowledge beyond producing maps.

The link between OpenStreetMap and the concept of Volunteered Geographical Information is also long-standing. Michael Goodchild, in his seminal paper from 2007 that defined Volunteered Geographic Information (VGI), mentioned OpenStreetMap as an example. Since then the literature frequently conflates OSM and VGI. In some recent papers statements such as 'OpenStreetMap is considered as one of the most successful and popular VGI projects' or 'the most prominent VGI project OpenStreetMap' are common[1] and, to some degree, the boundary between the two is being blurred. I also admit to be part of the problem—for example, with the title of my 2010 paper 'How good is volunteered geographical information? A comparative study of OpenStreetMap and Ordnance Survey datasets'. However, upon reflection on the characteristics of OpenStreetMap and other VGI projects, I became uncomfortable with the equivalence between OSM and VGI. The stance that Neis and Zielstra (2014) offer is, I suggest, more accurate: 'One of the most utilized, analyzed and cited VGI-platforms, with an increasing popularity over the past few years, is OpenStreetMap (OSM).'

The reason that it is valuable to differentiate between focusing on the OpenStreetMap project (what we can call OSM studies) and the more generic VGI research is partly due to the volume of papers specifically about the project, and what they reveal about the project. Over the years, several types of research papers that can be classified as OSM studies have emerged.

First, there is a whole set of research projects that use OSM data because it is easy to use and free to access (for example, in computer vision or even string theory). For these projects, OSM is just data to be used (see "Data Retrieval for Small Spatial Regions in OpenStreetMap" and "The Next Generation of Navigational Services Using OpenStreetMap Data: The Integration of Augmented Reality and Graph Databases", which arguably fall into this category). Second, there are studies of OSM data: quality, the history and evolution of objects in the database, what we can learn about the nature of the data and other aspects. The majority of this volume falls under this category (see "Assessment of Logical Consistency in OpenStreetMap Based on the Spatial Similarity Concept"–"Inferring the Scale of OpenStreetMap Features", "Route Choice Analysis of Urban Cycling Behaviors Using OpenStreetMap: Evidence from a British Urban Environment", "Building a Multimodal Urban Network Model Using OpenStreetMap Data for the Analysis of Sustainable Accessibility"–"Using Crowd-Sourced Data to Quantify the Complex Urban Fabric—OpenStreetMap and the Urban–Rural Index"). Third, there are studies that also look at the interactions between patterns of contribution and the data —for example, in trying to infer trustworthiness (see "Spatial Collaboration Networks of OpenStreetMap"). Fourth, there are studies that look at the wider

[1] These are deliberately unreferenced so as not to argue that specific authors are to blame.

societal aspects of OpenStreetMap—for example, what the spatial and social implications of data coverage are (see "Social and Political Dimensions of the OpenStreetMap Project: Towards a Critical Geographical Research Agenda"). Finally, there are studies of the social practices in OpenStreetMap as a project (see "The Impact of Society on Volunteered Geographic Information: The Case of OpenStreetMap").

In short, there is a significant body of knowledge regarding the nature of the project, the implications of what it produces, and ways to understand the information that emerges from it. Clearly, we now know that OSM produces good data and is aware of the patterns of contribution. What is also clear is that many of these patterns are specific to OSM. Because of the importance of OSM to so many application areas (including illustrative maps in string theory!), these insights are very important. Some of these insights are expected to be also present in other VGI projects but making such analogy needs to be done carefully, and only when there is evidence from other projects that this is the case. In short, we should avoid conflating VGI and OSM—and this volume provides a clear demonstration why this is the case.

November 2014

Prof. Mordechai (Muki) Haklay
Professor of Geographical Information Science
Department of Civil, Environment and Geomatic Engineering
University College London (UCL), UK

References

Goodchild M (2007) Citizens as sensors: the world of volunteered geography. GeoJournal 69(4):211–221

Haklay M, Weber P (2008) OpenStreetMap: user-generated street maps. IEEE Perv Comput 7(4):12–18

Haklay M (2010) How good is volunteered geographical information? A comparative study of OpenStreetMap and ordnance survey datasets. Environ Plan B 37(4):682–703

Neis P, Zielstra D (2014) Recent developments and future trends in volunteered geographic information research: the case of OpenStreetMap. Future Int 6, 1:76–106

Nelson A, de Sherbinin A, Pozzi F (2006) Towards development of a high quality public domain global roads database. Dat Sci J, 5:223–265

Taylor DRF, Caquard S (2006) Cybercartography: maps and mapping in the information era. Cartographica 41(1):1–6

Contents

Part II Social Context

Part III Network Modeling and Routing

Part IV Land Management and Urban Form

An Introduction to OpenStreetMap in Geographic Information Science: Experiences, Research, and Applications

Jamal Jokar Arsanjani, Alexander Zipf, Peter Mooney and Marco Helbich

Abstract Recent years have seen new ways of collecting geographic information via the crowd rather than organizations. OpenStreetMap (OSM) is a prime example of this approach and has brought free access to a wealth of geographic information—for many parts of the world, for the first time. The strong growth in the last few years made more and more people consider it as a potential alternative to commercial or authoritative data. The increasing availability of ever-richer data sets of freely available geographic information led to strong interest of researchers and practitioners in the usability of this data—both its limitations and potential. Both the unconventional way the data is being produced as well as its richness and heterogeneity have led to a range of different research questions on how we can assess, mine, enrich, or just use this data in different domains and for a wide range of applications. While this book cannot present all types of research around OpenStreetMap or even the broader category of User Generated Content (UGC) or Volunteered Geographic Information (VGI), it attempts to provide an overview of the current state of the art by presenting some typical and recent examples of work in GIScience on OSM. This chapter provides an introduction to the scholarly work on OpenStreetMap and its current state and summarizes the contributions to this book.

J. Jokar Arsanjani (✉) · A. Zipf
GIScience Research Group, Institute of Geography, Heidelberg University, 69120 Heidelberg, Germany
e-mail: jokar.arsanjani@geog.uni-heidelberg.de

A. Zipf
e-mail: zipf@uni-heidelberg.de

P. Mooney
Department of Computer Science, Maynooth University, Maynooth, Co. Kildare, Ireland
e-mail: peter.mooney@nuim.ie

M. Helbich
Department of Human Geography and Spatial Planning, Utrecht University, Heidelberglaan 2, 3584 CS Utrecht, The Netherlands

J. Jokar Arsanjani et al. (eds.), *OpenStreetMap in GIScience*,
Lecture Notes in Geoinformation and Cartography,
DOI 10.1007/978-3-319-14280-7_1

1 Introduction

Access to spatial data and cartographic products has changed radically over the last decade or so. Traditionally, governmental agencies, cartographic centers, and commercial agencies were the only sources for end-users seeking spatial data. One of the most formidable barriers to more widespread access to these geodata were created by often prohibitive high fees and license charges in combination with time- and purpose-limited copyright restrictions imposed. This business model was rather successful, but made access to high-quality geodata very difficult for all but a small number of end users. Changes in Information and Communication Technology (ICT) brought about by the Internet and social media and the availability of inexpensive portable satellite navigation devices has seen this traditional geodata business model challenged. One of the key driving forces in this change has been the OpenStreetMap (OSM) project. OSM was launched in 2004 with the mission of creating an editable map of the whole world and released with an open content license (http://wiki.openstreetmap.org/wiki/About). In general, OSM aims at building and maintaining a free editable map database of the world in a collaborative manner so that people and end-users are not forced to buy geodata in the traditional way and subsequently be subjected to restrictive copyright and license commitments. OSM started initially with a focus on mapping streets and roads. Since then it has moved far beyond these entities and it now contains a very rich variety of geographical objects (e.g., buildings, land use, Points of Interest) from all over the planet being mapped by thousands of volunteer contributors to the project. Aside from the obvious commercial benefits offered by OSM, the project has revolutionized the way in which geodata is collected. No longer are the collection of geodata and the development of cartographic products limited to specialists, geographic surveyors, or cartographers.

OSM is often referred to as the Wikipedia map of the world. As it is built on many of the same ICT structures as Wikipedia it offers its project contributors the possibility of (a) almost immediate updating of the map database as well as very frequent updating of associated editing software and other tools; (b) importing geodata recorded from Global Positioning System (GPS)-enabled devices, smartphones, and other digital maps tools; (c) access to the full history of mapping activities in OSM over its lifetime; and finally (d) collaboration with other OSM users and contributors through various communication channels including mailing lists, discussion forums, and physical meetings (Mooney and Corcoran 2013a). The gradual evolution of the OSM ecosystem has been very successful. The project got off to a slow start but since 2007 there has been an ever-increasing rate of people joining the project. In November 2014, OSM had approximately 1.85 million registered users and contributors (http://wiki.openstreetmap.org/wiki/Stats). As mentioned previously, the era of ubiquitous Internet, social media, open-source software, etc. has seen many citizen knowledge-based projects for a host of diverse purposes launched on the Internet over the last few years. OSM has been a unique case. The academic and industrial communities have recognized OSM not solely based on its rise to become an important distributor of geodata but its wider success in growing a global community of people willing to participate in the collection and maintenance of geodata. The OSM community is actively involved in much more than collecting geodata to build and maintain this global geodatabase. In addition, the community is involved in, for example, humanitarian work, open source software development to support OSM and the GIS community, and in building a network of support for those using and contributing to the OSM project.

In recent years, several scientific disciplines (e.g. geography, GIScience, spatial planning, cartography, computer science, and ecology) have realized the immense potential of OSM and it has become the subject of academic research. OSM offers researchers a unique dataset that is global in scale and a body of knowledge created and maintained by a very large collaborative network of volunteers. Research on OSM has shown that its geodata in some parts of the world are more complete and locationally and semantically more accurate than the corresponding proprietary datasets (e.g., Zielstra and Zipf 2010; Neis and Zipf 2012; Helbich et al. 2012), while being of high spatial heterogeneity. Skepticism amongst the GIS community and industry surrounding the quality of the geodata in OSM has seen a major effort being made on evaluating the quality of the OSM geodata. This has a led to the development of a number of software tools and methodologies for analyzing the quality (Roick et al. 2011; Helbich et al. 2012; Jokar Arsanjani et al. 2013a; Jokar Arsanjani and Vaz 2015c). Other approaches even try to improve the OSM data through algorithms dedicated to specific object types, such as addresses for geocoding (Amelunxen 2010). Investigation of the development and evolution of OSM across the globe over time has also emerged as a research topic for many academic studies (Mooney et al. 2012; Neis and Zipf 2012; Jokar Arsanjani et al. 2013c; Mooney and Corcoran 2013; Fan et al. 2014).

Extracting value-added information from the OSM database has become another emerging research topic for researchers to attempt to understand OSM better (Hagenauer and Helbich 2012; Mooney and Corcoran 2012; Mooney et al. 2013; Jokar Arsanjani et al. 2015). Hagenauer and Helbich (2012), for instance, predicted missing urban areas through artificial neural networks. Bakillah et al. (2014a) derived population estimations from OSM and an emerging important topic is land use maps that can be generated using OSM (Jokar Arsanjani et al. 2014; Jokar Arsanjani and Vaz 2015c). Klonner et al. (2014) investigated the updating of Digital Elevation Models and Fan et al. (2014) estimated building types from OSM.

Both inside and outside of the academic sphere, OSM is now being used increasingly in a variety of practical or scientific applications in different domains, which demonstrates the usability of the crowdsourced geodata in OSM. However, in all of these cases the characteristics of OSM must be considered. Because of the flexibility and open data-like structure of OSM, it is possible to use or even adapt and improve OSM for a large range of purpose-directed applications, as we will see below. As mentioned above, there are some data quality issues with the OSM database which can be mitigated against through specialized approaches to using the actual geodata (Goodchild 2013). This has brought about a host of examples of applications and domain-specific research. A first important category is the development of a set of different special routing and navigation systems that operate on a large scale. Examples include: routing for cars, bikes, and pedestrians such as in OpenRouteService, (Schmitz et al. 2008); emergency routing (Neis et al. 2010); wheelchair routing (Neis 2014); emergency response and evacuation simulation (Bakillah et al. 2012); indoor routing (Goetz 2012); or agricultural logistics (Lauer et al. 2014). Further typical uses of OSM include improving cartography (Rylov and Reimer 2014) or developing Location Based Services (LBS) (Schilling et al. 2009). Another innovation was the development of 3D city models from OSM (Over et al. 2010; Goetz 2013). Further research has focused on attempting to extend the current OSM spatial data model by working on extensions such as: 3D (Goetz and Zipf 2012), indoor mapping (Goetz and Zipf 2012), or wheelchair routing (Neis 2014) and using the results from this in a range of applications.

The relationship between OSM and open data standards, in particular Spatial Data Infrastructures (SDI) and the future direction of the Web 2.0 paradigm, is a question still requiring further discussion. In particular, the large volumes of data being updated by the minute that are now available pose challenges with regards to their handling and keeping them up to date on a global scale.

The discussion in the preceding paragraphs has shown that OSM has now emerged as a new research area. It has the potential to bring disparate research disciplines together and enhances interdisciplinary and multidisciplinary investigations. This interdisciplinary research collaboration can contribute to a more profound and cross-disciplinary understanding of citizens' knowledge-based efforts in projects such as OSM. It also provides an interesting platform for the academic research community to collaborate with these communities towards interactive collection of up-to-date geodata from citizens by means of novel computationally oriented methods such as network analysis, machine learning, and computer simulation models. As the examples above have demonstrated, these practical investigations on OSM provide a rich set of opportunities to discover novel and valuable patterns inherent in the geodata collected by citizens, to better understand the activities of contributors to open knowledge projects, the characteristics of their human–computer interactions, and the potential to tackle classical GIS research questions using this modern and revolutionary approach to the collection and distribution of geographic data.

2 A Short Overview of the OpenStreetMap Research Landscape

In this section, we present a brief overview of the OSM research landscape through a word cloud approach. To do so, a search query was applied on 16 August 2014 in Google Scholar looking for four terms "OpenStreetMap", "OSM", "VGI", and "Volunteered Geographic Information" either in the abstracts, titles or keywords. In total, 224 documents were collected. The collection of titles, abstracts, and keywords were explored by means of word clouds. Word clouds provide an intuitive

Fig. 1 Word cloud of the papers' titles

impression about common words and show the number of times a certain word appeared in the literature. This is expressed by varying font sizes. Larger font sizes refer to words that appear more often than smaller font sizes (Helbich et al. 2013). It should be noted that the aforementioned search terms were removed from the resulting word cloud as their usage frequencies were substantial and masked the other terms. We leave these figures for the readers to interpret them visually and gain some insights about the research on OSM so far.

As editors of this volume, we have been involved in research connected to OpenStreetMap for many years. From our own empirical experience, these word

Fig. 2 Word cloud of the papers' keywords

Fig. 3 Word cloud of the papers' abstracts

clouds capture the essence of academic research on OpenStreetMap from the past number of years. In Figs. 1 and 2 where the word clouds of paper title terms and keywords are presented we see a few dominant terms, particularly *data*, *quality*, *research*, *systems,* and *participation*. As we mentioned earlier, the issue of the quality of OSM data has been on the research agenda for many years now. It is likely to remain on the agenda for some time to come. However, these titles also reflect the expansion of OSM research to consider the citizen and volunteer participation, which drives the expansion of the project. The systems, models, sensors, and applications which collect, analyze, store, manage, transform, and distribute the data must now be studied and explored in more detail as OpenStreetMap grows in size potentially towards being considered as Geographic Big Data (Goodchild 2013).

The word cloud in Fig. 3 displays the most frequently occurring terms in the abstracts of the papers returned in our Google Scholar searches. Quality is a dominant term but we see the concepts of research, social, community, methods, and development emerging. There are obvious visual linkages to the word cloud in Fig. 4 where the abstracts of the chapters included in this volume are visualized.

Fig. 4 Word cloud of our chapters' abstracts

We can immediately see the same set of dominant terms. However, in our abstracts word cloud there is somewhat more diversity with urban modelling, navigation, modelling, and knowledge management related terms being highlighted.

3 Geography of OpenStreetMap

As already stated in the literature (e.g., Mooney and Corcoran 2012; Neis 2014), OpenStreetMap has its own geography across time and space. In other words, we rarely see identical patterns of contributions in two different regions/countries. When speaking of OpenStreetMap quality and contributions networks the importance of studying diverse case studies has been highlighted. Hence, in this section, two different maps are generated from the OSM statistics, which demonstrate the heterogeneity of OSM in different countries. Figure 5 displays the total number of created nodes in October 2014. This map displays a thematic categorization of created nodes, which is one of the key elements in measuring OSM contributions. It should be noted that in this comparative report, the size of the country, population, gross domestic product (GDP), and a number of other physical characteristics of the countries are not taken into consideration. However, they are of great importance in performing further in-depth analysis. For instance, the dominant land cover types in Canada and Australia should be excluded in considering the size of the country as apart from land cover there are no objects to be mapped and the contributed nodes have very likely occurred within urban areas. Besides, their populations are not comparable to the USA, China, and India.

Nonetheless, focusing on count gives an overall indication that the high number of node creation is not limited to European countries, but other countries are also

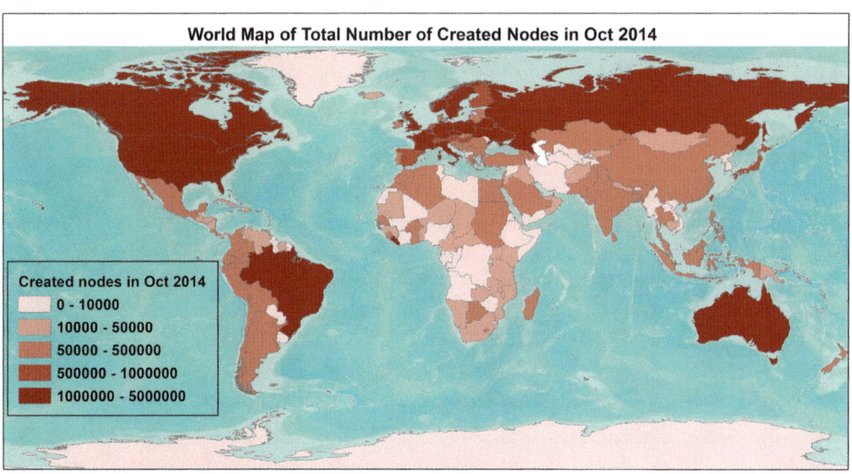

Fig. 5 A world map of the total created nodes in October 2014

emerging in OSM. Amongst these emerging countries, north America including USA and Canada, south American countries particularly Brazil, Australia, and some Asian and African countries can be named, which calls for further studies in these regions to find out how actively and accurately mapping in OSM is being undertaken. In terms of number of created nodes, in total over 46 million nodes were created in this month. In a number of countries, no nodes were created. However, Germany, United States, Russia, Czech Republic, Italy, Poland, France, Norway, Liberia, Canada, and Japan received the most created nodes, respectively.

In terms of total active members, i.e., mappers, in this period, while in total 3,048 members logged into OSM, the majority of them were from Germany (535), United States (215), Russia (212), France (195), Italy (156), Poland (155), UK (128), Spain (96), Austria (81), and Japan (55). In order to normalize the number of active members, the average number of active members per day in October 2014 is divided by the total population of the countries in terms of millions of people in 2010. Figure 6 shows the average number of active members per day for October 2014 per million people. This map helps to detect the countries that have a large portion of their population involved in the mapping process in OSM. Italy, the Netherlands, Kuwait, Croatia, and Liberia were at the top of the list. It is interesting to see that a number of countries from all continents have more than 1 member per million population active in mapping. On the contrary, a number of Asian and African countries have a very minor proportion of their population involved in mapping. This confirms the empirical findings that only a small portion of the population is mapping. It is worth mentioning that this finding is based on our analysis within the chosen timeframe for sharing general impressions about OSM and activities in OSM certainly also has a temporal pattern, which is an important indicator to be considered.

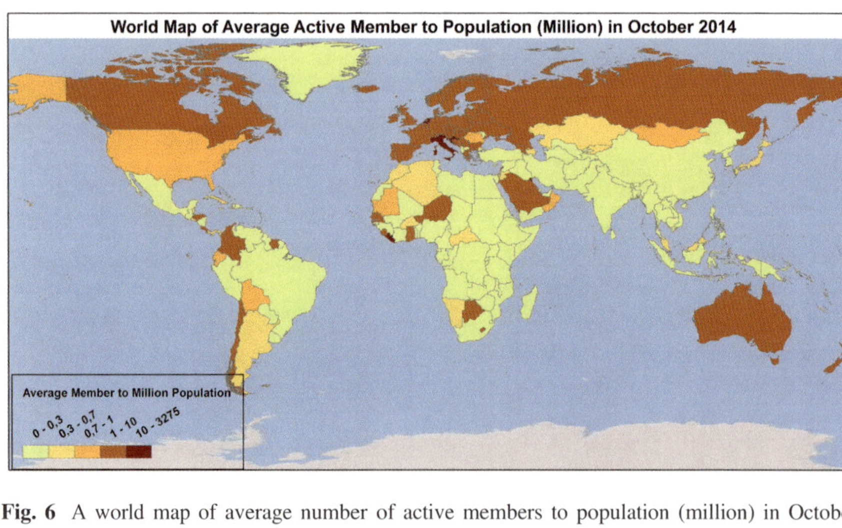

Fig. 6 A world map of average number of active members to population (million) in October 2014

It is interesting to note that contrary to Germany as an active country in OSM, some other countries are emerging in OSM and there are still large gaps in the OSM data from these countries. However, this can be viewed positively. Information dissemination about OSM within the last few years has promoted additional people to become members of OSM. Slowly but surely OSM is gaining popularity in these countries. Perhaps OpenStreetMap is helping address the participation inequality that is strongly represented in many types of User-Generated Content on the Internet today. This "Digital Divide" indicates that very small groups with specific demographic and geographical characteristics are responsible for production of most of the UGC we see on the Internet today. However, these map visualizations indicate that OpenStreetMap is reaching into countries and regions which heretofore would have felt the consequences of the digital divide (Graham et al. 2012). Improvements in ICT infrastructures and IT education for socially deprived groups such as women and children coupled with more ubiquitous access to smartphone technology has provided an environment where participation in OpenStreetMap can increase. Research will need to be undertaken to gain a better understanding of the social processes involved in these changes.

4 Objectives and Scope

The present volume entitled "OpenStreetMap in Geographic Information Science: experiences, research, and applications" presents a collection of experiences and research which has been carried out with OSM as the central and core theme. The book seeks to build a firm foundation for research work focused on integrating OSM. Chapters will address the following research topics:

(a) State-of-the-art and cutting-edge approaches for data quality analysis in OSM and VGI.
(b) Investigations on understanding OSM contributors and the nature of their contributions.
(c) Identification of patterns of contributions and contributors.
(d) Applications of OSM in different domains.
(e) Mining value-added knowledge and information from OSM.
(f) Limitations in the analysis of OSM data.
(g) Integration of OSM with commercial and non-commercial geodatasets.

We expect that this book will deliver significant scientific outcomes, which will further stimulate international research networking and collaboration. As outlined above, the inherent cross-disciplinary essence of OSM research combined with the emerging data quality, data mining, and patterns determination approaches to analysis of OSM means that contributing scholars for this book will be expected to have a diverse academic background not limited to geographic information science, cartography, computer science, statistics, and sociology. We feel that these trans- and inter-disciplinary contributions permit a deeper understanding of how the OSM project works and has become the phenomenal success that it is today. Last, but not least, the book will strive to bring OSM into the core of GIScience where the diverse worlds of new and classical geography and cartography will meet.

This book presents some cutting-edge developments and applications in the field of geography, spatial statistics, geographic information science, social science, and cartography. This collection of chapters is highly relevant for, but not limited to, the following potential audience and readership: researchers, postgraduates, and professionals.

The high response to our call for chapters shows that the intention of this book to be widely announced has been fulfilled. By the end of January 2014, a total of 34 chapter proposals were submitted and after an internal review by the editors, 30 authors of those originally 34 submitted chapter proposals were invited to submit a full chapter manuscript. After the final chapter submission deadline on 30 May 2014, a total of 29 manuscripts were submitted. Thereafter, each of the 29 chapter manuscripts was evaluated through a double-blind review process by at least two or three international experts in the respective field. For the review process, the standard Springer review guidelines were applied. Besides the innovative aspect of the research, the scientific quality of the research weighted heavily on the decision as to whether or not a manuscript was accepted or rejected. After two rounds of reviews conducted by international experts, the editors made the decision on whether or not the manuscript was fit for publication. In October 2014, 14 chapters were accepted and along with one introductory chapter and one conclusive chapter are now included in the present book.

5 Structure of the Book

The book covers several areas of OSM, each associated with a main theme of the book. The present volume has the following four sections: (1) Data management and quality, (2) Social context, (3) Network modeling and routing, and (4) Land management and Urban form. However, this structure should not be understood as fixed and definitive. Quite the contrary, the boundaries between these sections are partly fuzzy and overlap each other to some extent.

Section 1 on *Data Management and Quality* includes five chapters. In chapter "Assessment of Logical Consistency in OpenStreetMap Based on the Spatial Similarity Concept", Peyman Hashemi and Rahim Ali Abbaspour propose an approach for assessing the logical consistency in OSM based on the concept of spatial similarity in multi-representation considering three elements, i.e. directional relationships, topological relationships, and metric distance relationships. Jokar Arsanjani et al. in chapter "Quality Assessment of the Contributed Land Use Information from OpenStreetMap Versus Authoritative Datasets" attempt to comparatively assess the accuracy of the contributed OSM land use features in four German metropolitan areas versus the pan-European GMESUA dataset as a reference. Their empirical findings suggest OSM to be alternative complementary source for extracting LU information with over half of the selected cities mapped by mappers. Moreover, the results identify which land types preserve high/moderate/low accuracy across cities for urban land use mapping. The findings strengthen the potential of collaboratively collected LU features for providing temporal LU maps as well as updating or enriching existing inventories. Chapter "Improving Volunteered Geographic Information Quality Using a Tag Recommender System: The Case of OpenStreetMap" by Arnaud Vandecasteele and Rodolphe Devillers proposes an approach for both improving the semantic quality and reducing the semantic heterogeneity of VGI datasets through implementing a tag recommender system, called OSMantic plugin, that automatically suggests relevant tags to contributors during the editing process. Their approach helps contributors find the most appropriate tags for a given object, hence reducing the overall dataset semantic heterogeneity. This plugin is developed for the Java OSM editor (JOSM) and different examples illustrate how this plugin can be used to improve the quality of VGI data. In chapter "Inferring the Scale of OpenStreetMap Features", Guillaume Touya and Andreas Reimer propose and compare two concurrent approaches at automatically assigning scale to OSM objects. Their first approach is based on a multi-criteria decision making model, with a rationalist approach for defining and parameterizing the respective criteria, yielding five broad Level of Detail classes. Their second approach attempts to identify a single metric from an analysis process, which is then used to interpolate a scale equivalence. Both approaches are combined and tested against well-known CORINE data, resulting in an improvement of the scale inference process. The chapter closes with a presentation of the most pressing open problems. In chapter "Data Retrieval for Small Spatial Regions in OpenStreetMap", Roland M. Olbricht investigates what design choices are required to be able to answer almost any geographic query whilst serving common use cases fast enough such that the services based on this database are fast on affordable and standard sized hardware. He evaluates the usage patterns from the main instance of Overpass API on overpass-api.de by considering more than 40 million requests from 2012 and 2013.

Section 2 deals with *Social Context* and comprises of three chapters. In chapter "The Impact of Society on Volunteered Geographic Information: The Case of OpenStreetMap", Afra Mashhadi et al. address whether the society and its characteristics such as the socio-economic factors have an impact on what part of the physical world is being digitally mapped so that we can understand where crowd-sourced map information can be relied upon. They measure the positional and thematic accuracy as well as the completeness of OSM data and quantify the role of society on the state of this digital production and finally quantify the effect of social engagement as a method of intervention for improving user participation. Georg Glasze and Chris Perkins frame a research agenda in chapter "Social and Political Dimensions of the OpenStreetMap Project: Towards a Critical Geographical Research Agenda" that draws upon critical cartography, but widens the scope of analysis to the assemblages of practices, actors, technologies, and norms at work: an agenda which is inspired by the critical GIS literature, to take the specific social contexts and effects of technologies into account, but which deploys a processual view of mapping. They recognize that a fundamental transition in mapping is taking place, and that OSM may well be of central importance in this process. In chapter "Spatial Collaboration Networks of OpenStreetMap", Klaus Stein et al. describe a new type of spatial collaboration network that can be extracted from OSM edit history data and show how to apply the measurement of interlocking responses known from research on non-spatial collaboration in wikis to collaboration in OSM. Finally, they discuss the advantages of their approach by demonstrating an analysis of collaboration on OSM sample data.

Section 3 deals with *Network modeling and routing* and includes three chapters. In chapter "Route Choice Analysis of Urban Cycling Behaviors Using OpenStreetMap: Evidence from a British Urban Environment", Godwin Yeboah and Seraphim Alvanides undertake a route choice analysis using the cycling-friendly version of OSM as the transportation network for analysis, alongside GPS tracks and travel diary data for 79 Utility Cyclists around Newcastle upon Tyne in North East England. They suggest that OSM can provide a robust transportation network for cycling research, in particular when combined with GPS track data, and conclude that network restrictions for both observed and shortest paths are significant, suggesting that route directness is an important factor to be considered for restricted and unrestricted networks. Chapter "The Next Generation of Navigational Services Using OpenStreetMap Data: The Integration of Augmented Reality and Graph Databases" by Pouria Amirian et al. describes the implementation of a navigational application as part of the eCampus project in Maynooth University. The application provides users with several navigation services with navigational instructions through standard textual and cartographic interfaces and also through augmented images showing way-finding objects. Jorge Gil presents the process of building a multi-modal urban network model using OSM data in chapter "Building a Multimodal Urban Network Model Using OpenStreetMap Data for the Analysis of Sustainable Accessibility". He develops various algorithmic procedures to produce the network model, supporting the reproducibility of the process and addressing the challenges of using OSM data for this purpose and addresses the great potential of OSM for urban analysis, thanks to the detail of its attributes and its open and universal coverage.

Land management and urban form is the focus of the final Sect. 4, comprising of three chapters. Chapter "Assessing OpenStreetMap as an Open Property Map" by Mohsen Kalantari and Veha La provides an assessment of OSM as a crowdsourcing system in collecting and recording land tenure information using a case study in Victoria, Australia. Their chapter studies the completeness of the public property records in OSM, and the location, shape, area and description of the existing records, and finally discusses the potential of OSM as an Open Property Map. Jacinto Estima and Marco Painho in chapter "Investigating the Potential of OpenStreetMap for Land Use/Land Cover Production: A Case Study for Continental Portugal" review the existing literature on using OSM data for land use/cover database production and move this research forwards by exploring the suitability of the OSM Points of Interest. They conclude that OSM can greatly contribute to mapping specific land types. In chapter "Using Crowd-Sourced Data to Quantify the Complex Urban Fabric—OpenStreetMap and the Urban–Rural Index", Johannes Schlesinger presents an Urban–Rural Index (URI), which tackles the lack of any classification of the urban–rural continuum especially in regions of the world where accurate and up-to-date geodata is hardly available. His paper draws on the analysis of three study sites: Bamenda in Cameroon, Moshi in Tanzania, and Bangalore in India, and concludes that URI as a reproducible representation of the spatial complexity of the urban landscape and its surrounding areas has the potential to contribute to the understanding of urban development patterns.

In the final chapter (chapter "An Outlook for OpenStreetMap") by Peter Mooney, the future research perspective of research on OpenStreetMap is reviewed. In the chapter, he structures his future vision of OpenStreetMap research by using the content of this volume and other OSM literature as a basis for future work.

Last but not least, the editors express their appreciation to all reviewers for their kind support and their critical and constructive comments for each chapter. Undoubtedly, their efforts have significantly enriched the quality of the entire volume. We profoundly appreciate the efforts of all authors who submitted a full chapter manuscript and chose this book as their desired publication outlet for their research. Furthermore, we would like to thank Prof. Muki Haklay for his invited comment. Finally, Jamal Jokar Arsanjani thanks the Alexander von Humboldt foundation and the Institute of Geography at Heidelberg University, Germany, for providing the foundation for this book. Peter Mooney would like to thank the Environmental Protection Agency Ireland for funding support. Finally, we acknowledge the Springer team as well as the series editors William Cartwright, Georg Gartner, Liqiu Meng, and Michael Peterson for their great assistance throughout the whole publication process. Certainly, without all these helping hands, this volume would have never been published. The book is partially sponsored by the COST Action IC1203.

References

Amelunxen C (2010) On the suitability of volunteered geographic information for the purpose of geocoding. In: Car A, Griesebner G, Strobl J (eds) Proceedings of the geoinformatics forum salzburg, na. Retrieved http://koenigstuhl.geog.uniheidelberg.de/publications/2010/Amelunxen/GI_Forum_Amelunxen.pdf

Bakillah M, Andrés Domínguez J, Zipf A, Liang SHL, Mostafavi MA (2013) Multi-agent evacuation simulation data model with social considerations for disaster management context. In: Zlatanova S, Peters R, Dilo A, Scholten H (eds) Intelligent systems for crisis management, SE-1. Springer, Berlin, Heidelberg, pp 3–16. doi:10.1007/978-3-642-33218-0_1

Bakillah M, Liang S, Mobasheri A, Jokar Arsanjani J, Zipf A (2014a) Fine-resolution population mapping using OpenStreetMap points-of-interest. Int J Geogr Inf Sci 28:1940–1963

Bakillah M, Lauer J, Liang S, Zipf A, Jokar Arsanjani J, Loos L, Mobasheri A (2014b) Exploiting big VGI to improve routing and navigation services. In: Big data techniques and technologies in geoinformatics, pp 177–192

Fan H, Zipf A, Fu Q, Neis P (2014) Quality assessment for building footprints data on OpenStreetMap. Int J Geogr Inf Sci 28:700–719. doi: 10.1080/13658816.2013.867495

Goetz M (2013) Towards generating highly detailed 3D CityGML models from OpenStreetMap. Int J Geogr Inf Sci 27:845–865
Goetz M, Zipf A (2012) Using crowdsourced indoor geodata for agent-based indoor evacuation simulations. ISPRS Int J Geo-Inf 1(2):186–208. doi:10.3390/ijgi1020186
Goodchild MF (2013) The quality of big (geo) data. Dialogues Hum Geogr 3:280–284
Graham M, Hale S, Stephens M (2012) Digital divide: the geography of internet access. Environ Plan A 44:1009–1010
Hagenauer J, Helbich M (2012) Mining Urban land-use patterns from volunteered geographic information by means of genetic algorithms and artificial neural networks. Int J Geogr Inf Sci 26:963–982
Helbich M, Amelunxen C, Neis P (2012) Comparative spatial analysis of positional accuracy of OpenStreetMap and proprietary geodata. In: Jekel T, Car A, Strobl J, Griesebner G (eds) Geospatial crossroads @ GI_Forum 2012, Wichmann, Heidelberg, Salzburg, pp 24–33
Helbich M, Hagenauer J, Leitner M, Edwards R (2013) Exploration of unstructured narrative crime reports: an unsupervised neural network and point pattern analysis approach. Cartogr Geogr Inf Sci 40:326–336
Jokar Arsanjani J, Barron C, Bakillah M, Helbich M (2013a) Assessing the quality of OpenStreetMap contributors together with their contributions. In: 16th AGILE international conference on geographic information science, Leuven, Belgium, pp 14–17
Jokar Arsanjani J, Helbich M, Bakillah M, Hagenauer J, Zipf A (2013b) Toward mapping land-use patterns from volunteered geographic information. Int J Geogr Inf Sci 27: 2264–2278
Jokar Arsanjani J, Vaz E, Bakillah M, Mooney P (2014) Towards initiating OpenLandMap founded on citizens' science: the current status of land use features of OpenStreetMap in Europe. In: Huerta, Schade, Granll (eds) Proceedings of the AGILE'2014 international conference on geographic information science, Castellón, AGILE Digital Editions: Castellón, Spain, 3–6 June 2014
Jokar Arsanjani J, Helbich M, Bakillah M, Loos L (2015a) The emergence and evolution of OpenStreetMap: a cellular automata approach. Int J Digit Earth 8(1):74–88. doi:10.1080/17538947.2013.847125
Jokar Arsanjani J, Bakillah M, Arsanjani JJ (2015b) Understanding the potential relationship between the socio-economic variables and contributions to OpenStreetMap. Int J Digit Earth 0:1–16. doi:10.1080/17538947.2014.951081
Jokar Arsanjani J, Vaz E (2015c) An assessment of a collaborative mapping approach for exploring land use patterns for several European metropolises. Int J Appl Earth Obs Geoinf 35:329–337. doi:10.1016/j.jag.2014.09.009
Klonner C, Barron C, Neis P, Höfle B (2014) Updating digital elevation models via change detection and fusion of human and remote sensor data in Urban environments. Int J Digit Earth 1–19
Lauer J, Richter L, Ellersiek T, Zipf A (2014) TeleAgro+ - Analysis framework for agricultural telematics data, IWCTS '14, SIGSPATIAL'14, 4–7 Nov 2014, Dallas/Fort Worth, TX, USA
Mooney P, Corcoran P (2012) Characteristics of heavily edited objects in OpenStreetMap. Future Internet 4:285–305
Mooney P, Corcoran P (2013a) Analysis of interaction and co-editing patterns amongst OpenStreetMap contributors. Trans GIS
Mooney P, Corcoran P, Ciepluch B (2012) The potential for using volunteered geographic information in pervasive health computing applications. J Ambient Intell Humaniz Comput 1–15 (LA—English)
Mooney P, Rehrl K, Hochmair H (2013) Action and interaction in volunteered geographic information: a workshop review. J Locat Based Serv 7:291–311
Neis P (2014) Measuring the reliability of wheelchair user route planning based on volunteered geographic information. Trans GIS n/a–n/a, doi:10.1111/tgis.12087
Neis P, Zipf A (2012) Analyzing the contributor activity of a volunteered geographic information project—the case of OpenStreetMap. ISPRS Int J Geo-Inf 1:146–165

Neis P, Singler P, Zipf A (2010) Collaborative mapping and emergency routing for disaster logistics–case studies from the haiti earthquake and the UN Portal for Afrika. In: Proceedings of Geospatial Crossroads @ GI_Forum '10, Salzburg, Austria, 6–9 July 2010

Over M, Schilling A, Neubauer S, Zipf A (2010) Generating web-based 3D city models from OpenStreetMap: the current situation in Germany. Comput Environ Urban Syst 34:496–507

Roick O, Hagenauer J, Zipf A (2011) OSMatrix—grid-based analysis and visualization of OpenStreetMap. In: State of the Map EU 2011, Vienna, Austria

Rylov M, Reimer A (2014) A comprehensive multi-criteria model for high cartographic quality point-feature label placement cartographica. Int J Geogr Inform Geovis, doi:10.3138/carto.49.1.2137

Rylov M, Zipf A (2012) Solutions for limitations in label placement in OGC symbology encoding (SE) specification. In: Geoinformatik 2012, Heidelberg

Schilling A, Over M, Neubauer S, Neis P, Walenciak G, Zipf A (2009) Interoperable location based services for 3D cities on the Web using user generated content from OpenStreetMap. In: UDMS 2009. 27th Urban Data Management Symposium. Ljubljana , Slovenia

Schmitz S, Zipf A, Neis P (2008) New applications based on collaborative geodata—the case of routing. In: Proceedings of XXVIII INCA International Congress on Collaborative Mapping and Space Technology

Zielstra D, Zipf A (2010) Quantitative studies on the data quality of OpenStreetMap in Germany

Part I
Data Management and Quality

Assessment of Logical Consistency in OpenStreetMap Based on the Spatial Similarity Concept

Peyman Hashemi and Rahim Ali Abbaspour

Abstract The growth in the number of users and the volume of information in OpenStreetMap (OSM) indicate the success of this VGI-based project in attracting diverse sets of people from all over the world. A huge amount of information is generated daily by non-professional users and OSM faces the challenge of ensuring data quality. Spatial data quality comprises several basic elements; among them, logical consistency concerns the existence of logical contradictions within a dataset. It is one of the most important elements, but has not been studied much in VGI despite the key role in quality assurance. Because of the participatory nature of data collection and entry in OSM, the common consistency checking routines for spatial data should be revised. Since contributors have different views about objects, data integration in OSM may be considered as a form of multi-representation data combination. In this article, the concept of spatial similarity in multi-representation considering three elements, i.e. directional relationships, topological relationships, and metric distance relationships, is used to build a framework to determine the probable inconsistencies in OSM.

Keywords Volunteered geographic information (VGI) · Logical consistency · Spatial similarity · OSM · Topological relations

1 Introduction

The technological advancements in the geospatial domain such as developments in Web 2.0 mapping provide great opportunities to take advantage of a mass of volunteered users for geospatial data acquisition and enrichment. The well-known

P. Hashemi · R. Ali Abbaspour (✉)
Surveying and Geomatics Engineering Department, College of Engineering, University of Tehran, Tehran, Iran
e-mail: abaspour@ut.ac.ir

P. Hashemi
e-mail: hashemipeyman@ut.ac.ir

J. Jokar Arsanjani et al. (eds.), *OpenStreetMap in GIScience*, Lecture Notes in Geoinformation and Cartography,
DOI 10.1007/978-3-319-14280-7_2

projects such as OSM, Wikimapia, and Google Mapmaker, which are based on this idea, enable the users to play the role of geospatial data providers. The rising rate of users of such environments is an indication of the level of attraction. For example, the registered number of OSM members was 100,000 in 2009 while this reached over 1,500,000 by the beginning of 2014 (OSM statistics 2014).

Although VGI has proved a successful way to obtain detailed geographical information in a timely and low-cost manner, it suffers from some serious weaknesses (Goodchild and Li 2012). The Achilles' heel of VGI is the lack of metadata on spatial data quality parameters used for quality assurance mechanisms. Despite the professional surveyors, most of the users of VGI-based systems are unfamiliar with the standards on spatial data collection or they may feel no need to follow such rules. This leads to a serious problem when quality control procedures are employed to measure the quality level of spatial data. Thus, compatible, customized spatial data quality parameters and control procedures are required to insure the users to utilize those data by crossing the first step of collection of raw spatial data to spatial information production. The other reason to address the quality issues in VGI is the current movement from only data collection and enrichment towards volunteered geographic services (Sui et al. 2013).

There are several classifications of spatial data quality (Worboys and Duckham 2004; Morrison 1995; ISO 19157:2013), which are the results of efforts started in the 1980s in the geospatial information community. In almost all of them, five main elements of spatial accuracy, attribute accuracy, currency, logical consistency, and completeness are considered. They are more or less utilized in VGI studies to address the quality and uncertainty issues. Among the different elements of spatial data quality, logical consistency is studied less than the others in VGI research. Due to the numerous topological errors and loss of standards, which should be considered by users during data entry, different types of inconsistencies occur in VGI. Although there are mechanisms to discover some inconsistencies such as open polygons in some VGI projects like OSM, other basic errors such as overshoots and undershoots, which are of high frequency, remain unsolved. Therefore, it is necessary that the logical consistency, especially with a focus on topological aspects, be assessed in order to fulfill the spatial quality assurance in VGI.

Literature on the quality study in VGI, especially in OSM, may be generally divided into three categories. In the first category, the data in OSM is compared with reference data, which have higher precision and are generated by a mapping agency (Haklay 2010; Girres and Touya 2010; Zielstra and Zipf 2010; Neis et al. 2011). Comparisons of volunteered spatial data with reference data constitute a considerable portion of the research in this field. There are several reports of evaluations of this type which have been carried out in various countries such as the U.K. (Haklay 2010), France (Girres and Touya 2010), and Germany (Zielstra and Zipf 2010; Neis et al. 2011). Although this approach looks rational, using it in most cases in VGI, where there are no reference data, is impossible. Moreover, the costs and license limitations restrict access to high quality, commercial datasets. The research in the second category concentrates on user activities (Jokar Arsanjani et al. 2013), and the third category is comprised of research on the history of the

```
<node id="27128507" lat="50.0146754" lon="8.2429516" version="1" timestamp="2007-04-05T18:11:17Z"
  changeset="6911" uid="3609" user="seb"> <tag k="created_by" v="JOSM"/> </node>
<node id="27128507" lat="50.014677" lon="8.2429489" version="2" timestamp="2007-10-16T12:38:56Z"
  changeset="203178" uid="16643" user="Joh"/> </node>
<node id="27128507" lat="50.0144034" lon="8.2431315" version="7" timestamp="2013-03-16T23:41:50Z"
   changeset="15390280" uid="440308" user="spezialist"/> </node>
```


Fig. 1 A sample of the OSM history file

data (Keßler and de Groot 2013; Keßler et al. 2011). Data recorded as the history files in OSM data can play an alternative role as reference data. All OSM data such as nodes, ways, and relations as well as their previous versions are stored in the full history dump (OSM, full history dump 2014). These files contain information such as node position (latitude and longitude), object versions, modified times (timestamp), user identification (user name and id), and tags information (key and value). For instance, the submitted information for a node in the OSM history files, which is actually a gas station, is shown in Fig. 1. This file is in the XML format. This comprehensive history file allows researchers to conduct more statistical analysis on spatial aspects or a contributor's information. Thus, intrinsic quality indicators can be applied instead of reference datasets (Barron et al. 2013).

This paper addresses the logical, topological inconsistencies in VGI with a focus on the popular OSM project. A similarity-based framework is used to deal with this issue. The remainder of this paper is organized as follows: First, there is an introduction to logical consistency in general and then the focus is on the meaning of this concept in OSM. Afterwards, the proposed framework is introduced and its main components are described in detail. The last part of this section is the explanation of the proposed methodology and its evaluation. The final section summarizes the work.

2 Logical Consistency for OSM

The concept of logical consistency was initially introduced for database integration purposes (Kainz 1995). Logical constraints defined for a database are rules that adapt the data to the selected structure and provide the opportunity to optimize the storage/retrieval speed. This concept is then used by the geospatial community to address the different sorts of inconsistencies arising during data entry and analysis in GIS. Logical consistency is highly correlated with positional errors. Because there are numerous sorts of positional errors in VGI-based projects, the logical inconsistencies are of high frequency in these projects, which in turn could have a side effect on the analysis and usage of the information.

Based on ISO 19157:2013 standard, logical consistency can be defined as "degree of adherence to logical rules of data structure, attribution, and relationships". Four main sub-elements considered for logical consistency in this standard are conceptual consistency, domain consistency, format consistency, and topological consistency. Conceptual consistency monitors the adherence to the rules of the conceptual schema. Domain consistency checks the value ranges, which should be in certain value domains while format consistency controls the rate of stored data in accordance with the physical data structure. Although these components of logical consistency are meaningful when a standard geodatabase is discussed, they cannot be applied to OSM datasets. Topological consistency examines correctness of encoded topological characteristics. Despite the other components of logical consistency, those types of inconsistencies addressed in the latter one are of crucial importance in VGI.

There are several common errors such as undershoots, overshoots, and duplicate lines in volunteered data. The Quality Assurance page (OSM, Quality assurance 2014) serves as a directory for known error recognition services for OSM. These mechanisms provide some tools for users to recognize errors. Non-closed areas, dead-ended one-ways, broken polygons, and other similar errors can be detected by these tools. For instance, "keep right" is one of the error detection tools introduced in OSM (2014). Intersections of a highway with another highway and a street are shown in Fig. 2 as an example.

As shown in Fig. 2, many types of errors can be determined in this area using the keep right tool. Many of the recognized errors such as non-closed areas, dead-ended one-ways, floating islands, intersections without junctions, and overlapping ways, which are illustrated in the left part of Fig. 2, are related to inconsistencies in spatial data. If volunteered data is viewed in this way, it looks almost as if it were an incompatible, inconsistent set of data. While one of the main objectives of developing a volunteered geographic information environment is far beyond only achieving standard spatial data, considering its nature of being a social event, data consistency should be determined taking users' convenience and requirements into account.

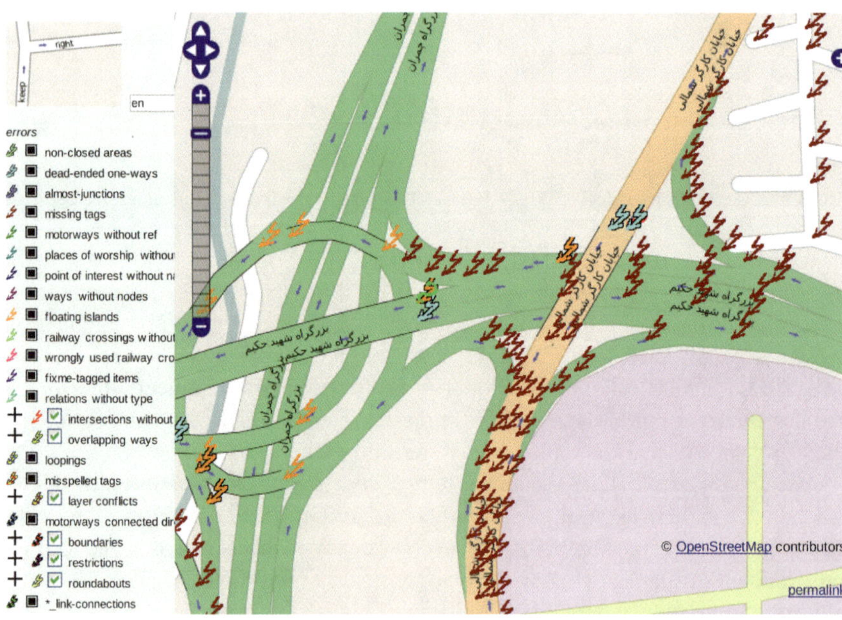

Fig. 2 Detected errors in an intersection using "Keep Right" tool (KeepRight 2014)

To illustrate this idea, a part of Wörrstadt city is presented from two different perspectives. The OSM website snapshot and a presentation of the history file in ESRI's ArcMap software are depicted in Figs. 3 and 4, respectively. The maps of OSM are frequently edited by users and the cartographic representations on the website may be very different from what users have provided. These maps are combinations of all the information provided by the users and some edits. For example, although there are no broken polygons or duplicate lines in Fig. 3, history files of the same area show numerous cases with gross errors (Fig. 4).

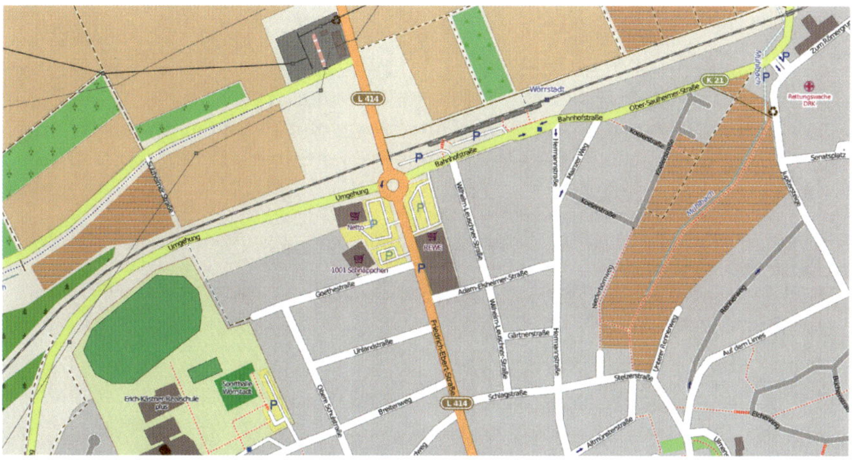

Fig. 3 Part of Wörrstadt city in OSM map (OSM 2014)

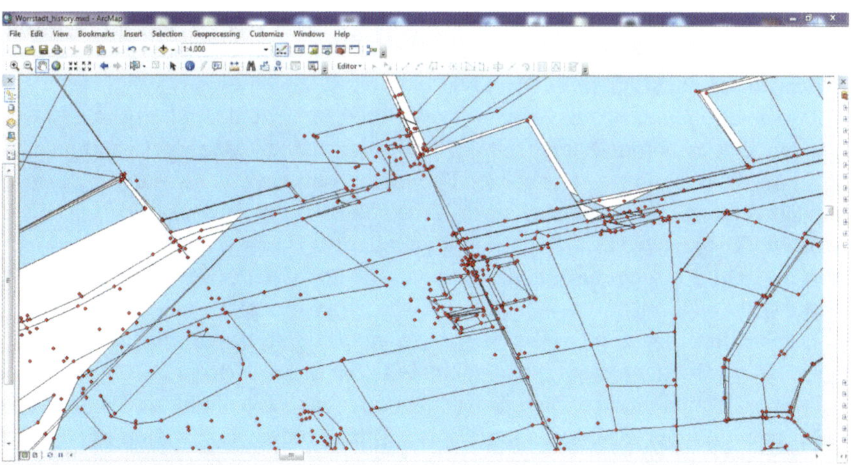

Fig. 4 A sample history file of Wörrstadt city opened in ESRI's ArcMap

While all of the users of OSM can change the objects in OSM, there are several versions of an object in the history file; hence, the consistency evaluation is similar to the assessment of different representations in multi-representations of objects (Egenhofer et al. 1994). In this study, multi-representation means different descriptions (spatial scenes) of the same object by different users, which in turn result in changes in properties of the object and its relationships with the other objects.

3 Proposed Framework

To model the logical consistency with a focus on the topological consistency of OSM data, a framework is developed in this paper based on the similarity measures in scenes according to three main spatial relationships. The proposed methodology is based on changes applied by users rather than comparisons with some reference datasets. The key concept in this methodology is the spatial scenes. Spatial scenes can be defined as follows (Li and Fonseca 2006):

> A spatial scene is a set of geospatial objects together with their spatial relations—topological relations, distance relations, and direction relations—and optionally other sorts of spatial properties, such as shape and relative sizes (areas and lengths), or attributes specifying the semantics of the spatial objects.

When a change is submitted by a user for a spatial object, two scenes of before and after the change are compared to assess their consistency. The spatial similarity is used as a criterion for comparison. This is selected because it is in a direct relation with the human cognition of a place. Similarity is defined as deviation from equivalence (Bruns and Egenhofer 1996). The analogies contribute to significant parts in human cognitions and are used as a general rule for classifications, generalization, and deductive inferences (Li and Fonseca 2006; Tversky 1977). Due to the spatial relations and characteristics in spatial scenes, several factors affect the evaluation of similarity measures. The plurality of these factors complicates the assessment of the similarity.

Two main approaches have been widely used for spatial similarity assessment: conceptual neighborhood approach and projection-based approach (Li and Fonseca 2006). In the conceptual neighborhood approach, similarity is measured by using a relevant network of concepts distances. The shortest path between two concepts in the network is selected as the criterion for similarity. The projection-based approach projects spatial objects and their relations onto the other space, which can be a vector space or a matrix space. This mapping transfers the problem of similarity assessment from the comparison of objects in spatial scenes to a vector or matrix space. There are some cases where a combination of the conceptual neighborhood and the projection-based approach is used (Goyal and Egenhofer 2001). The proposed methodology in this chapter also uses this combined approach to calculate the spatial similarity.

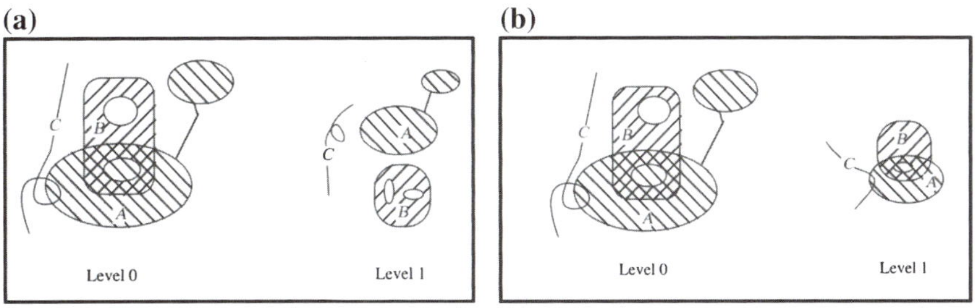

Fig. 5 **a** Two object-similar representations and **b** two relation-similar representations (Egenhofer et al. 1994)

Lower spatial similarity leads to different results in terms of analysis made on spatial data. The changes decrease the similarity rate and cause inconsistency within spatial objects. In other words, the changes in the shape of spatial objects and the relationships between them influence the logical consistency in the dataset. Each topological structure or space has some characteristics such that following them is a guarantee of consistency and validation for features defined by that structure. In multi-representation, topological relations are fundamental information for describing a scene, which should be free of conflict. Topological consistency constraints evaluate structures of the features and adapt to spatial feature relations and the combination of spatial relations (Kainz 1995). The topological relations between two regions are defined based on the well-known Egenhofer's 9-intersection matrix (Egenhofer and Sharma 1993).

There are two types of relations equivalences in multi-representation levels based on the comparison of topological invariants: object equivalence and relation equivalence (Egenhofer et al. 1994). Examples of these two types of equivalences are shown in Fig. 5. In the study of object similarity level, structural characteristics of each feature at different levels are analyzed together. For example, the number of arcs and nodes are counted, although node and edge equalities do not show structural consistency in two levels. In Fig. 6, the number of nodes and edges is equal for two objects, while holes are different; hence, they are partially inconsistent. To assess the relation similarity, relations between different components of an object and relations between an object and neighbor objects should be considered. Based on the possible

Fig. 6 Two representations of the same object

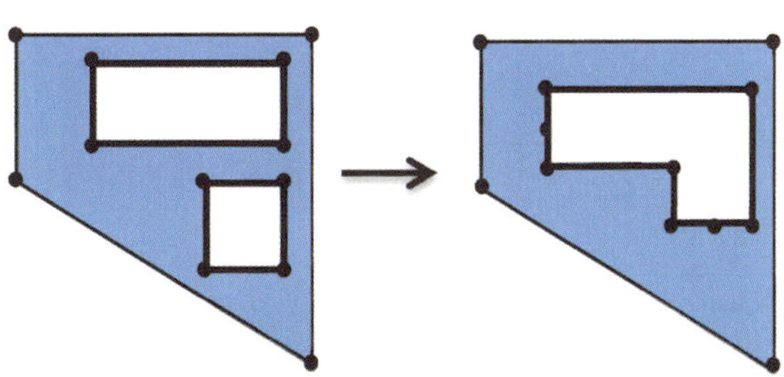

spatial relations, three types of relations are considered for evaluation in this paper: directional, topological, and metric distance relationships. The ways these concepts are employed are discussed in the following sections.

3.1 Directional Relationships

Cardinal directions have a considerable role in the description of spatial structures (Goyal and Egenhofer 2001). This role is highlighted when topological relations between objects are the same for two scenes. The evaluation method used for directional relationships in this study is based on the Goyal and Egenhofer method (Goyal and Egenhofer 2001). In this method, an object is selected as reference and nine geographical directions are used to define the direction-relation matrix (Fig. 7). The detailed direction-relation matrix is a 3 × 3 matrix (Eq. 1) which captures the neighborhoods around the reference object and shows for each tile how much of the target object falls into it (Goyal and Egenhofer 2001).

$$Direction(A,B) = \begin{pmatrix} \frac{area(B \cap NW_A)}{area(B)} & \frac{area(B \cap N_A)}{area(B)} & \frac{area(B \cap NE_A)}{area(B)} \\ \frac{area(B \cap W_A)}{area(B)} & \frac{area(B \cap 0_A)}{area(B)} & \frac{area(B \cap E_A)}{area(B)} \\ \frac{area(B \cap SW_A)}{area(B)} & \frac{area(B \cap S_A)}{area(B)} & \frac{area(B \cap SE_A)}{area(B)} \end{pmatrix} \quad (1)$$

Then a conceptual neighborhood graph (CNG) is used to compute the spatial similarity. The transfer cost from each node (i.e. NW, N, NE, E, SE, S, SW, and W) to its neighbors is assumed to be 1 on that edge. Transfer cost between two arbitrary nodes is the length of the shortest path between those nodes. For example, the cost of transfer between the NW node and the SE node is equal to 4.

Directional similarity between two scenes is defined as the minimum cost for transferring the elements with non-zero values of first direction matrix from their locations to the locations of the elements with non-zero values in the second direction matrix along the CNG to transform the first matrix to the second one (Goyal and Egenhofer 2001).

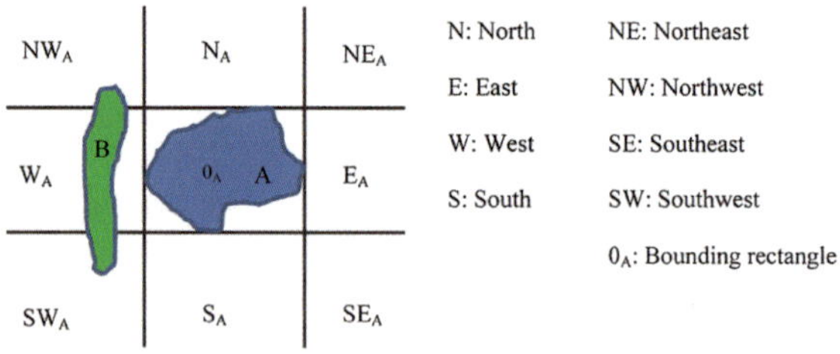

Fig. 7 The directional relation between two regions *A* and *B* using a projection-based system

3.2 Topological Relationships

To analyze the topological relations, topological distance based on 9-intersection matrices is employed (Egenhofer and Al-Taha 1992). Topological distance between two topological relations indicates the corresponding elements with different values. The difference between zero and non-zero elements of 9-intersection matrices is defined by Eq. 2.

$$\phi - \phi = 0;\ \neg\phi - \neg\phi = 0;\ \phi - \neg\phi = -1;\ \neg\phi - \phi = 1 \tag{2}$$

The topological distance between two topological relations, r_A and r_B is defined as the magnitude of the difference between two related intersection matrices (Egenhofer and Al-Taha 1992), M_A and M_B (Eq. 3).

$$\mathcal{T}_{r_A, r_B} = \|M_A - M_B\| \tag{3}$$

Topological distances between eight possible topological relations for two spatial objects are shown in Fig. 8, which can be used as a conceptual neighborhood graph between relations as well. For each change made on a feature, the cost for transferring the first topological relation to the second one is considered. Greater values of transfer cost or topological distance increase the chance for topological inconsistency. Computation of topological similarities using the concept of topology distance, the cost network (Li and Fonseca 2006), or five layers graph (Bruns and Egenhofer 1996) results in different consequences. Regardless of which cost graph has been used, it is essential for the model to have the ability to show the changes in topological relations. It is important since the changes in topological relations have a direct impact on the analysis results.

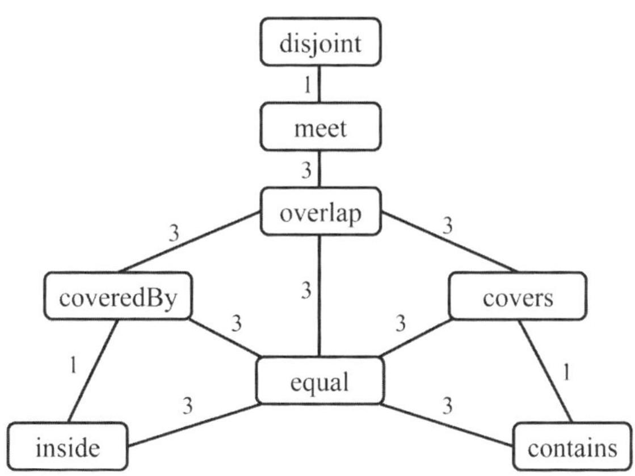

Fig. 8 The revised closest-topological-relationship-graph (Egenhofer and Al-Taha 1992)

3.3 Metric Distance Relationships

It is a challenging problem to define qualitative distance relationships for different types of spatial objects, because the concepts and conditions used in definition of these relations depend on the scale of spatial space. In this paper, the 4-granularity (equal, near, medium, far) metric distance network is used to determine the similarity of the metric distance relationship. There are four nodes and three edges on this network. The transformation cost is set as 1 on each edge. The rate of change depends on which distance for each qualitative distance parameter has been selected. According to our perception based on the different trial and error experiments, the value of 50-m distance is considered as near. For discovering whether the moved object is in the near area, it should be determined if the new object intersects with the 50-m buffer of the old object. In other words, in assessment of metric distance, the exact distance between objects is not calculated and checking to have an intersection with the buffer zone, which is around the reference object, is considered. This assumption is certainly with some approximations.

3.4 Proposed Methodology

The proposed methodology for assessment of logical consistency in OSM is shown in Fig. 9. When a spatial object is altered by a user, the spatial similarity of two scenes, i.e. before and after the change, is compared based on the mentioned relations. For computation of spatial similarity, a reference object must be selected. For this reason, the objects in a radius of 100 m from the changed object are considered as influential features in the selected scene. There is also a possibility to use only well-structured objects such as landmarks as the reference objects. The distance of 100 m is selected according to the density of urban features empirically. This distance can be changed based on the scale, type of region, and required accuracy. Considering the change, spatial similarity before and after alteration with each influential object as a reference object is computed. For greater spatial dissimilarity values, the possibility of inconsistency increases accordingly.

The effective methods and parameters in the calculation of spatial similarity are different. For example, if the changed object is a point and the reference object is a polygon, the most important relationship between these two object is if the point is inside of the polygon or not. Having such capabilities, the users are able to decide about the consequence of the changes.

One of the major affecting parameters on results is the selection of the reference object. In fact, on one hand, selection of all objects within a buffer is unnecessary

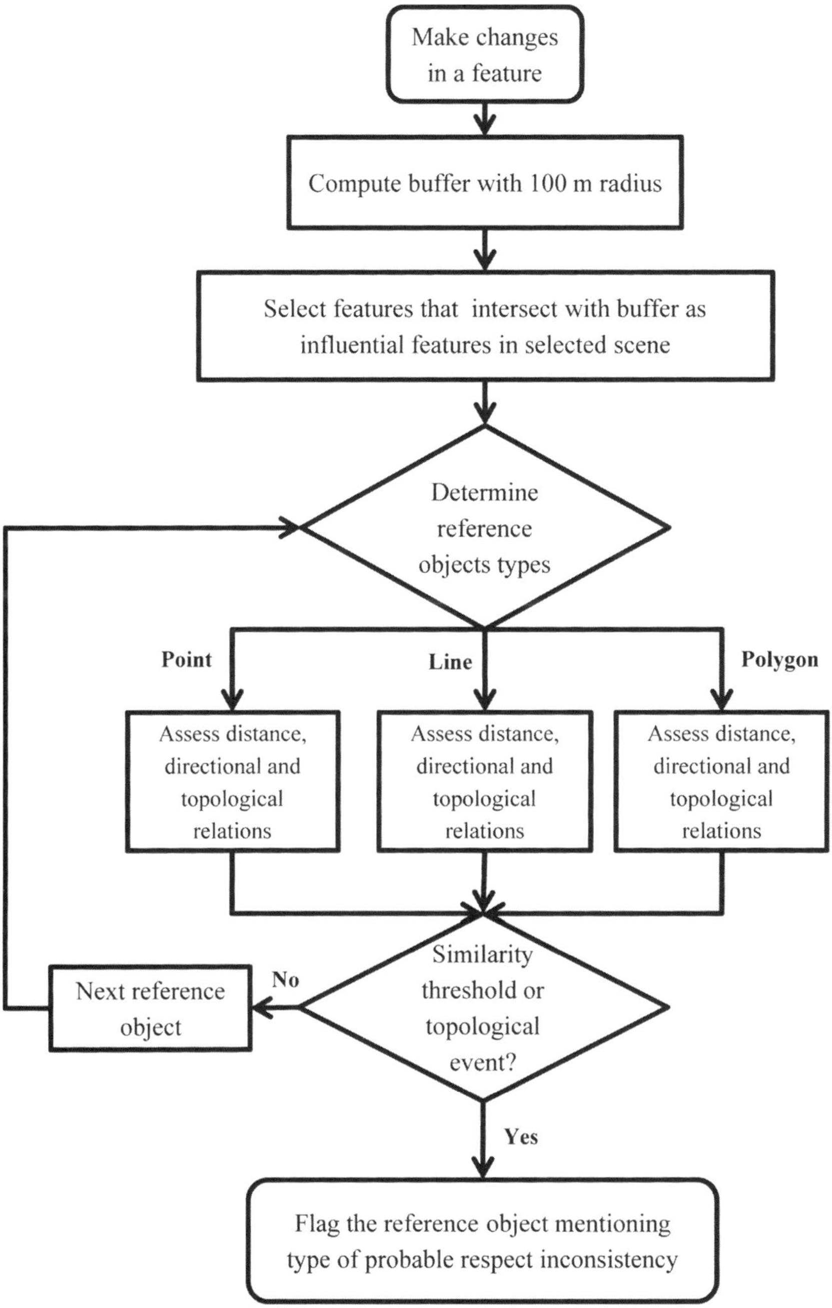

Fig. 9 The proposed methodology for detection of inconsistency in OSM

and on the other hand, selection of objects such as trees as the reference object is irrational, because these sorts of objects have no effects on the spatial cognition of people. Optimum selection of the reference object decreases the computational load of the program and increases its speed. Thus, it is recommended to choose the street crosses in road networks and existing landmarks as reference objects because the relationships between these objects are more accurate (Wang et al. 2011).

To illustrate the proposed methodology, an example of the implementation is described. Imagine the place of a gas station is changed from east of a highway, which is shown in Fig. 10a in the northeast to southwest direction, to the west side of it. Now it is desired to study the changes in the relations. In the first step, a buffer of 100 m distance is generated around the initial location of the gas station in order to find the influential objects (Fig. 10a). After selection of all influential objects (Fig. 10b), it is examined to discover which objects are affected due to the change in the location of the gas station. According to the introduced criteria for topologic situations for a point object, the program finds those lines and polygons for which the relationship type of them with the point is changed (Fig. 10c). The program alerts that the gas station exited from block A and entered block B. Moreover, this change results in the change in the direction of the gas station according to the highway.

To understand the effect of these parameters on the results better, the following example is explained. Imagine you are driving in north Kargar Street and while passing by the engineering college (blue balloon in Fig. 11) you are running out of gas. According to the map, the nearest gas station is located on the east side of Chamran Highway (green balloon in Fig. 11) and normally you need to travel

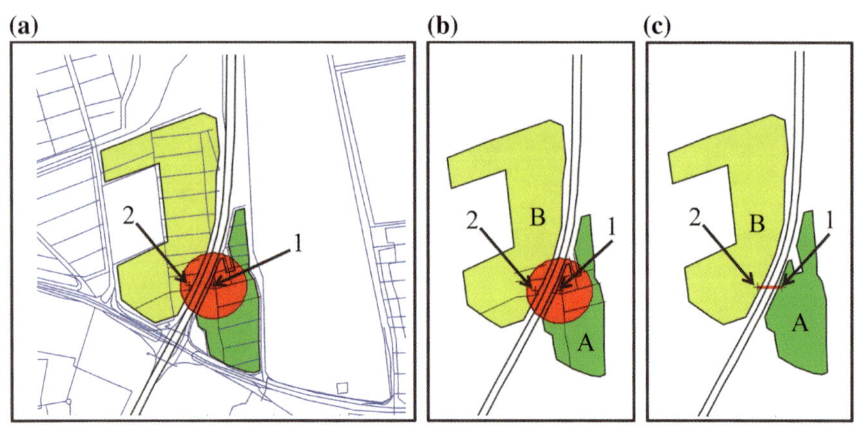

Fig. 10 **a** Change in the position of an object, **b** finding the influential objects, **c** the changed relationships due to the change in the position of an object, *arrow* (*1*) shows the first and *arrow* (*2*) shows the second position of the gas station

1.05 km to arrive (green line in Fig. 11). However, this distance is based on the assumption that the gas station is on the east side of the highway. Now assume that the position of the station is displaced a little on the map (yellow balloon in Fig. 11). Chamran Highway width is approximately 40 m. This error is trivial for a gas station position in a volunteered map. Now if the station is placed on the other side of the highway (west side) on the map, you need to travel the yellow line shown in Fig. 11 to arrive at the destination. If the gas station is really placed where it is shown on the map, someone needs to travel 1.64 km to arrive there. Considering the real distance of 1.05, 1.64 km is acceptable as well. While in reality, one does not arrive at the gas station by travelling a longer distance. The best scenario is to return by the same path (red line in Fig. 11) in which one needs to travel around 2.49 km. Therefore you should traverse 4.13 (1.64 + 2.49 = 4.13) km instead of 1.05 km, around four times more. Considering heavy traffic on the return back path, this case is not pleasant at all. This example indicates the importance of directional parameters beside the positional parameters.

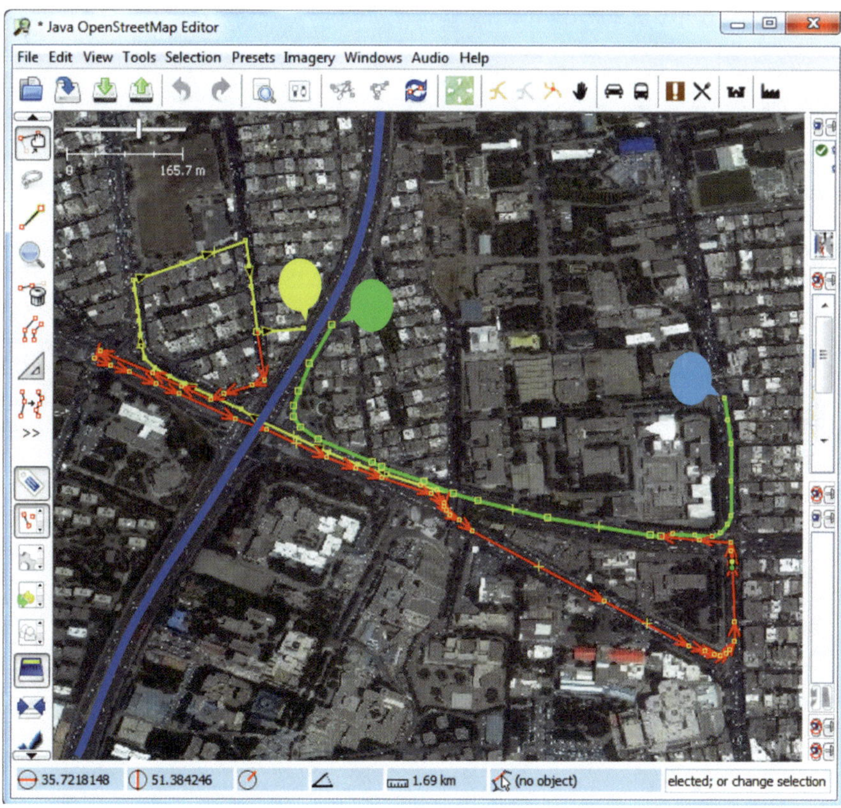

Fig. 11 Start point in north Kargar Street (*blue balloon*), real gas-station location (*green balloon*), imaginary gas-station location (*yellow balloon*), Chamran highway (*blue line*)

4 Implementation

To evaluate the proposed methodology, the dataset of the city of Wörrstadt in Germany in OSM is selected (Fig. 12). The dataset used in this study is based on the files extracted by Körner at extracts of the OSM full history dump (2014). In the first step, the data are unzipped using the osmconvert application. While the current version of the proposed framework is implemented offline in Matlab, results containing inconsistencies are displayed in ESRI's ArcMap. To demonstrate the functionality of the proposed methodology, two sample cases are presented, in which the changes in positions result in inconsistencies. In these cases, looking at the current versions of the maps in OSM shows none of the inconsistencies, but assessment of the history of the objects reveals these inconsistencies.

In the first sample, the change in direction and topological relationships are considered. The direction and topological relationships between Kleine Albanus street (id = 27022096) and the adjacent parking lot (id = 198260983) are changed. These changes cause a serious effect on finding the access way to the parking lot. Figures 13 and 14 show the current and past positions of the street and the parking lot, respectively. Moreover, Fig. 13 shows the current position of Adler pharmacy and Fig. 14 shows the changes of Adler pharmacy locations in time. The direction of the pharmacy is changed relative to both the adjacent parking lot and the Pariser highway (id = 23498981).

Fig. 12 A view of Wörrstadt region in OSM (OSM 2014)

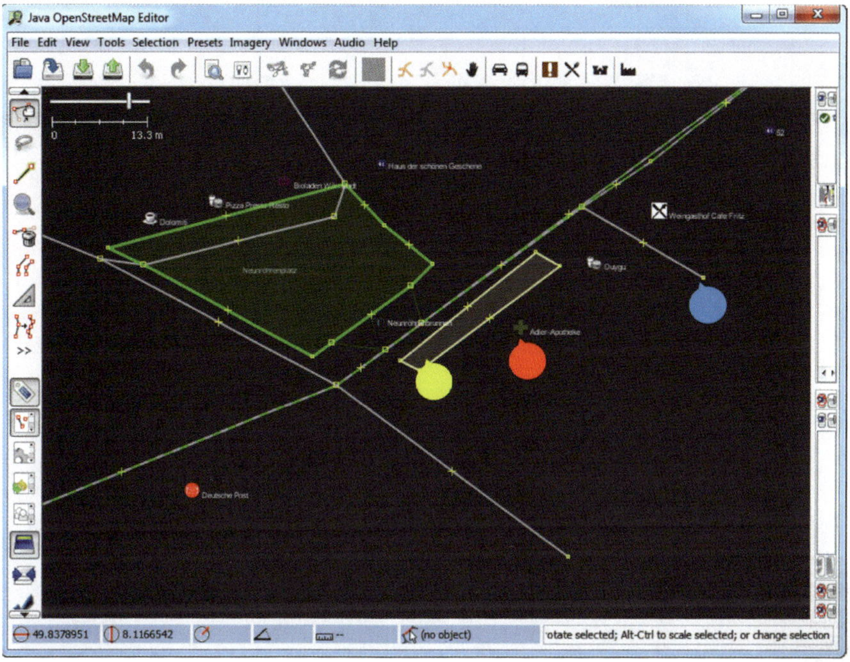

Fig. 13 Current location of Albanus street (*blue balloon*), parking lot (*yellow balloon*), and Adler pharmacy (*red balloon*) in OSM map shown with Java OpenStreetMap (JOSM) editor

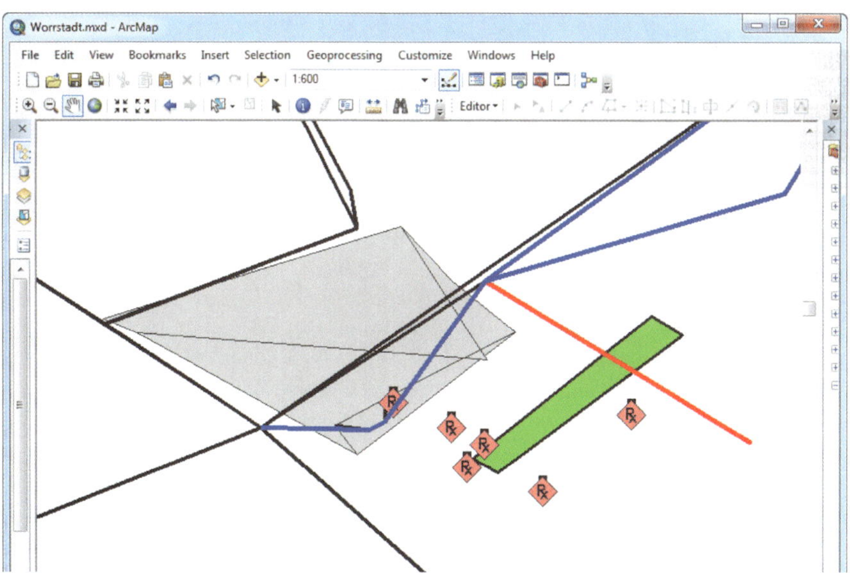

Fig. 14 Previous topological state of Albanus street (*red line*) and parking lot (*green rectangle*) are shown. In addition, changes of the direction of the pharmacy relative to the adjacent parking lot (*green rectangle*) and Pariser highway (*blue lines*) can be seen

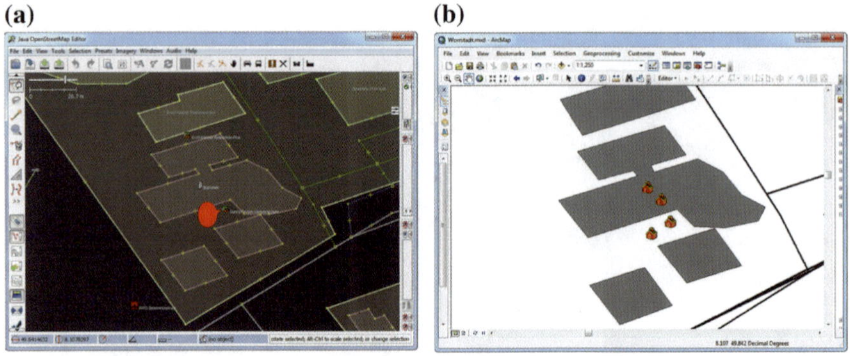

Fig. 15 **a** The current location of Forster school (*red balloon*) in JOSM and **b** changes in location of Forster shown with ArcMap

In the second case, a sample of exit from a polygon is considered. According to assessment of the history file, it is revealed that Forster (id = 280500466) school is outside the bounding polygon after changing. Figure 15 shows the current location and the past locations of this school. The mentioned cases are samples of inconsistencies, which are revealed after assessment of the similarity parameters in two scenes of before and after the changes. Generally, there are numerous inconsistencies like this in OSM. Finding the logical inconsistencies using the proposed framework and reporting them enables the users to be involved in the correction and resolution of inconsistencies more effectively.

5 Conclusion

Data provided by users in volunteered databases is increasing rapidly. While low cost and high update rates are the most important benefits provided by these data, there is a demand to develop methods and parameters for the quality evaluation of volunteer data. Among the different elements of spatial data quality, logical consistency is of crucial importance in quality assurance procedure. In this paper, logical consistency is studied from a different point of view. The idea is to import the features and qualitative relationships of objects with spatial positions to evaluate quality. It was demonstrated that the parameters such as direction, distance, and topological relationships between objects could directly affect human comprehension and analysis results. Because VGI in most cases reflects the cognitions of people from their surrounding environment, it is necessary to assess if the changes cause changes in spatial cognition. While the concept of similarity has a direct relation with spatial cognition, it has high capability to model a part of these changes. An important characteristic of this method is the ability to take advantage of both comparisons of data with reference and with itself. Given the potential of qualitative parameters, they can be added to factors of evaluation for volunteered data. These factors without adding much cost are valuable help to quality evaluation and vandalism recognition.

References

Barron C, Neis P, Zipf A (2013) A comprehensive framework for intrinsic OpenStreetMap quality analysis. Trans GIS 43–48. doi:10.1111/tgis.12073

Bruns T, Egenhofer MJ (1996) Similarity of spatial scenes. In: Kraak J-M, Molenaar M (eds) Seventh international symposium on spatial data handling. Delft, The Netherlands, pp 173–184

Egenhofer MJ, Al-Taha KK (1992) Reasoning about gradual changes of topological relationships. Theories and methods of spatio-temporal reasoning in geographic space. Springer, Berlin, pp 196–219

Egenhofer MJ, Sharma J (1993) Topological relations between regions in R^2 and Z^2. In: Abel D, Ooi BC (eds) Advances in spatial databases–3rd international symposium on large spatial databases, SSD 93, Singapore, Lecture notes in computer science. Springer, Berlin, pp 316–336

Egenhofer MJ, Clementini E, Di Felice P (1994) Evaluating inconsistencies among multiple representations. In: Proceedings of the 6th international symposium on spatial data handling, Edinburgh, Scotland, pp 901–920

Jokar Arsanjani J, Barron C, Bakillah M, Helbich M (2013) Assessing the Quality of OpenStreetMap contributors together with their contributions. Paper presented at the 16th AGILE international conference on geographic information science, Leuven, Belgium

OSM full history dump. http://osm.personalwerk.de/full-history-extracts/. Accessed 20 Sept 2014

Girres JF, Touya G (2010) Quality assessment of the French OpenStreetMap dataset. Trans GIS 14 (4):435–459

Goodchild MF, Li L (2012) Assuring the quality of volunteered geographic information. Spat Stat 1:110–120. doi:10.1016/j.spasta.2012.03.002

Goyal RK, Egenhofer MJ (2001) Similarity of cardinal directions. In: Jensen CS, Schneider M, Seeger B, Tsotras VJ (eds) Advances in spatial and temporal databases SE-3. Lecture notes in computer science, vol 2121, Springer, Berlin, pp 36–55

Haklay M (2010) How good is volunteered geographical information? A comparative study of OpenStreetMap and ordnance survey datasets. Environ Plan 37(4):682–703

ISO 19157:2013 Geographic information—data quality. ISO (International Organization of Standardization)

Kainz W (1995) Logical consistency. In: Guptill SC, Morrison JL (eds) Elements of spatial data quality, vol 202. Elsevier Science Ltd, Amsterdam, pp 109–137

Keßler C, de Groot RTA (2013) Trust as a proxy measure for the quality of volunteered geographic information in the case of OpenStreetMap. In: Geographic information science at the heart of Europe. Springer, Berlin, pp 21–37. doi:10.1007/978-3-319-00615-4_2

Keßler C, Trame J, Kauppinen T (2011) Tracking editing processes in volunteered geographic information: the case of OpenStreetMap. In: Duckham M, Galton A, Worboys M (eds) Identifying objects, processes and events in spatio-temporally distributed data (IOPE), workshop at conference on spatial information theory, Belfast, USA

KeepRight. http://keepright.at/. Accessed 20 Sept 2014

Li B, Fonseca F (2006) TDD: a comprehensive model for qualitative spatial similarity assessment. Spat Cogn Comput 6(1):31–62

Morrison JL (1995) Spatial data quality. In: Guptill SC, Morrison JL (eds) Elements of spatial data quality, vol 202. Elsevier Science Ltd, Amsterdam, pp 1–12

Neis P, Zielstra D, Zipf A (2011) The street network evolution of crowdsourced maps: OpenStreetMap in Germany 2007–2011. Future Internet 4(1):1–21. doi:10.3390/fi4010001

OpenStreetMap. http://openstreetmap.org. Accessed 20 Sept 2014

OpenStreetMap, full history dump. http://wiki.openstreetmap.org/wiki/Planet.osm/full. Accessed 20 Sept 2014

OpenStreetMap quality assurance. http://wiki.openstreetmap.org/wiki/Quality_assurance. Accessed 20 Sept 2014

OSM Statistics. http://wiki.openstreetmap.org/wiki/Statistics. Accessed 20 Sept 2014

Sui D, Elwood S, Goodchild MF (2013) Prospects for VGI research and the emerging fourth paradigm. In: Elwood S, Goodchild MF, Sui D (eds) Crowdsourcing geographic knowledge. Springer, The Netherlands, pp 361–375. doi:10.1007/978-94-007-4587-2

Tversky A (1977) Features of similarity. Psychol Rev 84(4):327–352

Wang J, Mülligann C, Schwering A (2011) An empirical study on relevant aspects for sketch map alignment. In: Advancing geoinformation science for a changing world. Springer, Berlin, pp 497–518

Worboys MF, Duckham M (2004) GIS: a computing perspective. CRC Press, Florida

Zielstra D, Zipf A (2010) A comparative study of proprietary geodata and volunteered geographic information for Germany. In: 13th AGILE international conference on geographic information science, Guimarães, Portugal

Quality Assessment of the Contributed Land Use Information from OpenStreetMap Versus Authoritative Datasets

Jamal Jokar Arsanjani, Peter Mooney, Alexander Zipf and Anne Schauss

Abstract Land use (LU) maps are an important source of information in academia and for policy-makers describing the usage of land parcels. A large amount of effort and monetary resources are spent on mapping LU features over time and at local, regional, and global scales. Remote sensing images and signal processing techniques, as well as land surveying are the prime sources to map LU features. However, both data gathering approaches are financially expensive and time consuming. But recently, Web 2.0 technologies and the wide dissemination of GPS-enabled devices boosted public participation in collaborative mapping projects (CMPs). In this regard, the OpenStreetMap (OSM) project has been one of the most successful representatives, providing LU features. The main objective of this paper is to comparatively assess the accuracy of the contributed OSM-LU features in four German metropolitan areas versus the pan-European GMESUA dataset as a reference. Kappa index analysis along with per-class user's and producers' accuracies are used for accuracy assessment. The empirical findings suggest OSM as an alternative complementary source for extracting LU information whereas exceeding 50 % of the selected cities are mapped by mappers. Moreover, the results identify which land types preserve high/moderate/low accuracy across cities for urban LU mapping. The findings strength the potential of collaboratively collected LU

J. Jokar Arsanjani (✉) · A. Zipf · A. Schauss
GIScience Research Group, Institute of Geography, Heidelberg University,
69120 Heidelberg, Germany
e-mail: jokar.arsanjani@geog.uni-heidelberg.de

A. Zipf
e-mail: zipf@uni-heidelberg.de

A. Schauss
e-mail: anneschauss@gmail.com

P. Mooney
Department of Computer Science, Maynooth University,
Maynooth, Co. Kildare, Ireland
e-mail: peter.mooney@nuim.ie

J. Jokar Arsanjani et al. (eds.), *OpenStreetMap in GIScience*,
Lecture Notes in Geoinformation and Cartography,
DOI 10.1007/978-3-319-14280-7_3

features for providing temporal LU maps as well as updating/enriching existing inventories. Furthermore, such a collaborative approach can be used for collecting a global coverage of LU information specifically in countries in which temporal and monetary efforts could be minimized.

Keywords Land use features · Comparative assessment · Global monitoring for environment and security urban atlas (GMESUA) · OpenStreetMap · Confusion matrix

1 Introduction

The process of mapping land is known as LU mapping (Thenkabail et al. 2005) and land cover (LC) mapping (Kasetkasem et al. 2005), and is reflected in LU/LC maps. LU and LC maps are of great importance for many purposes concerning urban and regional planning, LU policy making, etc. In fact, these two concepts are distinct in essence, as LU maps illustrate human activities, such as artificial surface construction, farming, and forestry that represent the usage of land (Ellis 2007; Wästfelt and Arnberg 2013), whilst LC maps display the physical and biological cover over the land surface regardless of the purpose for which they are used (De Sherbinin 2002; Ellis 2007; Vaz et al. 2012, 2013). In other words, LC maps identify which land types cover the land and LU maps classify the land based on the usage of the land (Paneque-Gálvez et al. 2013; Sexton et al. 2013). For instance, if a particular land parcel is covered by grass, its LC type is labeled as grassland, whilst this parcel might be a part of a meadow LU class.

Employing signal processing algorithms on remote sensing data coupled with in-field measurements and ancillary data have been the main source of collecting LU and LC features (Kandrika and Roy 2008; Pacifici et al. 2009; Saadat et al. 2011; Qi et al. 2012). Remote sensing images and techniques often require in-field surveying for the results' validation process, i.e., in situ measurements as ground-truth data play a great role in delivering final products. Within the field-surveying data collection, experts and native residents' knowledge of the environment are needed to minimize uncertainty of measurements (Cihlar and Jansen 2001; De Leeuw et al. 2011). Contrary to LC mapping, LU mapping requires in-field information collection on the status and current usage of each land parcel, which are scarcely achievable from remote sensing images. Therefore, investigators must collect ancillary data as well to label LU patterns appropriately. Thus, LU mapping is even more complicated than LC mapping, and the information collected by experts from local residents, land managers, and evidence sources plays a vital role for accurate LU mapping (Fritz et al. 2012).

So far, a noticeable amount of efforts have been spent on generating LU maps at global, regional, and local scales. Examples of global scale and coarse resolution datasets comprise Global Land Cover (GLC2000) (Fritz et al. 2003), Moderate-resolution Imaging Spectroradiometer [MODIS; (McIver and Friedl 2002)], and

GlobCover (Arino et al. 2012), among others. At a regional (e.g., European) scale, the CORINE 2000 (Büttner et al. 2002) and Global Monitoring for Environment and Security Urban Atlas (GMESUA; Seifert 2009) deliver LC and LU maps at continental and municipal levels, respectively. High-resolution images including SPOT, Rapideye, and ALOS images have been utilized to attain fine-scale maps of urban areas delivering GMESUA (Kong et al. 2012), while a European coverage of LC maps at a coarse spatial resolution of 100 and 250 m has been provided in the CORINE dataset. But, their accuracies have been a major concern as outlined by Mayaux et al. (2006), Strahler et al. (2006), Herold et al. (2008), Fritz et al. (2012). In terms of accuracy, the lack of sufficient accuracy is even more critical in the case of global LU mapping, because the collection of globally-covered in-field information of LU features is such a huge task (Foody 2002; Foody et al. 2013). Thus, the necessity of finding an alternative and complementary approach for mapping LU features becomes evident. This could presumably be responded by the Web 2.0 innovations.

Lately, the development of Web 2.0 technologies has resulted in the emergence of a large number of CMP projects, which collect information about geographical objects from citizens. The majority of these CMPs offer very high-resolution satellite and aerial images (even less than one meter spatial resolution) through image libraries (e.g., Google Maps, Bing Maps) in their interfaces, which enable people to visualize the whole globe by fine-resolution remote sensing images so that they can map any features and additionally attach respective attributes to them (Rouse et al. 2007). This simple and straightforward way of visual interpretation of remote sensing images can be considered as alternative approach for LU mapping and even achieving fine-resolution LU maps at a global scale. The CMPs as listed in Sester et al. (2014) also provide people with some basic mapping tools in order to mark and digitize the visible objects. Some examples of CMPs are: OSM (Ramm et al. 2011), Geo-wiki (Fritz et al. 2012; Comber et al. 2013; See et al. 2013), Eye into Earth (Birringer 2008), and Wikiloc (Castelein et al. 2010). The individuals are also capable of enriching the attributes of the objects with some personal knowledge about these objects. The capability of importing the recordings of GPS-enabled devices i.e., smart phones and GPS devices, is granted to enable anyone to contribute even if s/he has minimal mapping expertise.

In brief, people interpret and integrate remote sensing images along with their personal information and their GPS-enabled device records. This sort of information has been called Volunteered Geographic Information (VGI): (Goodchild 2007). Among the CMPs, OSM is a unique platform in collecting LU features, because OSM has been so far a pioneer CMP due to attracting a huge amount of public attention and contributions (Ramm et al. 2011) by having almost 1.9 million users until December, 2014 and continue to grow as outlined by Jokar Arsanjani et al. (2015a, 2015c). More interestingly, OSM is highly democratic in receiving contributions by enabling anyone to edit/modify the existing features and sharing the whole data history freely and openly with the public in a structured way (Flanagin and Metzger 2008; Koukoletsos et al. 2012). It should be noted that OSM collects spatial information in GIS vector formats such as points, polylines, and polygons depending on the type of objects and presents them through a number of organized tags as listed in Ramm (2014).

A literature review reveals that in contrast to extensive analysis of road networks in OSM (Ludwig et al. 2011; Mooney and Corcoran 2012), first attempts at analyzing LU features from the whole OSM datasets have been carried out by Hagenauer and Helbich (2012), Jokar Arsanjani et al. (2013) in which they tried to extract LU features from the shared objects. However, except (Jokar Arsanjani and Vaz 2015b) virtually no studies on comparative assessment of the OSM-LU features with other authoritative datasets have been ever published. Therefore, the idea of using the contributed LU features to OSM arises in order to see how suitably we can collect temporal LU features at a local scale from OSM or even exploit the contributed features for updating the current LU datasets. To do so, a comparative quality assessment analysis must be carried out to gain some insights about it. To conclude, statements about the suitability of voluntarily collected LU data still remain highly speculative and even less is known whether these data might identify mismatches or even complement authoritative LU datasets.

In addition to that, spatial heterogeneity in the data quality increases the complexity of comparative studies as proven by Haklay (2010), Helbich et al. (2012), Koukoletsos et al. (2012). Hence, this research intends to evaluate the quality of OSM-LU features compared to a recent pan-European LU dataset, namely GMESUA, as a reference, in order to find out how accurate LU features are attributed across four different German metropolitan areas. Besides preparing a LU dataset from OSM contributions, this study aims to cross-compare the degree of completeness and the attribute accuracy of the OSM-based LU features with the GMESUA data by means of a statistical assessment. To be more precise, this research seeks to find out: (a) how complete LU features are contributed to OSM, (b) how well OSM-LU features are attributed, (c) whether or not OSM-LU features are already usable for LU mapping, and (d) how effective the use of OSM data for questions in LU science would be.

The remainder of the paper is structured as follows: an overview of the utilized datasets and the chosen study sites are given in Sect. 2. Section 3 introduces the applied method while key results are presented in Sect. 4. Finally, Sect. 5 draws discussions and conclusions and Sect. 6 provides some recommendations.

2 Materials and Data Processing

2.1 OSM Dataset

The first datasets utilized in the present study is the OSM snapshot for November 5, 2013. The features tagged with “Natural” describe a wide variety of physical features, which are categorized into different categories such as water bodies, forest, etc. as described in Ramm (2014). “Land use” is the human use of land, which represents the purpose a land parcel is being used for (Ramm 2014). To extract relevant LU features, objects labeled with the tags “Land use” and “Natural” are exported from the OSM planet file into a uniform dataset.

2.2 GMESUA Dataset as a Reference Dataset

The second dataset, serving as reference data, is the pan-European GMESUA dataset, which comprises LU data for selected metropolitan areas exceeding 100,000 inhabitants. It is adapted to European needs, and contains information that can be derived chiefly from Earth Observation (EO) data supported by other reference data such as commercial-off-the-shelf (COTS) navigation data and topographic maps. Its minimum mapping unit (MMU) is between 0.25 and 1 ha, and a minimum width of linear elements of 100 m with ±5 m positional accuracy is applied (European Union 2011). Additionally, some complementary data are integrated to improve the accuracy of classification processes namely (a) COTS navigation data such as POIs, LU, LC, water bodies; (b) Google Earth for interpretation; (c) local city maps for certain classes; (d) local zoning data such as cadastral data; (e) field checks (on-site visits); and (f) high-resolution satellite images (finer than 1 m ground resolution) (European Union 2011). At the time of writing this paper, this dataset covers 305 urban regions within Europe. The thematic accuracy for all classes is above 80 %. For more details see the Urban Atlas mapping guide (European Union 2011). Table 1 represents the defined classes in GMESUA and in this article these classes will be recalled by their codes as well e.g., Isolated structures [113].

2.3 Study Areas

Four large metropolitan areas of Germany, from different regions, are selected: Berlin, Frankfurt am Main, Munich, and Hamburg. There are multiple reasons for choosing these areas in Germany. First, the OSM community in Germany is very active and dynamic, and therefore it is rational to begin from potentially well-mapped areas. Secondly, no bulk import of authoritative datasets for Germany into the OSM database has yet been reported. Thirdly, according to osmatrix.uni-hd.de (Roick et al. 2011) and (Jokar Arsanjani et al. 2014) these cities have received high rates of contributions. Fourthly, the reference dataset i.e., GMESUA for these cities are available. The selected areas cover approximately 35,000 km^2 and contain 15 major LU classes, so a wide variety of LU features from heterogeneous areas are identified. Both the selected areas and the input data are shown in Fig. 1.

3 Methods

In geodata quality analysis, the quality of geodata should be internally and externally considered (van Oort 2006; Gervais et al. 2009). Internal quality reflects the data production specifications, which recognizes errors in the data. The major standard organizations (e.g., ISO 19157, ICA, FGDC, and CEN) have introduced

Table 1 Classification scheme applied in the preparation of GMESUA datasets

Classification level	Level 1	Level 2	Level 3
Land	Artificial surfaces [100]	Urban fabrics [110]	Continuous urban fabrics [111]
			Discontinuous urban fabrics [112]
			Isolated structures [113]
		Industrial, commercial, public, military, private and transport units [120]	Industrial, commercial, public, military and public units [121]
			Road and rail network and associated lands [122]
			Port areas [123]
			Airports [124]
		Mine, dump and construction sites [130]	Mineral extraction and dump sites [131]
			Construction sites [132]
			Land without current use [133]
		Artificial nonagricultural vegetated areas [140]	Green urban areas [141]
			Sports and leisure facilities [142]
	Agricultural + seminatural areas + wetlands [200]	–	–
	Forests [300]	–	–
Water	Water [500]	–	–

their diverse quality criteria and the followings five are common amongst them: (1) thematic accuracy, (2) positional accuracy, (3) temporal accuracy, (4) logical consistency, and (5) completeness (Guptill and Morrison 1995). These data properties are introduced to the users through metadata files attached to datasets by producers (Devillers et al. 2007). On the other hand, external quality reflects the suitability of a dataset for a particular purpose and addresses the concept of "Fitness of Use" (FoU): (Guptill and Morrison 1995; Devillers et al. 2007). In this study, the internal aspects of data quality are considered.

The workflow of evaluating the OSM-LU dataset is summarized in Fig. 2 and described as follows. First, OSM features tagged with "Land use" and "Natural" are retrieved and merged together to result in a unique dataset. Second, overlaps and topological errors between the features are then resolved, which is described in details

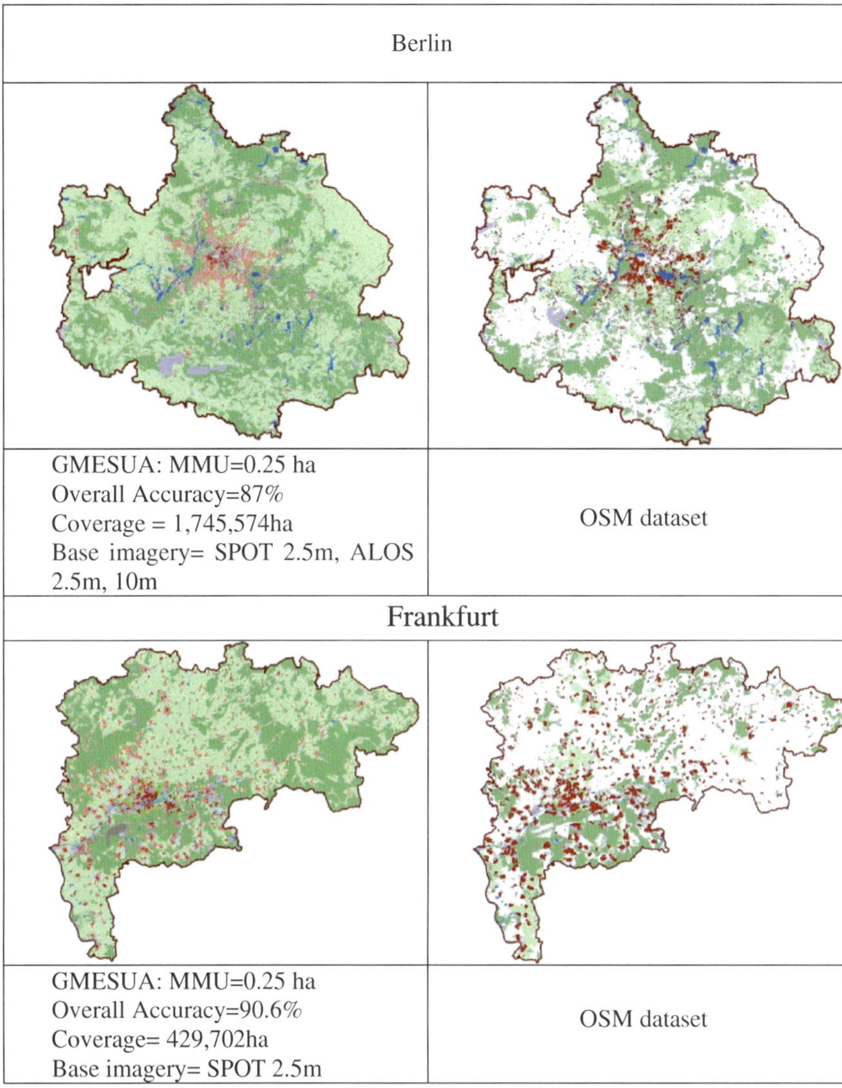

Fig. 1 The physical extent of the selected cities—Berlin, Frankfurt, Hamburg, and Munich, represented by the GMESUA datasets accompanied with their metadata (*left panels*) and contributed OSM features (*right panels*)

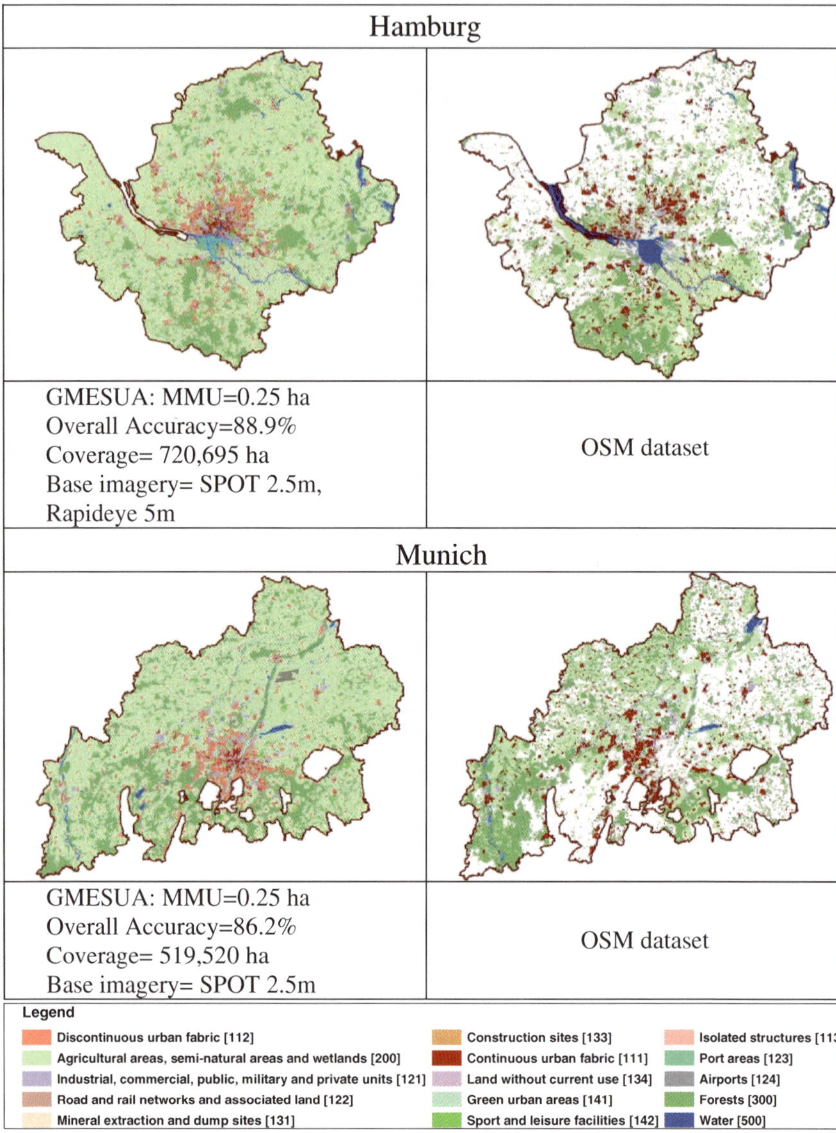

Fig. 1 (continued)

in Sect. 3.1. Third, the OSM features are re-labeled and matched according to the GMESUA nomenclature as will be explained in Sect. 3.2. Fourth, the degree of completeness for each city is determined to measure how much of the area is mapped. Finally, an error matrix between the OSM and GMESUA datasets is computed to measure the overall thematic accuracy of the OSM features along with a detailed per-class analysis accompanied with a map of agreement/disagreement values.

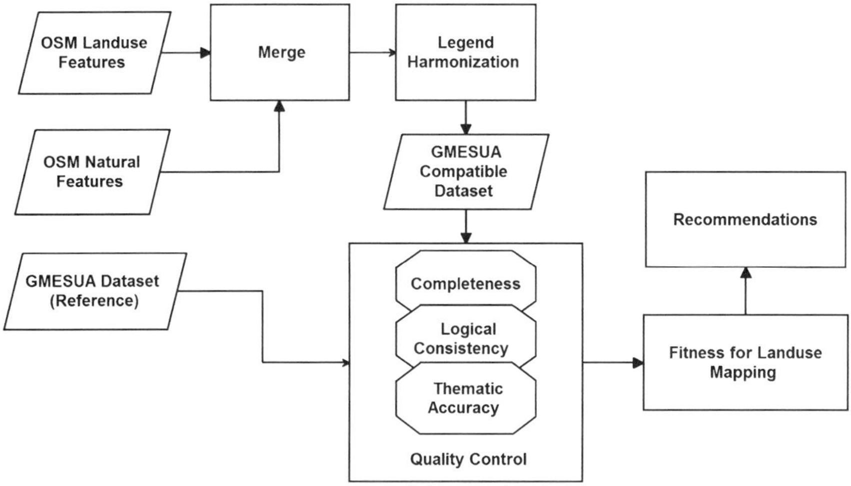

Fig. 2 The flowchart of evaluating OSM land use features

3.1 Logical Consistency and Topology

Logical consistency addresses how well logical relationships between the elements of the dataset are defined. If the objects are not topologically defined to each other, the dataset will fail in having proper internal relation between the objects (van Oort 2006). This issue is even more problematic when the features are polygons. There is no indicator to measure it quantitatively and in the metadata attached to the data is indicated with a Boolean value whether the data set has been cleaned from topological errors or not (Devillers et al. 2007). This concern is a challenge for the collected OSM polygon features, because the OSM contributions are mapped at different zoom levels, which result in dissimilar data scales and some features might overlap each other and consequently some areas possess more than one label (Sester et al. 2014).

However, depending on the type of data, the degree of goodness varies; for instance, this problem is not encountered for point datasets e.g., POIs. In the case of polyline datasets, such topological inconsistency can be observed at the road junctions as well as at the beginning and ending of the road segments. This issue is schematically illustrated in Fig. 3. Therefore, using this layer for any external application generally demands applying topological cleaning of features in order to clean them from overlaps and dangle errors. These issues are resolved automatically by applying topology for removing errors of our dataset including unclosed gaps, gaps between polygons, and overlapping polygons.

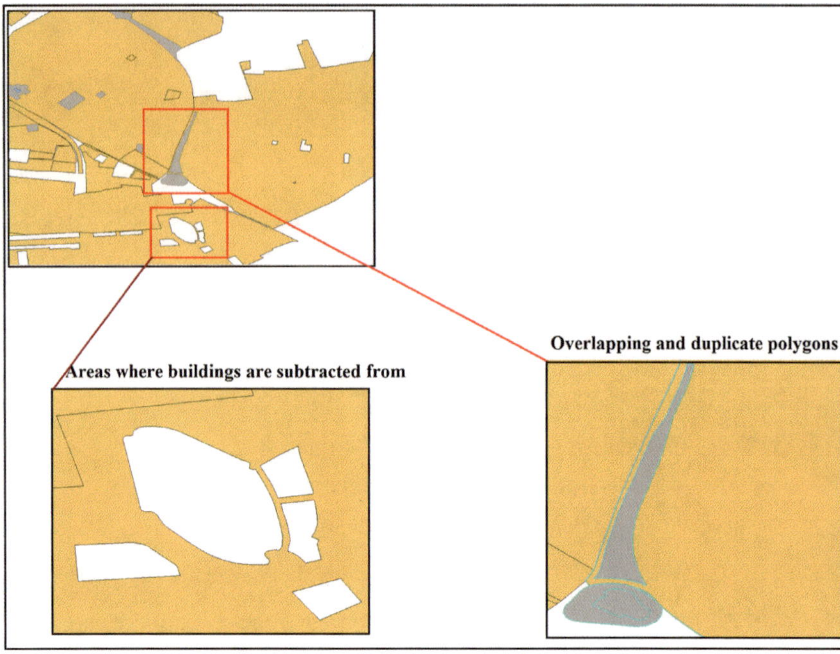

Fig. 3 A sample representation of duplicate polygons (*right-down*) and subtracted areas from land use dataset into building dataset (*left-down*)

3.2 Harmonization of the Datasets Nomenclatures

The OSM has its own classification nomenclature, which is designed for fine-scale LU classification, therefore, the OSM-LU features must be translated to globally known nomenclatures and in this case, the GMESUA nomenclature. This was carried out through adding an additional label compatible with the first and third level of classification of the GMESUA legend for artificial surfaces and non-artificial surfaces, respectively (see Estima and Painho 2013; Jokar Arsanjani et al. 2013). This helps to make a common LU nomenclature and also creates a dictionary for translating the contributions of individuals to the GMESUA adjusted LU types. For instance, water-related features are assigned as water [500]. Table 2 represents an exemplary dictionary of translating OSM features to GMESUA nomenclature.

Several difficulties were encountered such as (i) semantic understanding of terms in domestic language i.e., German terms with typos which had to be translated into English, (ii) the use of unidentified types of features by contributors and the incomplete attributes of features, which were between 10 and 15 % of area and removed from the analysis. This implies that the handling of the heterogeneous and unstructured contributions cannot be automatically handled.

Table 2 Translation of OSM tags (origin) to GMESUA nomenclature (target)

Origin: OSM tags	Target: GMESUA classes [CODE]
Wohngebiet, residential, apartments, residential, dorfgemeinschaft_breitenfurt	Continuous urban fabric [111]
House, hut, villa	Discontinuous urban fabric [112]
Isolated	Isolated structures [113]
Warehouse, university, social, school, sauna, retail, religious, public, power_station, power, place_of_worship, palace, office, museum, mosque, manufacture, kindergarten, Industrial, hotel, hospital, historic, greenhouse, glasshouse, factory, embassy, commercial, clubhouse, club, cinema, café, allotments, cemetery, fortress, greenhouse_horticulture, industrial, military, nursery, wayside_shrine, ruins, monument, monastery, memorial, industrial, grave, city_gate, castle, archaeological_site, tank, water_tower, warehouse, temple, storage_tank, library, church, chapel, cathedral, castle	Industrial, commercial, public, military and private units [121]
Bridge, railway, traffic_island, bicycle_parking, bus_station, fuel, motorcycle parking, parking entrance, parking space, taxi	Road and rail networks and associated land [122]
Ship	Port areas [123]
Airfield, airport	Airports [124]
Wastewater_plant, coal_heap, landfill	Mineral extraction and dump sites [131]
Construction	Construction sites [133]
Collapsed, greenfield, brownfield	Land without current use [134]
Park, grass, nature_reserve, recreation_ground, recreation_ground, zoo	Green urban areas [141]
Swimming_pool, leisure, alpine_hut, artwork, camp_site, caravan_site, information, picnic_site, theme_park, trail_riding_station, viewpoint, swimming_pool	Sport and leisure facilities [142]
Village_green, vineyard, scrub, orchard, meadow, green, grassland, grass, farmyard, farmland, farm, agriculture, agricultural, hühnerfarm	Agricultural areas, semi-natural areas and wetlands [200]
Forest, wood	Forests [300]
Water_basin, water_protected area, pond	Water [500]

3.3 Completeness

According to the published literature (Koukoletsos et al. 2012; Hecht et al. 2013), completeness is the major criterion for using OSM datasets as it is an indicator of how much of the whole has been mapped. Although measuring the degree of completeness for polyline and point features in OSM datasets requires a reference dataset, the completeness for LU features within a certain area can be measured even without having any reference dataset. Because every piece of the land should have an attribute, so the total coverage area is the maximum area to be mapped. The degree of completeness was measured by calculating a completeness index, which calculates the overall area (ha) mapped by contributors out of the whole area (ha) for each individual city.

3.4 Thematic Accuracy

The most important criterion to judge the quality of the contributed LU features is to find out how correctly the land parcels are attributed. This criterion i.e., thematic accuracy, is generally called "accuracy assessment" in the LU/LC classification literature (Congalton 1991; Foody 2002; Foody et al. 2013). The accuracy assessment reflects the difference between the target dataset and the reference dataset. The accuracy assessment process usually summarizes all data in a confusion matrix and reports several indicators such as "overall/per class accuracies", "Kappa index of agreement", "user's accuracy" and "producer's accuracy". In this study, a confusion matrix analysis is applied to achieve these measures. These measures have been the most straightforward and practical statistical tools for checking the degree of match between two thematic datasets as outlined in (Foody 2002; Herold et al. 2008). A measure for the overall accuracy is calculated by dividing the number of identical pixels by the total number of pixels. However, it does not identify how well individual classes between the two datasets match. Hence, the user's accuracy and producer's accuracy should be calculated to measure the accuracy of each class. The user's accuracy indicates the probability that a pixel from the OSM-LU map actually matches the GMESUA dataset, while the producer's accuracy refers to the probability that a specific LU type from the reference dataset is classified as such. These two measurements, typically for any given LU type, are not equal. For instance, if for a specific land type of 'A', with accuracies achieved of 89 and 78 % for user's accuracy and producer's accuracy respectively, it implies that as a user of the data, roughly 89 % of all the pixels classified as A are the same in the reference dataset and, as a producer, only 78 % of all A pixels are classified as such.

4 Results

4.1 Sensitivity to Pixel Size

The Kappa index proposed by Cohen (1960) intends to evaluate the degree of agreement between two or more datasets/observations and consequently provides an overall guide to quality of the map (see Landis and Koch 1977; Foody 2002).

Table 3 Kappa index analysis of the contributed land use features in comparison with GMESUA dataset at different pixel sizes for each city

City	Kappa index (pixel size in meter)				
	5	10	15	20	50
Frankfurt	0.36	0.362	0.361	0.361	0.36
Hamburg	0.402	0.402	0.408	0.403	0.402
Berlin	0.518	0.518	0.525	0.521	0.518
Munich	0.453	0.452	0.455	0.452	0.451

The datasets were converted from vector to raster format at different pixel sizes smaller than 50 m i.e., 5, 10, 15, 20, 30, and 50 m in order to find the most optimal pixel size. This analysis is done and represented in Table 3.

Table 3 demonstrates the computed Kappa indices of agreement between the two datasets at different pixel sizes. The pixel sizes between 10 and 15 m result in a slightly higher degree of match. Consequently, the confusion matrices for the selected cities were designed for the most optimal pixel size applicable for every city at 15 m.

4.2 Degree of Data Completeness

This measure for each city is shown in Fig. 4, which indicates that Berlin has reached the highest degree of completeness, and Frankfurt has the lowest degree of completeness. This value is much higher than the reported value for the Portugal continent at 3 % by Estima and Painho (2013) and confirms how greatly completeness index varies and how heterogeneous the quantity of contributions is. It should be noted that the selected areas consist of urban and rural areas, and the measured completeness index measures the degree of completeness for both regions in which the contributions in urban areas are relatively more than rural areas as illustrated in Fig. 4.

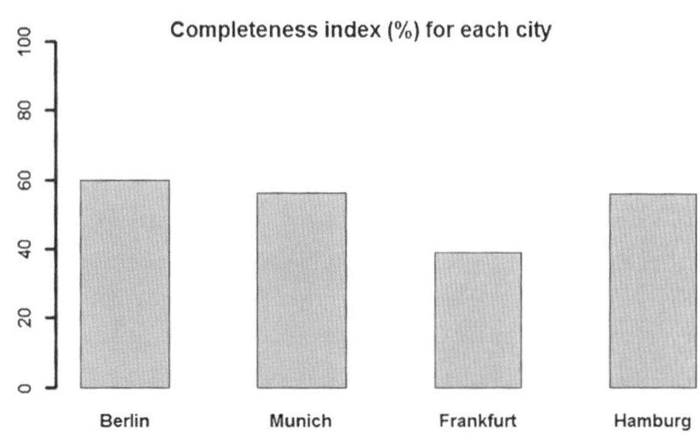

Fig. 4 Calculation of completeness index for each city

4.3 Overall and Per-class Analysis of Thematic Accuracy

As mentioned earlier, the GMESUA dataset is considered as the reference dataset and it will be cross compared with the OSM-LU dataset. In addition to calculating the Kappa index, the overall accuracy of the OSM-LU features, as well as user's and producer's accuracies, are calculated in order to discuss how LU types in each city are contributed as shown in Table 4. Due to heterogeneous accuracies across cities, interpretation of the confusion matrices is discussed for each city separately as follows.

4.3.1 Frankfurt

According to Table 4, among the selected cities the highest Kappa index of 56 % and overall accuracy of 76.5 %. This means that the OSM-LU map and GMESUA map match at a "moderately" rank according to Landis and Koch (1977). Per-class analysis of user's and producer's accuracies reveals that although roughly 98 % of Continuous urban fabric [111] and Water [500] classes have been correctly identified as such, only 20 % of the areas labelled as Continuous urban fabric [111] are actually Continuous urban fabric [111], while 88 % of the areas labelled as Water [500] are actually water [500]. Furthermore, while 71, 62, and 60 % of Forests [300], Industrial, commercial, public, military and private units [121], Agricultural areas + semi-natural areas + wetlands [200] classes have been correctly recognized as such, these classes have been mapped correctly at 94, 77, and 73 % rates. To sum up, the achieved user's accuracies confirm that the Forests [300] and Water [500] classes have a "very high" degree of accuracy with the reference data, while classes like Industrial, commercial, public, military and private units [121], Road and rail networks and associated land [122], and Agricultural areas + semi-natural areas + wetlands [200] are classified as "high" degree. Therefore, it could be concluded that these five classes could be of used for LU mapping purposes at a relatively good level of reliability. Despite the high value of producer's accuracy of Continuous urban fabrics [111] class, its low user's accuracy value (20.3 %) confirms that this class retains as those which are not reliable. On contrary, the contributions to the remaining classes confirm disagreements between the two data sources.

4.3.2 Munich

A Kappa index of 46 % ranked as "moderately" and overall accuracy of 67.1 % for contributed features in Munich are achieved according to Table 4. This means that 67.1 % of contributions are correctly classified. Analysis of the achieved per-class user's accuracies reveals that roughly 100 and 96 % of the Isolated structures [113] and Forests [300] classes have been correctly labelled by contributors, which are ranked as "very high" by Landis and Koch (1977). Moreover, classes such as Industrial, commercial, public, military and private units [121], Agricultural

Table 4 The values of confusion matrices between OSM and GMESUA land use features per city

Land class	City							
	Berlin		Munich		Frankfurt		Hamburg	
	Producer's accuracy (%)	Accuracy (%)	Producer's accuracy (%)	Accuracy (%)	Producer's accuracy (%)	Accuracy (%)	Producer's accuracy (%)	Accuracy (%)
Continuous urban fabric [111]	93.7	5.9	93.2	7.9	98	20	66	11
Discontinuous urban fabric [112]	0	0	0	0	0	0	0	0
Isolated structures [113]	0.1	52.4	0	100	0	0	0	0
Industrial, commercial, public, military and private units [121]	40.8	50.8	56.6	69	62	78	51	66
Road and rail networks and associated land [122]	4.2	66.3	6.5	66	11	72	7.4	79
Port areas [123]	0.0	77	0.0	0.0	0.0	0.0	0.0	0.0
Mineral extraction and dump sites [131]	19.1	0	68.3	0	57	0	61	66
Construction sites [133]	4.4	5.9	0.0	0.0	0.0	0.0	12	22
Land without current use [134]	1.7	0.1	2.2	5.3	0.7	1.2	2.3	4.2
Green urban areas [141]	6.3	41.3	0.8	2	3	4.2	52	7.1
Sport and leisure facilities [142]	54.3	68.9	8.5	49	11	40	36	83
Agricultural areas, semi-natural areas and wetlands [200]	76.3	94.6	45	68	61	74	70	95
Forests [300]	94.2	81.1	69.1	96	71	94	90	71
Water [500]	84.5	86.1	97.7	60	98	89	87	71
Overall accuracy (%)	75.9		67.1		76.5		63.9	
Index (%)	52.3		46		56.2		41	

areas + semi-natural areas + wetlands [200], Road and rail networks and associated land [122], and Water [500] are positioned at the second rank i.e., "high" in this city with values of roughly 69, 67, 66, and 60 %, respectively. The remaining classes possess a low level of reliability to be used for LU mapping, so that they are not recommended for usage.

4.3.3 Berlin

Based on the presented values in Table 4 for Berlin, a Kappa index of 52 % ranked as "moderate" and an overall accuracy of 75.9 % is measured. Per-class analysis of the user's accuracies shows that roughly 94, 86, and 81 % of the Agricultural areas + semi-natural areas + wetlands [200], Water [500], and Forests [300] classes have been correctly labelled by contributors, while classes such as Port areas [123], Sport and leisure facilities [142], and Road and rail networks and associated land [122] have been mapped at 77, 69, and 66 % rates of accuracy, respectively. Therefore, the remaining classes retain at moderate to low level of agreement with the reference dataset.

4.3.4 Hamburg

According to Table 4, among the selected cities the lowest Kappa index of 41 % and overall accuracy of 63.9 % are achieved, which means the mapped LU features to OSM for Hamburg match at a "moderately" rank with the GMESUA dataset. Analysis of the achieved per-class user's accuracies reveals that Agricultural areas + semi-natural areas + wetlands [200], Sport and leisure facilities [142] classes have been mapped at approximately 94 and 83 % rate of match i.e., "very high" with the reference dataset, which are followed by Road and rail networks and associated land [122], Water [500], Forests [300], Industrial, commercial, public, military and private units [121], and Mineral extraction and dump sites [131] classes ranked as "high". The remaining classes lack of sufficient accuracy.

4.4 Spatial Distribution of Agreements and Disagreements

Spatial distribution of agreement and disagreement between the OSM and reference datasets is visualized in Fig. 5. Pink pixels identify areas where the two datasets agree on having the same land type whilst blue pixels indicate areas where the two datasets represent dissimilar LU types. In each city map, a black bounding box marks the clustered urban areas of each city. Generally speaking, depending on the city the Isolated structures [113], Industrial, commercial, public, military and private units [121], Road and rail networks and associated land [122], Sport and leisure facilities [142], Agricultural + semi-natural + wetlands [200], Forests [300], and Water [500] contain the highest level of agreement between the two datasets for

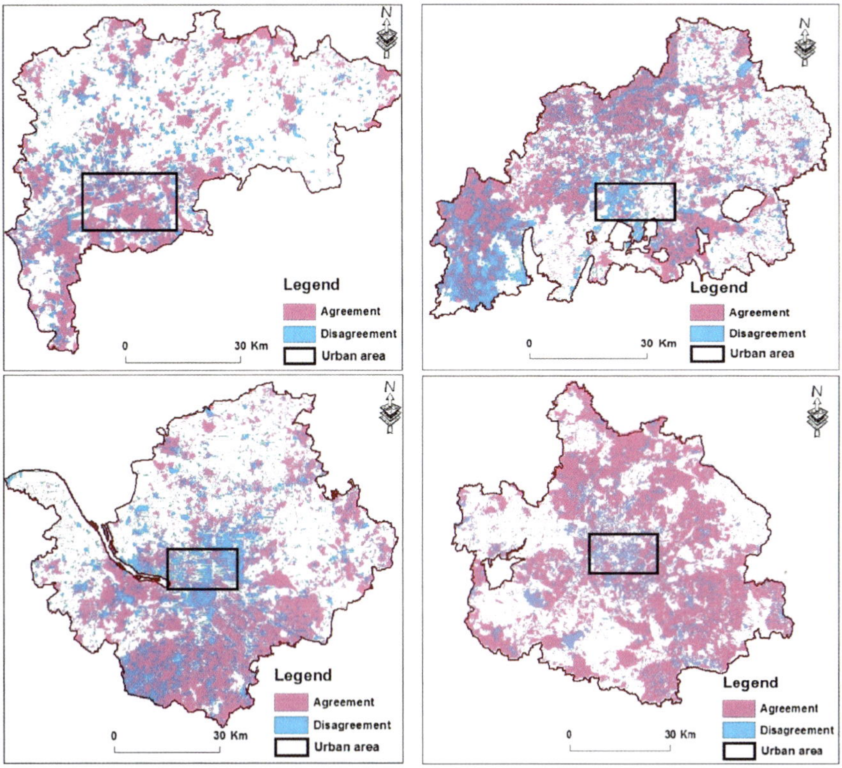

Fig. 5 Spatial distribution of agreement and disagreement between OSM land use features and GMESUA dataset for Frankfurt (*top-left*), Munich (*top-right*), Hamburg (*down-left*), and Berlin (*down-right*). *White* areas indicate unmapped areas

all cities. In contrast, the remaining classes have a large lack of agreement, assuming that they are correctly reflected in the reference dataset. However, it is an invalid statement to relate these mismatches with the certainty of OSM features as they might represent the correct information. Visual analysis of Fig. 5 reveals that in urban areas of Hamburg and Munich the amount of disagreement increases, while in Berlin and Frankfurt a mixture of both are existent. Furthermore, it is concluded that the contributed OSM-LU features are heterogeneously distributed over inside/outside urban areas, however, the density in urban areas is higher.

5 Discussions and Conclusions

The recent rapid emergence of online CMPs, such as OSM, has attracted large numbers of individuals to share their personal knowledge, as well as records from their GPS-enabled devices, with the public. This bottom-up process of collecting individuals' contributions has resulted in the generation of a tremendous amount of geolocated information. This new type of user generated geoinformation has been leveraged new applications such as indoor mapping (Goetz and Zipf 2010), routing applications (Bakillah et al. 2014a) and tourism recommendations. Environmental monitoring through extracting LU features from the VGI are still few. While (Hagenauer and Helbich 2012; Jokar Arsanjani et al. 2013) extracted LU information from VGI at acceptable level of accuracy, the capacities to integrate VGI into LU science have not yet been fully discovered. The findings from the Geo-wiki project in terms of users engagement and reliability of crowdsourced information reveal that although the crowdsourced information from users are reliable enough, the number of involved users is not significant. Although the question on how to attract users and how to keep them active in the crowdsourcing process is still unanswered, OSM has so far attracted around 1.7 million users. More importantly, quality analysis of the contributed streets network (Hochmair et al. 2014), buildings footprints (Fan et al. 2014), and POIs (Bakillah et al. 2014b) in OSM proves the promising avenues of using OSM features for multiple applications. Therefore, a huge potential in OSM exists to be explored. Thus, in this study, we comparatively evaluated the completeness and thematic accuracy aspects of the contributed OSM-LU features in four large German metropolitan areas (Berlin, Frankfurt, Hamburg, and Munich) and see how reliable we could start using them.

Some of the lessons learned from this investigation are as follows. The analysis of completeness index reveals that, between 40 and 60 % of the selected areas are mapped. From a logical consistency perspective, in some cases overlapping polygons and topological issues are found that cause additional data processing. Furthermore, the existing differences in LU legends highlight an obvious inconsistency that makes harmonizing LU datasets difficult. From a thematic accuracy viewpoint, the contributed features have, in general, a "moderate" rank of Kappa indices and their overall accuracies are between 63 and 77 %. It must be noted that the overall accuracy of GMESUA datasets barely exceeds 90 % and, therefore, the computed accuracies are noticeable at the current stage of OSM. Per-class analysis of the LU types show that, depending on the city, Isolated structures [113], Industrial, commercial, public, military and private units [121], Road and rail networks and associated land [122], Sport and leisure facilities [142], Agricultural + semi-natural + wetlands [200], Forests [300], and Water [500] reach substantial level of accuracies, which implies that these classes are highly exploitable. It is worth mentioning that integrating ground-truth information with other reference data for accuracy assessment could be highly beneficial for producing hybrid LU products.

According to European Union (2011), archived images from 2005 until 2010 have been used for LU mapping and this could have caused a major source of disagreement, whereas the OSM-LU contributions have mainly been uploaded after 2009, and therefore, some information from OSM might be even more accurate than the reference data. Moreover, the MMU of the GMESUA datasets is 0.25–1 ha and, therefore, land parcels smaller than this MMU are neglected from mapping, while in OSM even smaller parcels are mapped, i.e., a smaller MMU in OSM is achieved. This means that in some parts while a polygon in GMESUA dataset is representing a specific LU type, the same area in OSM-LU dataset is covered by multiple small polygons representing different polygons. From the mappers' viewpoint, the citizens' perception of LU types should be further investigated to see how people visually interpret LU types.

As a final conclusion, the OSM-LU features suggest a promising input source of updating LU inventories and also LU mapping. Certainly, the longer OSM is in existence, more contributions will be collected and the higher accuracy LU maps can be retrieved.

6 Recommendations

This study attempts to draw some recommendations that will lead future research possibly in the interesting directions. The following recommendations are proposed to the environmentalists and LU researchers based on the findings in this paper. According to the indicated completeness indices of the four cities, as well as the accuracy statistics, the contributed LU features to OSM in the selected areas account for a potential alternative data source for mapping LU features. However, further studies on other areas must be conducted to see the heterogeneity of completeness and thematic accuracy across space. Furthermore, applying data mining techniques as well as data fusion with other available datasets GMESUA for extracting the LU features of incomplete areas are of high importance. Additionally, the land types with the highest accuracies can be separately incorporated for respective applications.

This investigation of the OSM-LU features will be posted on the OSM blogs, and mailing and discussion lists, to inform the OSM community about current academic progress in the area of LU features. This enables experts to: (a) possibly find ways to draw the attention of mappers to LU features, (b) determine the possible existing problems with the OSM ontology of the LU dataset, or (c) it is the case that users are not able to add further features in the urban areas, because the volume of mapped objects (e.g., POIs, roads, building, etc.) do not let users to add new features including LU polygons, or (d) ultimately the LU attributes are not easily understandable for people and they have insufficient interest in mapping them.

Acknowledgments Jamal Jokar Arsanjani acknowledges the funding of the Alexander von Humboldt foundation.

References

Arino O, Ramos Perez JJ, Kalogirou V, Bontemps S, Defourny P, Van Bogaert E (2012) Global land cover map for 2009 (GlobCover 2009)

Bakillah M, Lauer J, Liang SHL, Zipf A, Jokar Arsanjani J, Mobasheri A, Loos L (2014a) Exploiting big VGI to improve routing and navigation services. Big Data Tech Technol Geoinformatics 177–192

Bakillah M, Liang S, Mobasheri A, Jokar Arsanjani J, Zipf A (2014b) Fine-resolution population mapping using OpenStreetMap points-of-interest. Int J Geogr Inf Sci 28: 1940–1963

Birringer J (2008) Eye into Earth. Space Culture 11:59

Büttner G, Feranec J, Gabriel J (2002) Corine land cover update 2000

Castelein W, Grus Ł, Crompvoets J, Bregt AA (2010) Characterization of volunteered geographic information. In: 13th AGILE international conference on geographic information science 2010, Guimarães, pp 1–10

Cihlar J, Jansen LJM (2001) From land cover to land use: a methodology for efficient land use mapping over large areas. Prof Geogr 53:275–289

Cohen J (1960) A coefficient of agreement for nominal scales. Educ Psychol Meas 20:37–46

Comber A, See L, Fritz S, Van der Velde M, Perger C, Foody G (2013) Using control data to determine the reliability of volunteered geographic information about land cover. Int J Appl Earth Obs Geoinf 23:37–48

Congalton RG (1991) A review of assessing the accuracy of classifications of remotely sensed data. Remote Sens Environ 37:35–46

De Leeuw J, Said M, Ortegah L, Nagda S, Georgiadou Y, DeBlois M (2011) An assessment of the accuracy of volunteered road map production in western Kenya. Remote Sens 3:247–256

De Sherbinin A (2002) A CIESIN thematic guide to land land-use and land land-cover change (LUCC), NY, pp 10–20

Devillers R, Bédard Y, Jeansoulin R, Moulin B (2007) Towards spatial data quality information analysis tools for experts assessing the fitness for use of spatial data. Int J Geogr Inf Sci 21:261–282

Ellis E (2007) Land-use and land-cover change. Earth

Estima J, Painho M (2013) Exploratory analysis of OpenStreetMap for land use classification. In: Proceedings of the second ACM SIGSPATIAL international workshop on crowdsourced and volunteered geographic information GEOCROWD'13, ACM, New York pp 39–46

European Union (2011) Mapping guide for a european urban atlas

Fan H, Zipf A, Fu Q, Neis P (2014) Quality assessment for building footprints data on OpenStreetMap. Int J Geogr Inf Sci 28:700–719

Flanagin AJ, Metzger MJ (2008) The credibility of volunteered geographic information. GeoJournal 72:137–148

Foody GM (2002) Status of land cover classification accuracy assessment. Remote Sens Environ 80:185–201

Foody GM, See L, Fritz S, Van der Velde M, Perger C, Schill C, Boyd DS (2013) Assessing the accuracy of volunteered geographic information arising from multiple contributors to an internet based collaborative project. Trans GIS 17(6):847–860

Fritz S, Bartholomé E, Belward A, Hartley A, Stibig HJ, Eva H, Mayaux P, Bartalev S, Latifovic R, Kolmert S et al (2003) Harmonisation, mosaicing and production of the global land cover 2000 database (beta version). Office for Official Publications of the European Communities Luxembourg, Luxembourg

Fritz S, Mccallum I, Schill C, Perger C, See L, Schepaschenko D, van der Velde M, Kraxner F, Obersteiner M (2012) Geo-Wiki: an online platform for improving global land cover. Environ Model Softw 31:110–123

Gervais M, Bédard Y, Levesque M, Bernier E, Devillers R (2009) Data quality issues and geographic knowledge discovery. Geogr Data Min Knowl Discov pp 99–115

Goetz M, Zipf A (2010) Extending OpenStreetMap to indoor environments : bringing volunteered geographic information to the next level. CRC Press, Delft

Goodchild MF (2007) Editorial: citizens as voluntary sensors: spatial data infrastructure in the world of web 2.0 2: 24–32

Guptill SC, Morrison JL (1995) Elements of spatial data quality. Elsevier Science, Oxford

Hagenauer J, Helbich M (2012) Mining urban land-use patterns from volunteered geographic information by means of genetic algorithms and artificial neural networks. Int J Geogr Inf Sci 26:963–982

Haklay M (2010) How good is volunteered geographical information? A comparative study of OpenStreetMap and ordnance survey datasets. Environ Plan B Plan Des 37:682–703
Hecht R, Kunze C, Hahmann S (2013) Measuring completeness of building footprints in OpenStreetMap over space and time. ISPRS Int J Geo-Information 2:1066–1091
Helbich M, Amelunxen C, Neis P (2012) Comparative spatial analysis of positional accuracy of OpenStreetMap and proprietary Geodata. In: International GI_Forum, Salzburg
Herold M, Mayaux P, Woodcock CE, Baccini A, Schmullius C (2008) Some challenges in global land cover mapping : an assessment of agreement and accuracy in existing 1 km datasets. Remote Sens Environ 112:2538–2556
Hochmair HH, Zielstra D, Neis P (2014) Assessing the completeness of bicycle trail and lane features in OpenStreetMap for the United States. Trans GIS. doi:10.1111/tgis.12081
Jokar Arsanjani J, Helbich M, Bakillah M, Loos L (2015a) The emergence and evolution of OpenStreetMap: a cellular automata approach. Int J Digital Earth 8(1):74–88. http://www.tandfonline.com/doi/abs/10.1080/17538947.2013.847125
Jokar Arsanjani J, Vaz E (2015b) An assessment of a collaborative mapping approach for exploring land use patterns for several European metropolises. Int J Appl Earth Obs Geoinf 35:329–337
Jokar Arsanjani J, Helbich M, Bakillah M, Hagenauer J, Zipf A (2013) Toward mapping land-use patterns from volunteered geographic information. Int J Geogr Inf Sci 27:2264–2278
Jokar Arsanjani J, Mooney P, Helbich M, Zipf A (2015c) An exploration of future patterns of the contributions to OpenStreetMap and development of a Contribution Index, Trans GIS
Jokar Arsanjani J, Vaz E, Bakillah M, Mooney P (2014) Towards initiating OpenLandMap founded on citizens' science: the current status of land use features of OpenStreetMap in Europe. In: Huerta Schade G (ed) Proceedings of the AGILE'2014 international conference on geographic information science, 3–6 June 2014, AGILE digital editions, Castellón
Kandrika S, Roy PSS (2008) Land use land cover classification of Orissa using multi-temporal IRS-P6 Awifs data: a decision tree approach. Int J Appl Earth Obs Geoinf 10:186–193
Kasetkasem T, Arora MK, Varshney PK (2005) Super-resolution land cover mapping using a markov random field based approach. Remote Sens Environ 96:302–314
Kong F, Yin H, Nakagoshi N, James P (2012) Simulating urban growth processes incorporating a potential model with spatial metrics. Ecol Indic 20:82–91
Koukoletsos T, Haklay M, Ellul C (2012) Assessing data completeness of VGI through an automated matching procedure for linear data. Trans GIS 16(4):477–498
Landis JR, Koch GG (1977) The measurement of observer agreement for categorical data. Biometrics 33:159–174
Ludwig I, Voss A, Krause-Traudes M (2011) A comparison of the street networks of Navteq and OSM in Germany. In: Geertman S, Reinhardt W, Toppen F (eds) Advancing geoinformation science for a changing world SE-4 lecture notes in Geoinformation and cartography. Springer, Berlin, pp 65–84
Mayaux P, Eva H, Gallego J, Strahler AH, Herold M, Member S, Agrawal S, Naumov S, De Miranda EE, Di Bella CM et al (2006) Validation of the global land cover 2000 map. IEEE Trans Geosci Remote Sens 44:1728–1739
McIver D, Friedl M (2002) Using prior probabilities in decision-tree classification of remotely sensed data. Remote Sens Environ 81:253–261
Mooney P, Corcoran P (2012) The Annotation Process in OpenStreetMap. Trans GIS 16:561–579
Pacifici F, Chini M, Emery WJ (2009) A neural network approach using multi-scale textural metrics from very high-resolution panchromatic imagery for urban land-use classification. Remote Sens Environ 113:1276–1292
Paneque-Gálvez J, Mas J-F, Moré G, Cristóbal J, Orta-Martínez M, Luz AC, Guèze M, Macía MJ, Reyes-García V (2013) Enhanced land use/cover classification of heterogeneous tropical landscapes using support vector machines and textural homogeneity. Int J Appl Earth Obs Geoinf 23:372–383
Qi Z, Yeh AG-O, Li X, Lin Z (2012) A novel algorithm for land use and land cover classification using RADARSAT-2 polarimetric SAR data. Remote Sens Environ 118:21–39

Ramm F (2014) OpenStreetMap data in layered GIS format http://www.geofabrik.de/data/geofabrik-osm-gis-standard-0.6.pdf

Ramm F, Names I, Files SS, Catalogue F, Features P, Features N, Related T, Infrastructure T, Generation P, Features L, et al (2011) OpenStreetMap data in layered GIS format pp 1–21

Roick O, Hagenauer J, Zipf A (2011) OSMatrix—grid-based analysis and visualization of OpenStreetMap. In: State of the map EU 2011, Vienna, Austria

Rouse LJ, Bergeron SJ, Harris TM (2007) Participating in the geospatial web: collaborative mapping, social networks and participatory GIS. In: Scharl A, Tochtermann K (eds) The geospatial web advanced information and knowledge processing. Springer, London, pp 153–158

Saadat H, Adamowski J, Bonnell R, Sharifi F, Namdar M, Ale-Ebrahim S (2011) Land use and land cover classification over a large area in iran based on single date analysis of satellite imagery. ISPRS J Photogramm Remote Sens 66:608–619

See L, Comber A, Salk C, Fritz S, van der Velde M, Perger C, Schill C, McCallum I, Kraxner F, Obersteiner M (2013) Comparing the quality of crowdsourced data contributed by expert and non-experts. PLoS ONE 8:e69958

Seifert F (2009) Improving urban monitoring toward a European urban atlas. In: Global mapping of human settlement; remote sensing applications series. CRC Press, USA

Sester M, Jokar Arsanjani J, Klammer R, Burghardt D, Haunert J-H (2014) Integrating and generalising volunteered geographic information. In: Burghardt D, Duchêne C, Mackaness W (eds) Abstracting geographic information in a data rich world, in series: lecture notes in geoinformation and cartography. Springer, Berlin, pp 119–155

Sexton JO, Urban DL, Donohue MJ, Song C (2013) Long-term land cover dynamics by multi-temporal classification across the landsat-5 record. Remote Sens Environ 128:246–258

Strahler AH, Boschetti L, Foody GM, Friedl MA, Hansen MC, Herold M, Mayaux P, Morisette JT, Stehman SV, Woodcock CE (2006) Global land cover validation: recommendations for evaluation and accuracy assessment of global land cover maps. Office for Official Publications of the European Communities, Luxemburg

Thenkabail PS, Schull M, Turral H (2005) Ganges and Indus river basin land use/land cover (LULC) and irrigated area mapping using continuous streams of MODIS data. Remote Sens Environ 95:317–341

Van Oort P (2006) Spatial data quality: from description to application. Wageningen University

Vaz E, Nijkamp P, Painho M, Caetano M (2012) A multi-scenario forecast of urban change: a study on urban growth in the Algarve. Landsc Urban Plan 104:201–211

Vaz E, Walczynska A, Nijkamp P (2013) Regional challenges in tourist wetland systems: an integrated approach to the Ria Formosa in the Algarve, Portugal. Reg Environ Change 13:33–42

Wästfelt A, Arnberg W (2013) Local spatial context measurements used to explore the relationship between land cover and land use functions. Int J Appl Earth Obs Geoinf 23:234–244

Improving Volunteered Geographic Information Quality Using a Tag Recommender System: The Case of OpenStreetMap

Arnaud Vandecasteele and Rodolphe Devillers

Abstract Studies have analyzed the quality of volunteered geographic information (VGI) datasets, assessing the positional accuracy of features and the semantic accuracy of the attributes. While it has been shown that VGI can, in some contexts, reach a high positional accuracy, these studies have also highlighted a large spatial heterogeneity in positional accuracy and completeness, but also concerning the semantics of the objects. Such high semantic heterogeneity of VGI datasets becomes a significant obstacle to a number of possible uses that could be made of the data. This paper proposes an approach for both improving the semantic quality and reducing the semantic heterogeneity of VGI datasets. The improvement of the semantic quality is achieved by using a tag recommender system, called OSMantic, which automatically suggests relevant tags to contributors during the editing process. Such an approach helps contributors find the most appropriate tags for a given object, hence reducing the overall dataset semantic heterogeneity. The approach was implemented into a plugin for the Java OpenStreetMap editor (JOSM) and different examples illustrate how this plugin can be used to improve the quality of VGI data. This plugin has been tested by OSM contributors and evaluated using an online questionnaire. Results of the evaluation suggest a high level of satisfaction from users and are discussed.

Keywords Volunteered geographic information (VGI) · Semantic similarity · Data quality · OpenStreetMap (OSM)

A. Vandecasteele (✉) · R. Devillers
Department of Geography, Memorial University of Newfoundland,
St. John's, NL A1B 3X9, Canada
e-mail: a.vandecasteele@mun.ca

R. Devillers
e-mail: rdeville@mun.ca

J. Jokar Arsanjani et al. (eds.), *OpenStreetMap in GIScience*,
Lecture Notes in Geoinformation and Cartography,
DOI 10.1007/978-3-319-14280-7_4

1 Introduction

With the increasing availability of geolocation devices and the development of more user-friendly software, professional expertise in cartography is no longer necessary to collect, publish and share geographic information (GI) (Goodchild and Li 2012; Haklay 2013). These developments have led to the emergence of Volunteered Geographic Information (VGI), geographic user-generated content described by Goodchild (2007) as resulting from "the widespread engagement of large numbers of private citizens, often with little in the way of formal qualifications, in the creation of geographic information." VGI opened the door to an alternative to traditional data producers by providing citizens with tools and web services to describe and share their geographic knowledge (Flanagin and Metzger 2008).

VGI datasets are increasingly used as basemaps by different online mapping products and Web mapping applications. However, VGI data are still far being used at their real potential by mapping companies, national mapping agencies and many users due, amongst other things, to the absence of detailed information about data quality and the difficulty of assessing quality using traditional geographic informatics systems approaches (Elwood et al. 2012; Flanagin and Metzger 2008). Whilst quality evaluations made by several studies (Girres and Touya 2010; Haklay 2010; Ludwig et al. 2011) have documented relatively high overall positional accuracy of VGI data, they have also identified challenges. In terms of coverage, some areas can be very well mapped while others miss most features (Goodchild 2007). In terms of attributes, also called "tags" in some VGI projects, their semantic can be unclear and their quality can vary spatially and temporally (Ballatore et al. 2013). Different names (e.g., "forest" and "wood") can be used to represent similar geographic phenomena (synonymy) while a same name can be used to describe different geographic phenomena (polysemy) (Ballatore et al. 2013; Mooney and Corcoran 2012a). The nature of VGI compared to typical cartographic products makes the use of traditional data quality evaluation methods difficult, paving the way to new data quality metrics (Keßler and de Groot 2013; van Exel and Dias 2011). While positional accuracy assessments of VGI have received significant attention, fewer efforts have looked at the semantic quality of VGI (Mooney and Corcoran 2012a). Moreover, no studies have attempted to improve the quality of VGI during the editing process.

While most studies looked at measuring the quality of VGI data in comparison to authoritative datasets, this paper proposes to enhance the quality of VGI data by improving the user experience during the contributing process. This is done using a recommender system that automatically suggests tags that could be added to better describe map features, and by detecting tags associated to the same feature that appear too dissimilar. Relationships between tags are computed based on the semantic similarity between the tags and the number of times a tag has been used in the database. This approach has been implemented and tested by developing a plugin for an editing tool for the OpenStreetMap[1] (OSM) project. Started in 2004,

[1] OpenStreetMap Website: http://openstreetmap.org/ (Accessed 3 October 2014).

OpenStreetMap is the most popular example of a VGI project, aiming to create a free geospatial database of the world crowdsourced by volunteers. Since its beginning, OSM has grown rapidly. At the time of writing this paper, the OSM project involved more than 1.8 million volunteers, has more than 4.2 billion global positioning system (GPS) points, 2.5 billion nodes and 254 million ways (i.e., polylines and polygons),[2] making it the largest global crowdsourced geographic vector dataset ever created.

This paper is structured as follows: Sect. 2 presents relevant work about quality evaluation of VGI datasets. Section 3 describes sources of the semantic heterogeneity. Section 4 presents the prototype and some examples of its use. In Sect. 5, results of the evaluation of the plugin are discussed. The last section outlines our conclusions and future plans.

2 Volunteered Geographic Information Quality

VGI allows people with little knowledge in geographic information to contribute to the creation of maps that are made publicly available. In a context where mapping was traditionally conducted by professional cartographers, such a change has been met with significant initial skepticism, leading to a number of academic studies analyzing VGI data quality. However, the nature of VGI makes classical data quality elements (e.g., completeness and positional accuracy) and their measurements, as defined by the ISO/TC 211 standards on data quality (ISO/TC 211 2002), of limited use in some cases. New metrics for assessing external quality of VGI may have to be proposed to capture specific properties of VGI data. Some of the possible missing quality elements include measurements of the spatial and semantic data heterogeneity, a problem that did not exist in such extent with most traditional maps that had precise specifications allowing the production of relatively homogeneous data products.

Despite the relative novelty of VGI, many research studies have analyzed the quality of OSM datasets in different countries and contexts. One of the popular and classical approaches used to assess the overall quality of VGI datasets is to compare VGI to an authoritative dataset of the same area that acts as ground-truthing data. These comparisons used different measurements, such as differences in features' length or attribute values. These analyses have been conducted in different countries such as England (Haklay 2010), France (Girres and Touya 2010), and Germany (Ludwig et al. 2011). Results indicate that populated places can reach higher positional accuracy and data completeness than the authoritative dataset, while less populated places tend to be characterized by lower accuracy and completeness. A challenge with the use of such traditional data quality assessment methods is that VGI datasets are now, in many parts of the world, more complete and accurate than

[2] OpenStreetMap Statistics: http://wiki.openstreetmap.org/wiki/Stats (Accessed 3 October 2014).

authoritative datasets, violating the basic assumptions of those quality assessment methods. To overcome these challenges, recent approaches explored the possibility of assessing the quality of VGI data using intrinsic VGI properties such as the number of edits, the number of contributors, or a combination of them (Haklay et al. 2010; Keßler and de Groot 2013; Mooney and Corcoran 2012b). By taking into account the evolving nature and the collaborative aspects of VGI, these approaches allow for a better assessment of VGI data quality. Based on our literature review and on the work of Goodchild and Li (2012), these newest quality evaluation methods can be summarized in three main approaches: data-centric, user-centric, and context-centric.

2.1 Data-Centric Approach

Unlike traditional mapping activities, VGI results from collaborative mapping processes. To manage the collaborative aspect of maps' edition, most VGI projects use a version control system (VCS) that records all the changes (e.g., creation, edition, deletion) made to geographic objects. VCS allows retrieving and if necessary reverting back to a previous version of the objects. The data-centric approach takes advantage of this VCS to compute quality metrics.

Data-centric quality assessment methods were at first focused on the analysis of a single parameter. For example, Haklay et al. (2010) analyzed the relationship between the number of contributors and the positional accuracy of the data, as objects in the database can be updated by many contributors. Results showed that high accuracy is achieved when there are at least 15 contributors per square kilometer. While positional accuracy is important in a map, having accurate and complete attributes is also of major importance for users. Mooney and Corcoran (2012b) analyzed how heavily edited object (i.e., more than 15 versions) are tagged in the OSM database. As suggested by the authors, such heavily edited objects could gain the status of 'featured' objects and could be recognized for their quality.

Recent studies used a combination of indicators to provide a better estimation of the quality of VGI datasets. For example, Keßler and de Groot (2013) assessed the quality of OSM objects using a combination of five main parameters. Three of these parameters (i.e., number of versions, number of users, and confirmations) have a positive influence on trustworthiness, while the other two (i.e., tags corrections and rollbacks) have a negative influence. Results from this analysis showed that more than 75 % of the selected features were correct when compared to ground-truthing data. More recently, Barron et al. (2014) created an extensible Open Source framework for evaluating the quality of OSM data without relying on any other authoritative dataset. This Open Source framework, called iOSMAnalyzer, uses the intrinsic characteristics of VGI data to compute more than 25 data quality indicators.

2.2 User-Centric Approach

Instead of deriving quality metrics from VGI data, the user-centric approach assesses the quality of VGI by evaluating contributors using different parameters such as contributors' motivations, experience, or recognition (Rehrl et al. 2013). For example, if a contributor already has edits that have resulted in no or few modifications by other contributors in an active contributing area, this contributor could be identified as trustable. Such assessments could be easily applied to OSM, due to the fact that more than 80 % of edits appear to be made by about 10 % of contributors (Bégin et al. 2013; Mooney and Corcoran 2012b).

Several studies analyzed the motivations of VGI contributors using different parameters, such as their goals, or their relations with other users (Budhathoki et al. 2010; Coleman et al. 2009; Haklay 2013; Neis and Zipf 2012), but these studies are often too generic to assess VGI data quality. One of the most interesting approaches is the one proposed by van Exel and Dias (2011) who coined the concept of 'crowd quality'. The crowd quality indicator is based both on an analysis of the users (e.g., experience, recognition) and on the data. Nevertheless, the relationship between the results of these analyses and VGI data quality has not yet been explored.

2.3 Context-Centric Approach

Studying the geographic context in which map elements are found can be used to enhance the quality of VGI data, although little work has been done using such an approach. For example, adding a pub in a place surrounded by nursing homes or in the middle of a park could be automatically identified as a potential error (Goodchild and Li 2012; Mülligann et al. 2011). Such an approach relies on the knowledge of the context in which geographic information exists, being related to Tobler's first law of geography stating that "all things are related, but nearby things are more related than distant things" (Tobler 1970). Nevertheless, this approach could require in practice an explicit schema that links the different geographical concepts. These links and the distance between concepts would allow measuring relationships between sets of geographical concepts (e.g., pubs and nursing homes) and possibly help identify potential errors.

In OpenStreetMap, two main projects have started exploring the possibility of creating a semantic schema of the OSM concepts by organizing them into an ontology. The first ontology, LinkedGeoData, aims to provide a geographic dataset for the semantic web. LinkedGeoData enriches OSM data with other semantic datasets such as DBPedia[3] or GeoNames[4] and then publishes them in a semantic form. LinkedGeoData contains more than 1 billion nodes, 100 million ways and

[3] DBPedia Website: http://dbpedia.org/ (Accessed 3 October 2014).

[4] GeoNames Website: http://www.geonames.org/ (Accessed 3 October 2014).

approximately 20 billion triples (Auer et al. 2009). Despite the advantages of LinkedGeoData, this ontology is a simple tree structure where objects are linked using relationships of the form "is_a" (Ballatore et al. 2013). Moreover, due to the size of the OSM database, keeping good scalability is still a major challenge (Auer et al. 2009). The second ontology, called OSMonto, has been designed for use in a navigation web service (Codescu et al. 2011). Compared to LinkedGeoData, OSMonto provides a richer description of relationships between the OSM tags (e.g., "hasCuisine" relationships for a seafood restaurant). However, not all OSM concepts have been implemented for performance and stability reasons. Moreover, as OSMonto has been designed for a specific application, reusing it in a generic context can be difficult. Despite the potential of these approaches, none has been officially integrated into the OSM infrastructure.

In addition to these ontological approaches, two main projects have looked at the problem of semantics in VGI. The first project conducted by Mülligann et al. (2011) combined semantic similarities with a point pattern analysis in a two-step method to study the geography of OSM semantics. First, the semantic similarity is computed from the OSM dataset by looking at the spatial co-occurrence of features. Second, spatial–semantic patterns are identified indicating potential correlations between two different types of geographic objects or between two geographic objects of the same type. For example, the addition of a new fire station close to an existing one (i.e., average distance between these two features below the spatial–semantic pattern of this feature) could be detected as a potential error. The second project proposed by Ballatore et al. (2013) used a semantic network created from the extraction of the relationships between OSM concepts in the OSM Wiki. This extraction is performed using the OSM Wiki Crawler developed by the authors. In this semantic network, OSM concepts are represented as vertices and edges between these concepts are computed by measuring the relatedness between the Wiki pages. Using this semantic network, different similarity measures (e.g., SimRank, P-Rank) based on the co-citation algorithm are tested and a semantic similarity score between two concepts was computed. This semantic network has been used for the development of the tag recommender system presented in this paper.

3 Semantic Heterogeneity of OpenStreetMap Dataset

3.1 Sources of Semantic Heterogeneities

In OpenStreetMap, real-world phenomena are described by associating a set of attributes to geographic primitives (Fig. 1). In the current OSM application programming interface (API 0.6), four types of geometric primitives can be created: point (often described as "node" in the OSM vocabulary), line (often described as "way"), polygon and relation. Attributes are described using structured key/value pairs called "tags", where the "key" is similar to an attribute (e.g., "amenity") and "value" is the value of this attribute for the given geometric object (e.g., "cafe").

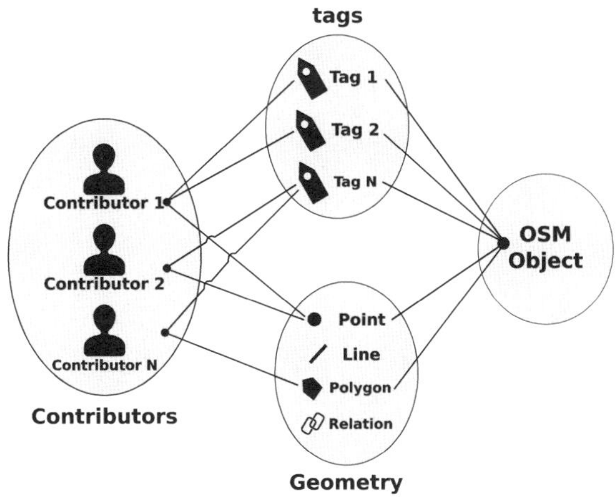

Fig. 1 Example of a schematic depiction of the OSM collaborative contributing process

To be valid, an OpenStreetMap object must have at least one tag, although in practice, precisely describing an OSM object often requires using several tags. For example, a traffic signal can be generally described by the tag "highway = traffic_signals". However, other tags could be added to extend the description of the object, such as a tag "tactile_paving = yes/no" for traffic signals equipped with textured ground surface indicators to assist visually impaired people. These tags are the result of informal and continuous discussions within the community and the most popular tags are described in the OSM Wiki.[5] However, there is no restriction on the use of tags, and contributors can freely create their own tags when necessary (Mooney and Corcoran 2012a).

While such a flexible collaborative approach allows for a rich description of geographic objects that can capture local specificities, it also creates significant geometric and semantic heterogeneities. Semantic heterogeneity refers in this paper to the diversity and amount of attributes/tags used to describe a same object in a VGI dataset. For example, for the geometric component, many objects (e.g., a "university") can be represented by either a point or a polygon. Polygons boundaries can themselves be subject to interpretation depending on how features' boundaries are easy to identify on the ground. On the semantic side, with more than 40,000 tags currently recorded in the OSM database, the sources of semantic heterogeneity are possibly more complex. Contributors cannot possibly know all of the tags and knowing a large number of common tags requires a certain experience.

From our analyses, three main sources of semantic heterogeneity can be identified and were conceptualized in Fig. 2. The first one relates to the nature of the concept itself. Some geographic concepts can result in more ambiguity in their classification than others, being more difficult to conceptualize and hence possibly leading to a lack of consensus in the VGI community. In OpenStreetMap, this is the case for

[5] OSM Wiki: http://wiki.openstreetmap.org/wiki/Map_Features (Accessed 3 October 2014).

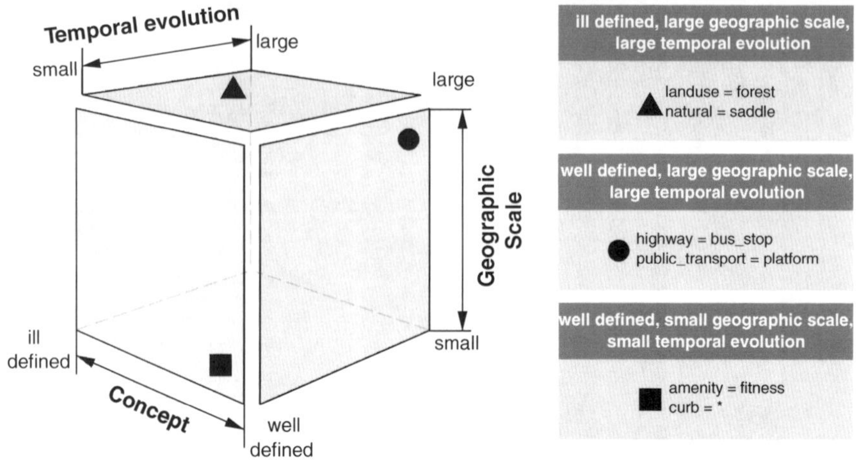

Fig. 2 Framework capturing three main sources of heterogeneity in VGI projects

example for the tags "landuse = forest" and "natural = wood" (Ballatore et al. 2013) or the tag "amenity = fitness" that have more than 10 other similar concepts (e.g., "amenity = fitness_center", "amenity = gym", "amenity = sports_center", "leisure = fitness"). The result of this semantic ambiguity is that contributors tend to use tags interchangeably depending on their mapping habits. The second main source of semantic heterogeneity comes from the geographic scale a concept is used at. Some concepts are only used in specific places (or used in a specific way in some places), while others are used all around the world. Depending on the conceptual representation of a geographic object, the meaning of those concepts or the way people describe them can vary (Smith and Mark 2003). For example, in Newfoundland, Canada, most natural water areas are named as "ponds" and not "lakes", while "ponds" in OSM refer to areas of water created by human activity. A local contributor could misuse such a tag based on its local meaning. The same applies for the OSM key "kerb" that people sometimes describe as "curb". While such a difference in the way contributors describe features locally can be a source of semantic heterogeneity globally, it can enrich semantics locally. The third main source of semantic heterogeneity (i.e. third axis on Fig. 2) comes from the temporal evolution of tags use and definition. VGI projects can exist for years (e.g., OSM has existed since 2004) and the use of tags can change over the years. A given tag use can decrease when replaced by a more appropriate tag, tag definitions can evolve and tags can even be deleted.[6] For example, while the tag "highway = bus_stop" has been used to describe places where passengers wait for a bus, new OSM

[6] OSM Wiki Proposed Features "Abandoned": http://wiki.openstreetmap.org/wiki/Category:Proposed_features_%22Abandoned%22 (Accessed 3 October 2014).

recommendations advocate for the use of "public_transport = platform". As OSM map objects have been created since its inception, the present OSM dataset can include objects that are described using tags not used in the same way nowadays.

3.2 OpenStreetMap Tag Distribution

Analyzing the distribution of tags can provide useful information on the semantic heterogeneity of the OSM database. The tagging system in OpenStreetMap is based on human input where contributors can freely add new tags or modify/delete existing ones. According to Zipf's "Principle of Least Effort" (Zipf 1949) and since power laws tend to be standard signatures of self-organized human activities, a relation between the distribution of OSM tags and the general shape of the distribution can be expected.

To test this hypothesis, data from the TagInfo database were analyzed. TagInfo provides a wide range of information such as how many times a tag or a key was used in the entire OpenStreetMap database. More complex data are also available, such as the number of times one specific key or tag is used together with another key or tag. For example, using the "tag_combinations" TagInfo's table allows knowing that the OpenStreetMap tags "access = private" and "highway = service" have been used together more than 2 million times. The entire TagInfo dataset (green dots in Fig. 3) and two subsets of the 1,000 and 500 most used tags (blue and red dots in Fig. 3, respectively) were used for this analysis. Results show a strong

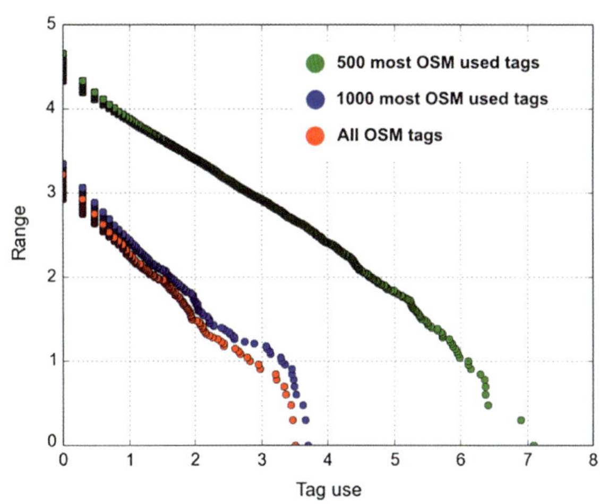

Fig. 3 Cumulative number of tag usage frequency based on their relative position. The plot is on a log–log scale: the *vertical* axis shows the logarithm of the relative rank position, while the *horizontal* axis shows the logarithm of the cumulative frequency of tags in that relative position. The singular shape of the distribution is representative of a Zipf Law

relationship between the distribution of tags and the frequency of their use (Fig. 3). Tags on the left part of the graph can be identified as “Power tags” (Peters and Stock 2010), being OSM tags shared and used by a broad spectrum of contributors. The right part of the graph represents tags that are used by only a minority of contributors. These tags can be related to a specific context and then are valuable for those minorities or can simply result from errors made by contributors.

4 Improving VGI Dataset Quality Using Semantic Measurements and Folksonomy of Tags

To reduce the semantic heterogeneity of the OSM dataset while at the same time helping contributors during the editing process, an Open Source plugin called OSMantic[7] (OSM Semantic) has been developed for the Java OpenStreetMap (JOSM) editor. This plugin automatically suggests related tags to the contributors to enhance the user experience and automatically display warnings in case of possible tag misuse. Two main components compose this plugin. The first component is the database that contains all the semantic relationships between the tags, being used to provide relevant information to the users. The second component is a user interface where the information from the database is displayed.

4.1 Data Sources

Two main data sources were used for this project: the TagInfo Database and the OpenStreetMap Semantic Network created by Ballatore et al. (2013). This semantic network measures the semantic similarity between concepts. It was created using the OSM Wiki Crawler that extracts OSM tags and their relationships from the OSM Wiki website and computes the semantic similarity score of each pair of tags. The P-Rank (Penetrating Rank) co-citation algorithm (Zhao et al. 2009) was chosen to compute the semantic similarity score between sets of tags. Compared to other algorithms (e.g., SimRank) that only take into account incoming links, P-Rank is a recursive algorithm based on both the incoming and outgoing links of the networks. In other words, in P-Rank “two entities are similar if (1) they are referenced by similar entities and (2) they reference similar entities” (Zhao et al. 2009). In P-Rank, the semantic similarity is measured between two vertices a and b on iteration k and assuming that $Rk(a, b) = 1$ if $a = b$ and $a \neq b$. The P-Rank similarity score is updated at each iteration until the maximum number of iterations (k) is reached. At the end of the iterations, the P-Rank similarity score is a value ranging from zero

[7] OSMantic Wiki page: https://wiki.openstreetmap.org/wiki/JOSM/Plugins/OSMantic (Accessed 3 October 2014).

(i.e., no semantic similarity) to one (i.e., tags are similar). The OSMantic Plugin currently works with tags that have points, lines, and polygons as OSM primitives. Further work will have to be done to handle the more recent OSM "relation" primitives.

As mentioned previously, an OpenStreetMap object is composed of a geographic primitive and a set of tags. Analyzing the relationships between the tags used to describe an OSM object can provide valuable information during the OSM editing process. This is why a new dataset that combines the OpenStreetMap Semantic Network and the TagInfo database has been created. The result of this combination was named "semantic specificity" (Table 1), a value used by the plugin to identify related tags. For performance reasons and to keep a semantic consistency, only the tags that were used more than 1,000 times in the entire OSM database were used. A second filtering process has been applied to only keep the relevant tags needed to help the contributors during the editing. For example, the tags that contain the key "source", "note", or "attribution" were deleted. The end result is a table with 4,765 rows. Table 1 presents the most popular tags ordered by the number of time they were used. The information in this table indicates how many times the OSM tags that contain the Key1 and the Value1 has been used at the same time with the tags composed by the Key2 and the Value2. Such information is available for each different OSM primitive (node, ways, relations) associated with the semantic specificity. For example, the first line must be read as: the tags "highway = service" and the tags "service = parking_aisle" were used together more than 1 million times. OSM objects that used this tag pair are mainly represented as a way (or polygon) and the semantic specificity value between these tags is 0.0517.

The count (all, node, ways and relations) value and the semantic specificity value are used by the plugin to compute how a given tag is related to another tag. First, the plugin identifies potential relationships between OSM tags using the count value, and then results are ordered in descending order using the semantic specificity value as a weight.

Information from this table can be used to guide OSM contributors during the editing process. For example if a contributor wants to label a road as "highway = service", OSMantic will automatically suggest adding the tag "service = parking_aisle" as the number of times these have been used together and the relation between these two tags are high.

4.2 User Interface and Prototype Functionalities

The main goal of this plugin is to provide OSM users with useful information during the editing process. To reach this goal, two main functionalities have been designed. The first one is the automatic suggestion of tags depending on the tags already specified for the selected OSM object. The second functionality is the possibility for the contributors to receive a notification when the similarity between the tags of a selected OSM object is too low.

Table 1 Most used tags in the OSM database after the filtering process has been applied

Key1	Value1	Key2	Value2	Count all	Count node	Count ways	Count relations	Semantic specificity
Highway	Service	Service	parking_aisle	1,334,683	77	1,334,581	25	0.0517
Bicycle	Yes	Foot	Yes	831,685	182,701	648,538	446	0.0712
Intermittent	Yes	Waterway	Stream	635,821	4	635,812	5	0.0785
admin_level	8	Boundary	Administrative	482,822	211	352,483	130,128	0.0516

In order to select the best OSM editor to implement these functionalities, an evaluation of the most popular editors has been realized. Two mains categories of OSM editor can be distinguished. The first category, like Potlatch or iD editors, is based on a Web architecture and is accessible through a Web browser. The second category, like Merkaartor or Java OpenStreetMap (JOSM), includes desktop applications. While iD tends to be the most popular editor, there is currently no direct way to add new functionalities. Due to these limitations, OSMantic has been implemented into the most commonly used desktop OSM editor: JOSM. During the last years, usage statistics show[8] that JOSM has been used by over 40 % of the contributors. JOSM is a powerful free and open-source standalone desktop application that allows contributors to create, edit, or delete data from OSM. JOSM has a flexible API that allows creating plugins that extend JOSM basic functionalities. This capability is one of the main reasons why JOSM was used to develop the user interface of the OSMantic plugin.

4.2.1 Automatic Suggestions of Tags

To help contributors during the editing process, JOSM has a menu called "presets" that lists the most common tags used. This list of tags is organized by categories such as highways, sports, or facilities. After creating or selecting a geographic object, contributors can choose the appropriate ones within the list of tags. By default, JOSM only contains a limited number of tags, but this list can be freely extended by adding manually the desired tags or by downloading presets already created by other contributors.[9] While "presets" provides a good reminder if a contributor does not remember exactly the exact tags for a geographic object, it implies that the contributor already knows the tags that have to be used. In addition, in the absence of knowledge of how tags relate to each other, selecting a tag does not allow the suggestion of other related tags. For example, if a contributor wants to tag a road, the following tags would probably be the surface of the road or the number of lanes. Finally, due to the high number of existing tags, listing all of them or even the most important ones is impossible and listing only a limited portion of the tags could lead to an overall semantic impoverishment of the database.

Improving the contribution process by automatically suggesting related tags can help contributors provide a richer, more accurate and more consistent description of OSM objects, hence allowing a reduction of the current overall database semantic heterogeneity. When editing an object that already has a set of tags, the plugin analyzes the existing tags to propose potential related tags. For contributors, the benefits of such tag recommender functionality include a higher efficiency in

[8] Editor usage statistics: http://wiki.openstreetmap.org/wiki/Editor_usage_stats (Accessed 3 October 2014).

[9] OpenStreetMap Presets: https://josm.openstreetmap.de/wiki/TaggingPresets (Accessed 3 October 2014).

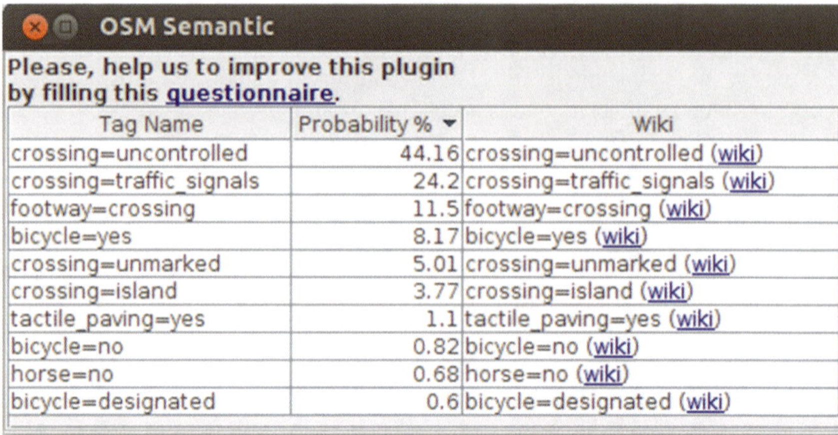

Tag Name	Probability %	Wiki
crossing=uncontrolled	44.16	crossing=uncontrolled (wiki)
crossing=traffic_signals	24.2	crossing=traffic_signals (wiki)
footway=crossing	11.5	footway=crossing (wiki)
bicycle=yes	8.17	bicycle=yes (wiki)
crossing=unmarked	5.01	crossing=unmarked (wiki)
crossing=island	3.77	crossing=island (wiki)
tactile_paving=yes	1.1	tactile_paving=yes (wiki)
bicycle=no	0.82	bicycle=no (wiki)
horse=no	0.68	horse=no (wiki)
bicycle=designated	0.6	bicycle=designated (wiki)

Fig. 4 Example of suggested tags provided to the contributor in the OSMantic plugin. This example is for an object "highway = traffic_signals" being edited by the contributor

finding appropriate complementary tags, the potential to discover new tags, building an expertise in OSM, and gaining more confidence in selecting a specific tag (Pu et al. 2011). For each attribute of the selected object, the similarity of being related to other OSM tags is computed. The similarity is computed by aggregating the "Count all" value and the "Semantic specificity" value (Table 1) for each object. Results are then compared to calculate the final similarity for a given OSM object and are displayed in the JOSM interface in the form of a table structured into three columns (Fig. 4). The first column shows potentially related tags, the second column shows the likelihood that the given tag is related to the selected OSM object and the last column is a direct hyperlink to the description of this tag in the OSM Wiki, allowing contributors to get more information on the tag use. Figure 4 shows a concrete example of the use of this functionality for an OSM object only described by the tag "highway = traffic_signals". The JOSM interface displays others relevant tags, such as "crossing = uncontrolled" or "tactile_paving = yes".

4.2.2 Notification of Unrelated Tags

As mentioned earlier, precisely describing an object in OpenStreetMap usually requires using several tags. While some of them are well described in the OSM Wiki, others are still poorly described and some of them do not have a description at all. This situation can lead new contributors to combine tags that have little or no relationship.

To prevent contributors from adding tags that are too dissimilar, the relationship between new tags being added and the existing tags is computed. If a low or lack of relationship is detected then a notification is automatically displayed to the user. Figure 5 illustrates such a case with a contributor willing to add the new tag

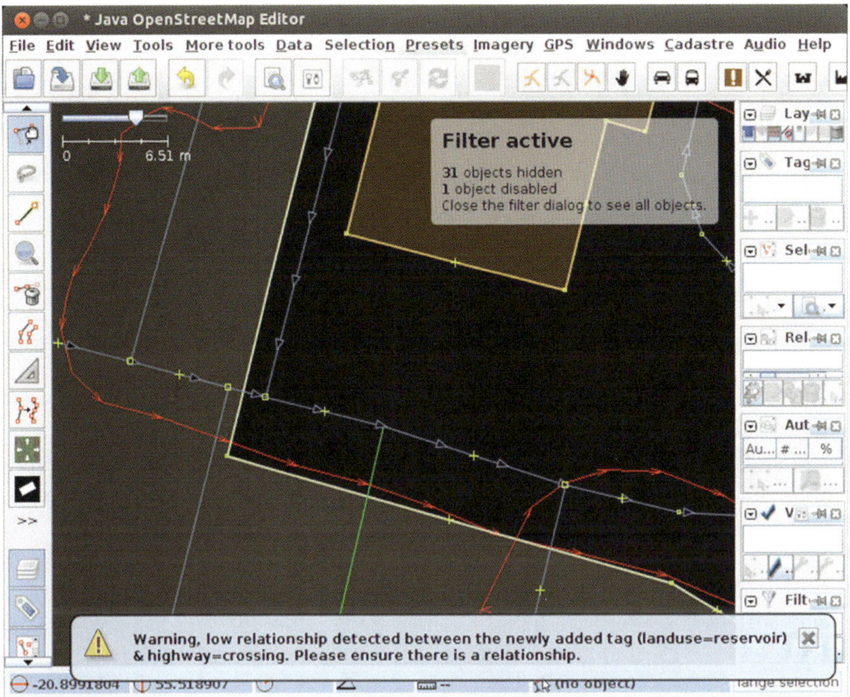

Fig. 5 Warning (at the *bottom*) of a low relationship between tags

"landuse = reservoir" to an OSM object that already has a "highway = crossing" tag. As the similarity between these two tags is low, a notification is displayed in the system.

5 Evaluations and Results

5.1 Experimental Setup

A formal evaluation of the plugin was designed based on the unifying evaluation framework ResQue (REcommender Systems' Quality of User Experience) of Pu et al. (2011). ResQue is specifically dedicated to the evaluation of recommender systems and is structured around four main dimensions: perceived system qualities, users' beliefs, users' subjective attitudes, and users' behavioral intentions. The four dimensions are subdivided into 15 constructs, each addressed by one or several questions. The complete ResQue model suggests asking 32 questions to the users. Out of the 15 constructs, several of them were assessed as not being relevant to this study, being mostly related to commercial recommender systems (e.g., users' purchase intentions). No questions on these specific constructs were hence asked. As suggested by Pu et al. (2011) and to shorten the response time, a shorter questionnaire was used to assess the quality of the OSMantic plugin (Table 2). The questionnaire that includes 15 questions begins with three general questions characterizing the respondent's profile, then the other questions focus on the evaluation of the plugin. Answers to questions used a five-point Likert scale ranging from "strongly disagree" (1) to "strongly agree" (5).

5.2 User Evaluation Results

A total of 30 people, mainly recruited via the OpenStreetMap mailing list and social media, participated in the survey. Answer consistency and reliability was assessed using the Cronbach's alpha measure (Bland and Altman 1997). The measure provided a value of 0.907, showing a high reliability between the survey responses. Tests identified one respondent as an outlier. After further verification, the respondent gave arbitrary weights to the questions to be able to leave a comment on

Table 2 ResQue constructs and questions of the survey used to evaluate the OSMantic plugin

Id construct	Construct	Question
1	General information	What type of OpenStreetMap (OSM) contributor are you?
		How often do you contribute to OSM?
		How often do you contribute to JOSM?
2	Recommendation accuracy	The proposed tags were relevant
3	Recommendation novelty	The tags recommender system helped me discover new tags
4	Recommendation diversity	The proposed tags were diverse
5	Interface adequacy	The information provided by the tag recommender system was clear and adequate
6	Information sufficiency	The information provided by the tag recommender system was sufficient to select new tags
7	Perceived usefulness	The recommender system helped me find appropriate tags
8	Interaction adequacy	I found the hyperlink to the OSM Wiki useful
		I found the ability to add tags by doing a double click useful
		I found the probability display useful for tag selection
9	Perceived ease of use	I became quickly familiar with the recommender system
10	Overall satisfaction	Overall, I am satisfied with the tag recommender system
11	Use intentions	I will use this tag recommender system again

Table 3 Contribution frequency to OSM of surveyed contributors

Type of contributor	OSM contribution frequency		
	Rarely	At least once a month	At least once a week
Novice	2	0	0
Intermediate	0	2	7
Expert	0	2	16

the system at the end of the survey. This response was then removed from further analysis but the comment was considered for future improvements to the plugin.

As indicated in Table 3, most survey participants were expert or intermediate contributors contributing at least once a week.

Table 4 represents how often contributors use JOSM for editing.

Figures 6, 7 and 8 present the result of the Likert scale for the functionality of the plugin. For each of them, results have been grouped by contributor type (i.e. novice, intermediate and expert).

Figure 6 presents the survey results related to the perceived usefulness of the proposed plugin (constructs 2–7 in Table 2). While all the answers agree on a general usefulness of the plugin, differences can be observed depending on the type of contributor. Globally, the more a contributor knew about OpenStreetMap, the less the plugin was helpful. This can be seen from the answers to "The tags recommender system helped to discover new tags" and "The recommender system helped me find appropriate tags". While a large majority of the novice and intermediate contributors agreed on the usefulness of the plugin, this value was lower for the expert contributors. These results are not surprising as the plugin was mainly designed to help novice and intermediate contributors. While the usefulness of this plugin for expert contributors is not as high, experts largely agreed on the relevancy, diversity, and sufficiency of the proposed tags. Such results confirm that the tags suggested by the plugin are often relevant.

Figure 7 presents the survey results related to the Interaction Adequacy construct that characterizes how effective some specific plugin functionalities are. Three questions were asked of the participants. The first one was related to the hyperlink that links the tags displayed by the plugin to the OSM Wiki website. The second question related to the ability to add a tag to an OSM object by a simple double-click on the cell of the plugin. The last question was associated to the display of the similarity between the selected OSM object and the tags proposed by the plugin.

Table 4 Use of JOSM by surveyed contributors

Type of contributor	JOSM usage frequency		
	Rarely	At least once a month	At least once a week
Novice	1	1	0
Intermediate	0	2	7
Expert	1	2	15

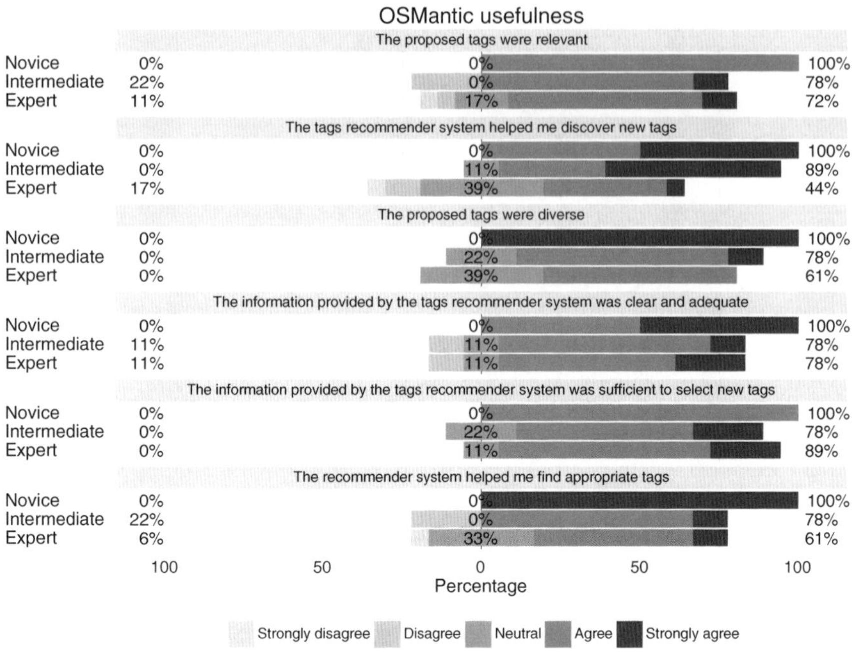

Fig. 6 Perceived usefulness of the OSMantic plugin expressed using a Likert scale. Percentages on the *left* indicate the proportion of negative assessments (i.e. strongly disagree or disagree), percentages in the *middle* represent the proportion of neutral responses and percentages on the *right* indicate the proportion of positive assessments (i.e. agree or strongly agree)

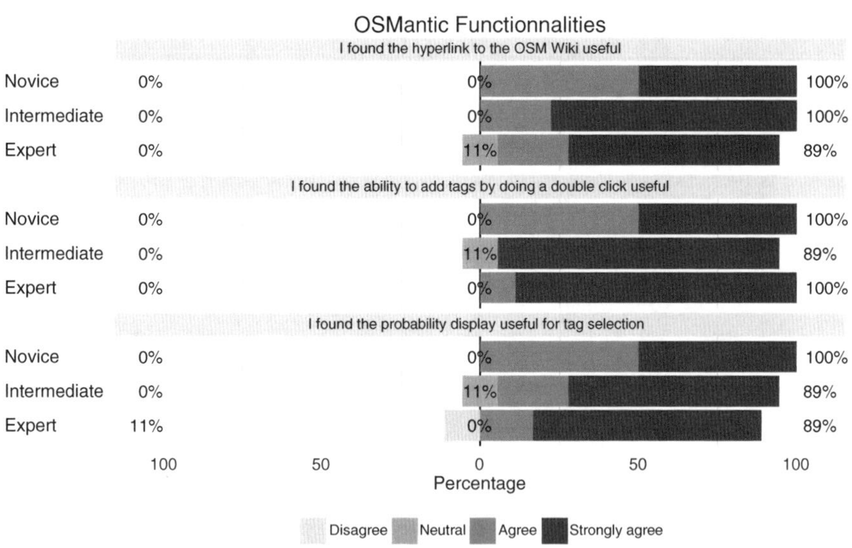

Fig. 7 Interaction adequacy of the main functionalities of the OSMantic plugin

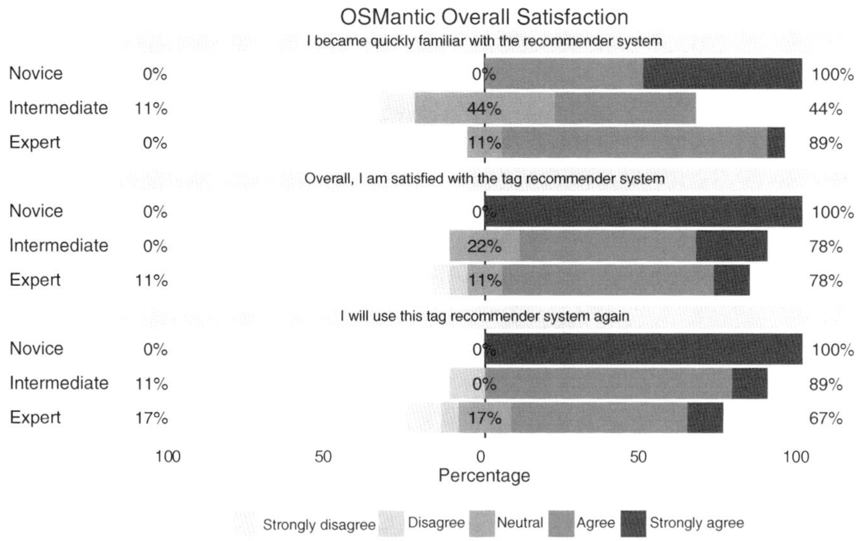

Fig. 8 Overall satisfaction with the OSMantic plugin

In general, results indicate a strong agreement of users, independently of the contributor type. The survey also allowed participants to leave comments and a lot of positive comments were made about the functionalities. While the display of a warning in case of a low similarity between the tags has not been specifically evaluated in this survey, this functionality received several negative comments. The main reason appears to be the high number of false-positives (i.e. warning on things are that not actual problems), the time allowed to read the warning before it disappears that was too short, and the number of warnings that was too high. For the false-positive results, two main causes can be identified. The first one is that by only using a subset of all the tags (only 1,000 tags have been kept), some potentially relevant relationships can be missing. The second reason is that some of the relationships between OSM tags may not be detected by the semantic network.

Figure 8 presents the survey results related to the overall satisfaction of the participants with the plugin. In general, results are positive, showing good satisfaction regarding the use of the plugin.

6 Discussion and Conclusions

The growing popularity of VGI projects such as OpenStreetMap has radically changed geographic information production, dissemination, and use. While such free and up-to-date data can often be a valuable alternative to authoritative datasets, assessing and enhancing the quality of VGI datasets remains a significant challenge.

Specifically, the high semantic heterogeneity of VGI datasets can be a major obstacle to the use of such data for diverse data analysis. For these reasons, this paper has investigated the use of semantic similarity to support OSM contributors' training and editing, at the same time enhancing the quality of VGI data. Such an approach allows reducing the overall semantic heterogeneity within a VGI dataset. Semantic similarity measurements were used to provide contributors with suggestions of other tags that could be used to describe an object being edited, leading both to a richer semantic and a more consistent use of tags. To test our approach, a plugin integrated into the JOSM editor was developed and examples of its use were presented. These examples illustrated potential situations where the use of semantic similarity measurements could reduce the semantic heterogeneity. To validate our approach, a formal evaluation of the plugin was conducted using a survey completed by 30 OSM contributors. Results show a good overall perception of the plugin and its different functionalities. The participants of the survey also suggested improvements that could be integrated in the next version of the plugin. These improvements are, for example, the possibility to handle the recent OSM primitive 'Relation' that is increasingly popular amongst contributors, the possibility to hide the warning in cases of low similarity between tags, and the enhancement of the algorithm that computes the similarity. Generally, the development of tools that could help novice and intermediate contributors develop an expertise could play a significant role in the improvement of VGI data quality and efforts to improve VGI semantics more specifically.

While results and comments from the evaluation are positive, potential improvements to the approach have been identified. One of the main improvements is related to the semantic network used to compute the semantic similarity between the tags. While it is a good indicator of a potential link between a pair of tags, it does not provide enough semantic information on how tags are related to each other. For example, a semantic network does not allow computing semantic distances between OSM tags as all the relations are at the same level. Providing a formal ontology of the OSM concepts could help solve this limitation.

Acknowledgments This research was funded by the Natural Science and Engineering Research Council of Canada (NSERC) through the second author's NSERC Discovery Accelerator Supplement Program. Authors also thank Dr. Andrea Ballatore for sharing his results, Vincent Privat for his helpful feedback on JOSM, Daniel Bégin for his early tests of the plugin and the 30 OSM contributors that participated in the evaluation of the plugin.

References

Auer S, Lehmann J, Hellmann S (2009) LinkedGeoData: adding a spatial dimension to the web of data. In: Proceedings of the 8th international semantic web conference, ISWC'09, Washington, DC. Lecture notes in computer science, vol 5823, Springer, Berlin, pp 731–746

Ballatore A, Bertolotto M, Wilson DC (2013) Geographic knowledge extraction and semantic similarity in OpenStreetMap. Knowl Inf Syst (KAIS) 37(1):61–81

Barron C, Neis P, Zipf A (2014) A comprehensive framework for intrinsic OpenStreetMap quality analysis. Trans GIS 18(6):877–895. doi:10.1111/tgis.12073
Bégin D, Devillers R, Roche S (2013) Assessing volunteered geographic information (VGI) quality based on contributors' mapping behaviours. In: Proceedings of the 8th international symposium on spatial data quality ISSDQ 2013, Hong Kong, China, pp 149–154
Bland JM, Altman DG (1997) Statistics notes: Cronbach's alpha. BMJ 314:572
Budhathoki NR, Nedović-Budić Z, Bertram B (2010) An interdisciplinary frame for understanding volunteered geographic information. J Geospat Inf Sci Technol Pract 64(1):11–26
Codescu M, Horsinka G, Kutz O, Mossakowski T, Rau R (2011) DO-ROAM: activity-oriented search and navigation with OpenStreetMap. In: Claramunt C, Levashkin S, Bertolotto M (eds) Fourth international conference on geospatial semantics, vol 6631, Brest, France. Lecture notes in computer science, Springer, Berlin, pp 88–107
Coleman DJ, Georgiadou Y, Labonte J (2009) Volunteered geographic information: the nature and motivation of produsers. Int J Spat Data Infrastruct Res 4:332–358
Elwood S, Goodchild MF, Sui DZ (2012) Researching volunteered geographic information: spatial data, geographic research, and new social practice. Ann Assoc Am Geogr 103(3):571–590
Flanagin AJ, Metzger MJ (2008) The credibility of volunteered geographic information. GeoJournal 72:137–148
Girres J-F, Touya G (2010) Quality assessment of the French OpenStreetMap dataset. Trans GIS 14(4):435–459
Goodchild MF (2007) Citizens as sensors: the world of volunteered geography. GeoJournal 69:211–221
Goodchild MF, Li L (2012) Assuring the quality of volunteered geographic information. Spat Stat 1:110–120
Haklay M (2010) How good is volunteered geographical information? A comparative study of OpenStreetMap and Ordnance Survey datasets. Environ Plann B: Plann Des 37:682–703
Haklay M (2013) Citizen science and volunteered geographic information: overview and typology of participation. In: Sui D, Elwood S, Goodchild M (eds) Crowdsourcing geographic knowledge. Springer, The Netherlands, pp 105–122
Haklay M, Basiouka S, Antoniou V, Ather A (2010) How many volunteers does it take to map an area well? The validity of Linus' law to volunteered geographic information. Cartographic J 47 (4):315–322
ISO/TC 211 (2002) 19113 Geographic information—quality principles. International Organization for Standardization (No. ISO 19113:2002). International Organization for Standardization (ISO)
Keßler C, de Groot RTA (2013) Trust as a proxy measure for the quality of volunteered geographic information in the case of OpenStreetMap. In: Association of geographic information laboratories for Europe. Presented at the Agile 2013, Leuven, Belgium, pp 21–37
Ludwig I, Voss A, Krause-Traudes M (2011) A comparison of the street networks of Navteq and OSM in Germany. In: Geertman S, Reinhardt W, Toppen F (eds) Advancing geoinformation science for a changing world. Lecture notes in geoinformation and cartography. Springer, Berlin, pp 65–84
Mooney P, Corcoran P (2012a) The annotation process in OpenStreetMap. Trans GIS 16(4):561–579
Mooney P, Corcoran P (2012b) Characteristics of heavily edited objects in OpenStreetMap. Future Internet 4(1):285–305
Mülligann C, Janowicz K, Ye M, Lee W-C (2011) Analyzing the spatial-semantic interaction of points of interest in volunteered geographic information. In: Egenhofer M, Giudice N, Moratz R, Worboys M (eds) Spatial information theory. Lecture notes in computer science. Springer, Berlin, pp 350–370
Neis P, Zipf A (2012) Analyzing the contributor activity of a volunteered geographic information project—the case of OpenStreetMap. ISPRS Int J Geo-Inf 1(2):146–165
Peters I, Stock WG (2010) "Power tags" in information retrieval. Libr Hi Tech 28(1):81–93

Pu P, Chen L, Hu R (2011) A user-centric evaluation framework for recommender systems. In: Proceedings of the 5th ACM conference on recommender systems, RecSys '11. ACM, New York, USA, pp 157–164

Rehrl K, Gröechenig S, Hochmair H, Leitinger S, Steinmann R, Wagner A (2013) A conceptual model for analyzing contribution patterns in the context of VGI. In: Krisp JM (ed) Progress in location-based services. Lecture notes in geoinformation and cartography. Springer, Berlin, pp 373–388

Smith B, Mark DM (2003) Do mountains exist? Towards an ontology of landforms. Environ Plan 30(3):411–427

Tobler W (1970) A computer movie simulating urban growth in the detroit region. Econ Geogr 46:234–240

van Exel M, Dias E (2011) Towards a methodology for trust stratification in VGI. In: Volunteered geographic information (VGI)—research progress and new developments. Presented at the association of American geographers annual meeting 2011, Seattle, Washington, p 4

Zhao P, Han J, Sun Y (2009) PRank: a comprehensive structural similarity measure over information networks. In: Proceedings of the 18th ACM conference in information and knowledge management. Presented at the CIKM, 09, ACM Press, New York, pp 553–562

Zipf GK (1949) Human behavior and the principle of least effort: an introduction to human ecology. Martino fine books, USA. p 588

Inferring the Scale of OpenStreetMap Features

Guillaume Touya and Andreas Reimer

Abstract Traditionally, national mapping agencies produced datasets and map products for a low number of specified and internally consistent scales, i.e. at a common level of detail (LoD). With the advent of projects like OpenStreetMap, data users are increasingly confronted with the task of dealing with heterogeneously detailed and scaled geodata. Knowing the scale of geodata is very important for mapping processes such as for generalization of label placement or land-cover studies for instance. In the following chapter, we review and compare two concurrent approaches at automatically assigning scale to OSM objects. The first approach is based on a multi-criteria decision making model, with a rationalist approach for defining and parameterizing the respective criteria, yielding five broad LoD classes. The second approach attempts to identify a single metric from an analysis process, which is then used to interpolate a scale equivalence. Both approaches are combined and tested against well-known Corine data, resulting in an improvement of the scale inference process. The chapter closes with a presentation of the most pressing open problems.

Keywords Generalization · Scale · Data quality · Level of detail · OpenStreetMap

---

G. Touya (✉)
COGIT, Institut national de l'information géographique et forestière, Paris, France
e-mail: guillaume.touya@ign.fr

A. Reimer
Department of Geography, Heidelberg University, Heidelberg, Germany
e-mail: andreas.reimer@geog.uni-heidelberg.de

J. Jokar Arsanjani et al. (eds.), *OpenStreetMap in GIScience*,
Lecture Notes in Geoinformation and Cartography,
DOI 10.1007/978-3-319-14280-7_5

1 Introduction

Studying the quality of OpenStreetMap (OSM) data has been a hot research topic in recent years, as OSM has grown and applications flourished. OSM quality can be assessed by comparisons with a reference dataset (Haklay 2010; Girres and Touya 2010; Zielstra and Zipf 2010), with intrinsic measurements (Barron et al. 2013), with contributor trust inference (Skarlatidou et al. 2011; Keßler and de Groot 2013), or in a fitness for use context, like pedestrian routing (Mondzech and Sester 2011). The latter method is the concern of this chapter, particularly the assessment of the fitness of OSM data for high-quality cartography (Sester et al. 2014). Besides being crucial for automated derivation of cartographic products, scale evaluation also touches issues for the appraisal of OSM data for geographic analyses, linking the issue with the bigger problem of scale in Volunteered Geographic Information (VGI) (Sester et al. 2014; Feick and Robertson 2014).

OSM data can be very rich in the regions with good completeness, but the level of detail of objects is very heterogeneous (Touya 2012). For instance, Fig. 1 shows three very complex religious buildings: the first one is captured with many details while the others are captured coarsely. This heterogeneity is a major obstacle for automated cartography, as existing automatic processes, like map generalization processes, are parameterized for a given homogeneous input scale. As a consequence, producing smaller scale maps by automatic generalization requires the inference of the initial scale or level of detail for every object as it can be different for each. Even for the production of large-scale maps (e.g. city plans) where map generalization is not necessary, level of detail is important. By introducing confusion in the spatial relations between detailed and less-detailed objects, such inconsistency may mislead the map reader. Whether a building is actually in a clearing or in a patch of woodland that was imported from a generalized source or even incompletely mapped is undecidable without some information on scale (Touya and Brando 2013). Apart from cartographic communication, investigations into land cover/land use obviously face similar problems. Two concurrent methods have been proposed to infer the scale/LoD of OSM objects individually (Touya and Brando 2013; Reimer et al. 2014). In this chapter, we review and compare both methods to improve the quality of scale/LoD inference for OSM.

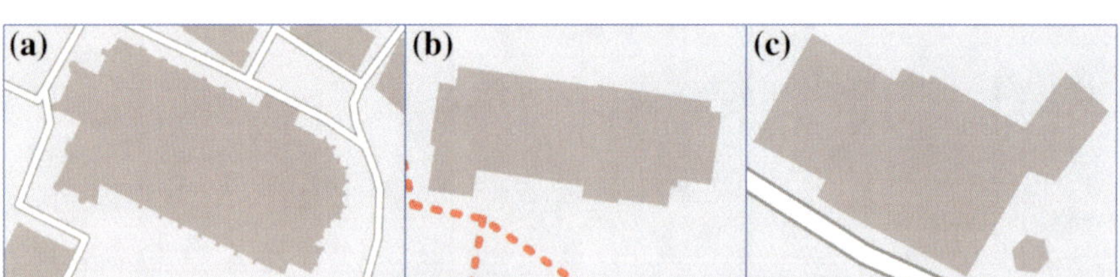

Fig. 1 Different levels of detail for three very complex religious buildings with similar granularity. **a** Is captured from cadastre data while **b** and **c** are captured from Bing images

The second section discusses the cartographic and geographic notions of scale and level of detail. The third section briefly presents both scale inference methods, while Sect. 4 explores how these methods can be combined to improve the scale inference of OSM data. Section 5 describes how scale inference can be used to enhance the automated mapping techniques for OSM data, to illustrate the usefulness of the research. After a discussion in the sixth section, the chapter is concluded and further research is proposed.

2 Scale and Level of Detail

All scientific modelling approaches implicitly or explicitly must deal with the question of the conceptual and numerical relationships between reality and the model being created. In the most straightforward interpretation, a numerical relationship between reality and the model is the ratio of values in reality to values in the model. For all spatial sciences and endeavors, be they model railroads or traffic simulations, informative numerical ratios as an expression of the reality–model relationship are based on geometries. For maps as graphical models of geographic phenomena, the ratio between distances as measured on the map and as measured in reality is called "scale" (International Cartographic Association 1972). As it has so far been inefficient to provide maps at an arbitrary scale, mapping agencies have prepared maps and map series at certain scale groups. Furthermore, in cartography, scale has become shorthand for usage environment for the map and conceptual organization of the subject matter depicted. Freitag (1962) writes: "The scale of a map is one of its constitutive elements; it determines information density, readability, significant contents and area of application." Due to this historical development, the seemingly objective and quantifiable numerical scale has become attached to precepts about the conceptual relations between geographic reality and its graphical model (Freitag 1962). This was driven by and is congruent with developments in geography (Sudgen and Hamilton 1971) and other geosciences such as ecology (Steinhardt 1999; Levin 1992; Chave 2013). In all cases, ranges of numerical scales are grouped into conceptual hierarchies. Consequentially, each level of the hierarchy is understood to be the realm of certain phenomena and specialized research methods only applicable within the bounds of that level. Haggett et al. (1965) proposed normalization over the surface of the earth as the basis for a logarithmic subdivision as a "yardstick" (G-scale) for a continuous hierarchy of modelling perspectives. Without any direct connection to the geosciences or cartography, computer graphics research saw itself confronted with the problem of "enhancing performance and realism" of "computer produced pictures" for graphical models of reality (Clark 1976). It is not surprising that one of the most enduring solutions was to devise hierarchical geometric models, which came to be known as "Levels of Detail" (LoD) (Clark 1976). As such, LoD and the hierarchical models of space in the geosciences are quite similar in purpose and argumentation. Both approaches also share the need for a concretization for specific use cases.

Touya and Brando (2013) define the LoD of a geographical dataset as the conjunction of several factors, namely the conceptual schema of the data, the semantic resolution, the geometric resolution, the geometric precision, and the granularity. The *conceptual schema* component is the way ground truth is represented in the geographical database: polygonal features representing forests or point features representing individual trees are conceptual schemas that correspond to different LoDs. The *semantic resolution* is the quantity of details in the attribute data attached to geographical features. By analogy with raster resolution, the *geometric resolution* of vector features is approximately the minimum distance between two vertices of the geometry. The *geometric precision* is simply the positional shift between ground truth and the represented feature. Finally, *granularity* describes the size of the smallest shapes of features, such as the smallest protrusions of buildings or the smallest width for sharp bends in a road. Biljecki et al. (2014) proposed a formal framework to measure LoD, but it is applicable for 3D city models.

As CityGML standard did for 3D city models (Kolbe 2009), LoD categories can be defined for maps. Touya and Brando (2013) proposed five LoD categories that would be used in this chapter, from the more detailed to the less detailed: street, city, county, region, and country. The *street* LoD contains features represented for parcel management or street orientation. For example, the British Ordnance Survey MasterMap can be considered as a street LoD dataset. The *city* LoD contains features to describe what is visible on the ground (buildings, roads, rivers, forests, etc.). It is the LoD of classical 1:25k topographic maps. The *county* LoD contains features that represent a small region to allow tourist-like displacements (e.g., visits, hiking, cycle rides). The *regional* LoD is related to the representation of a large region and only contains important roads and geographic features. Finally, the *country* LoD is even less detailed, for the representation of countries or very large regions.

In order to compare LoD and scale inferences, we have to match the LoD categories used in this chapter with scale values. Table 1 summarizes an attempt at such a matching. Although this matching is somewhat subjective, it was used in the remainder of the chapter to compare scale and LoD inferences.

3 Two Methods for the Automatic Inference of Scale

As we have noted before, information on the scale/LoD of a given dataset is crucial for further geoprocessing operations such as automated generalization. Before the rise of datasets like OSM, information on scale/LoD was just part of the metadata,

Table 1 Matching the LoD categories from Touya and Brando (2013) with map scale ranges, i.e. the objects with a given LoD category can be mapped within this scale range, without significant generalization

LoD category	Street	City	County	Region	Country
Scale range	<1:15k	1:15k–1:50k	1:50k–1:150k	1:150k–1:750k	>1:750k

but with user generated geodata, capture or intended scale/LoD are not documented. To address these problems, we present two different data enrichment approaches that attempt to infer the scale/LoD of OSM-data. The first approach can be understood as a rationalist (Touya and Brando 2013), the second as an empiricist approach (Reimer et al. 2014).

3.1 Scale Inference with a Multiple Criteria Decision Technique

3.1.1 Measuring Level of Detail

The definition of LoD presented in Sect. 2 showed that it is more complex than geometric resolution alone, as many aspects are involved in its characterization. Assuming that LoD inference is the aim, rather than scale, several measures have to be used to properly infer LoD. Following the definition given in Sect. 2, measures can be used to infer LoD while covering all its aspects (Touya and Brando 2013).

Measuring the conceptual schema:
The feature type of the object, in the Open Geospatial Consortium sense, can help us to measure the schema aspect of the level of detail. For instance, buildings or points of interest have a higher LoD than built-up areas. Although the Resource Description Framework (RDF) structure of OSM data prevents deriving of the feature type from classes, it can be derived from the main tags of objects like highway, building, or amenity.

Measuring the semantic resolution:
The annotation process to add semantic information on objects improves the level of detail of OSM objects by giving specifications on the object, but OSM is sparsely tagged (Mooney and Corcoran 2012) and only a few objects contain several tags. A very simple measure that counts tags is used. Metadata-like tags, such as source, or created_by, are counted as well as property tags like name, and language specific tags, such as species: fr that gives the French name of a tree species while the generic tag species gives the Latin name.

Measuring the geometric resolution:
Geometric resolution can be measured by assessing the density of vertices in relation to the length of objects. Indeed, we assume that a higher frequency of vertices denotes a digitization of objects at a larger scale or level of detail. The vertex density measure has been empirically normalized between 0 and 1 using the maximum and minimum values found in French authoritative datasets at different LoDs (1:25k, 1:100k and 1:250k). However, such a measure penalizes curvy lines that require more vertices to digitize than a straight line. So, curved objects (e.g. rivers) may be artificially considered as more detailed. So, we added a measure, the median of edge lengths (Fig. 2), to better capture the resolution of curved objects.

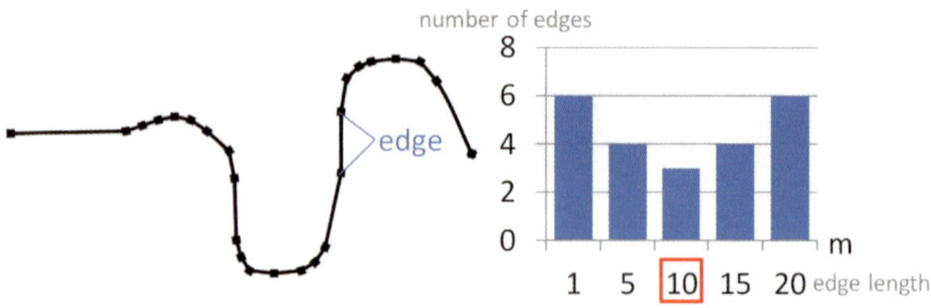

Fig. 2 The median of edge lengths gives a clearer idea of geometric resolution for curved objects like mountain roads or rivers

It has been proved more effective than vertices density on mountain roads by Girres (2011). This measure is also empirically normalized using the same authoritative datasets.

Measuring the geometric precision:
Intrinsically assessing the geometric precision is a complex task and previous research giving a geometric precision for OSM objects used a reference dataset (Haklay 2010; Girres and Touya 2010). However, the source tag in OSM informs on where the contribution comes from (e.g. GPS survey, Bing imagery, imported from open datasets). Knowledge on the rough precision of such sources can give us a vague but useful insight on geometric precision. The main possible values for the source tag have been listed from the TagInfo tool, and a rough LoD equivalency was derived when possible: a GPS survey is usually quite precise (value set to city) while digitization from Landsat creates imprecise objects (value set to region).

Measuring granularity:
Granularity captures the minimum size of details in an object shape. The detail can be the whole shape, a simple protrusion, or a bend in a linear object. To capture granularity in all three cases, three measures are proposed: size, smallest edge, and coalescence. First, small areas indicate a high granularity for small-sized objects but do not indicate anything about the granularity of large objects. The measure was calibrated with the smallest sizes found in the authoritative datasets at several LoDs used before. The size criterion is only applied to polygons. Then, the smallest edge criterion analyzes the length of the shortest edge between two vertices to infer granularity. This criterion is a classical measure to assess building granularity in cartographic generalization (Stoter et al. 2009). Note that when used on raw data, shortest edge values can become infinitesimally small due to nigh-double clicks and other digitization errors. A clean-up step is suggested when used on OSM data, for example. Finally, the coalescence (i.e. symbol overlap in a curve) criterion is used for the inference of linear features' granularity. This criterion is based on the principle that if a linear feature symbol coalesces at a given symbol width, the feature cannot be displayed at a scale that requires such a width for eye perception (Girres 2011). It could also be used on polygons but it has not been tested.

To summarize, eight criteria can be used to infer LoD in OpenStreetMap:

- Feature type (conceptual schema).
- Number of tags (semantic resolution).
- Vertex density (geometric resolution).
- Median edge length (geometric resolution).
- Capture source (geometric precision).
- Size (granularity).
- Smallest edge (granularity).
- Coalescence (granularity).

3.1.2 Combining Criteria to Infer a LoD Category

The eight proposed criteria are quite diverse, and hardly comparable, so their aggregation to infer a LoD category is challenging. Multiple-criteria decision-making is a computer science domain that researches methods to that are able to cope with diverse and intuitively hard-to-compare measurements (Figueira et al. 2005a). Such techniques can help us to infer LoD from the eight proposed criteria. Among the large number of existing techniques, we chose the ELECTRE TRI method (Figueira et al. 2005b) as its properties match our needs:

- The decision is a classification into categories.
- There are more than three criteria.
- The criteria are hardly comparable (i.e. how to compare the vertex density measure with the capture source criterion?).
- The criteria give fuzzy results that may be insignificant taken individually.

In ELECTRE TRI, the comparison between two vectors of measures (that gather the measures for all criteria) is not made by aggregating the measures. Comparisons are made criterion by criterion, and according to this comparison, each criterion votes for or against the assertion that "vector u outranks vector v". To make a classification, each category is assigned a lower-bound vector and an upper-bound vector to which the vectors to classify are compared. Therefore, we had to define, the lower-bound and upper-bound vectors for each category: e.g., the feature type "building" is the feature type criterion value for the lower-bound vector of the category City LoD. The bounds definition is based on the authoritative datasets. The criteria are able to vote indifference when the criterion alone is not sufficient to decide what outranks what. For instance, the criterion "feature type" is indifferent when the values up for comparison are "footpath" and "residential road". The criteria are also able to vote a veto, i.e. the outranking decision is based only on this criteria. For instance, the granularity criterion will put a veto if a 1-m smallest edge is compared to a 500-m smallest edge.

As the importance of any single criterion in the global method is hard to assess, a sensitivity analysis was conducted. We compared the global result (i.e. with all criteria) with results derived from only a subset of the criteria. The sensitivity analysis shows stability for some objects (e.g. building and forest), i.e. whatever the criteria used, the LoD category is the same. Other objects show more variability: for the tested streets or rivers, the choice of the criteria among the eight available significantly impacts the LoD category that is inferred.

As a strategy to make the method more robust against such feature type-dependent sensitivity, we compute the LoD for all the possible permutations of criteria (with at least four criteria to all criteria). The geometric mean of all the inferred categories is used as the new LoD value, robust against outliers and changing sensitivity.

The multiple criteria decision method and all the criteria have been implemented in the open source research platform GeOxygene (Grosso et al. 2012), which is able to load OSM datasets. Some results automatically computed on built-up areas, like in Reimer et al. (2014), are presented in Fig. 3.

3.2 Empiric Scale Inference

The empiric approach as presented in Reimer et al. (2014) is based on an inversion of Töpfer's radical law (Töpfer and Pillewizer 1966). The basic idea is to create a specific radical law applicable to a specific class of objects by empirically determining a measurable constant and inverting it, to determine the scale equivalency of

Fig. 3 LoD inference results for built-up areas in France: LoD is globally low because built-up areas are imported from Corine land cover

a given object. The scale equivalency is then expressed as if it was a cartographic scale. This empiric scale inference method is thus divided into three phases:

1. Identification of a suitable measure.
2. Determination of the specific constant.
3. Automated application.

Phases 1 and 2 are currently not fully automated, but conceivably could be automated, assuming some further work in defining criteria for selecting suitable measures. Note that phase 2 needs to be informed by the relation of the selected measure to the object, potentially modifying the degree of power of the equation. Area-based measures will most likely be of third degree and linear measures of second-degree complexity, disregarding other factors such as a drop or increase in apparent density. For a detailed discussion of the determination of the exponent x in the equation:

$$n_F = n_A \cdot \sqrt{\left(\frac{M_A}{M_F}\right)^x},$$

where n_F is a measure in the follow-up scale, n_A is a measure at the original scale and M_A and M_F are the scale factors at original and follow-up scale respectively, please see Töpfer (1979), pp. 43.

For the case of urban area polygons, the vertex frequency was identified from many other potential measures generated for urban area polygons for 1:250k scale map products from three different national mapping agencies. Note that vertex frequency was used for a related purpose by Dutton (1999). Other potential measures such as minimum edge length, area, vertex per area, angular resolution, minimum angle, etc. were tested and discarded either due to being too different across the tested maps or due to being statistically dependent on the area of the polygon. The vertex frequency tested negatively for dependence on the area of the polygon, i.e. there is no statistically significant correlation between frequency and polygon size. Furthermore, the measure showed very stable distributions with low standard deviations around a concrete value of 1 vertex per millimeter drawing space across a wide range of maps of the same scale produced by different organizations. We thus set up the scale equivalency S_E as:

$$S_E = 250{,}000 \cdot \left(\frac{1.0 \frac{v}{mm}}{n_F}\right)^2$$

Note how this is equivalent to the 1st degree (x = 1) selection equation based on Töpfer (1979):

$$M_F = M_A \cdot \left(\frac{n_A}{n_F}\right)^2.$$

Contrary to common experience, no significant numerical difference could be detected whether curvy rivers and coastlines were part of the urban area polygons or not. We highlight this as an example of the conceptual difference compared to the multi-criteria analysis, where the choice of criteria is made a priori, i.e. without a preceding statistical–empirical analysis. Whereas intuition and expert knowledge suggest that there is a strong influence on the artificial/natural form dichotomy of polygons on vertex frequency, these effects are not measurable for this specific object type and scale range.

4 Combining Both Methods to Improve Scale Inference

4.1 Compared Evaluation of Both Inference Methods

To compare both methods, scale was used and the improved LoD inference (the mean of category inferences when criteria vary) was interpolated to scale, using the scale ranges from Table 1, and a linear interpolation inside each category (i.e. a 2.5 LoD mean corresponds to a 1:50,000 scale). Then, five test areas were selected with only built-up areas as the empiric scale inference is only tuned for such objects. Four of the test areas come from areas of interest highlighted in Reimer et al. (2014): Africa, western Ukraine, Australia, and Belgium. A test area in France, where the multiple criteria method was initially tested, was also added. Figure 4 clearly highlights the general trend on all tested datasets, that both methods do not infer scale the same way. The scale equivalency (SE) tends to infer larger scale than the multiple criteria method (SL). The results are summarized for all built-up areas of all five test datasets in Table 2.

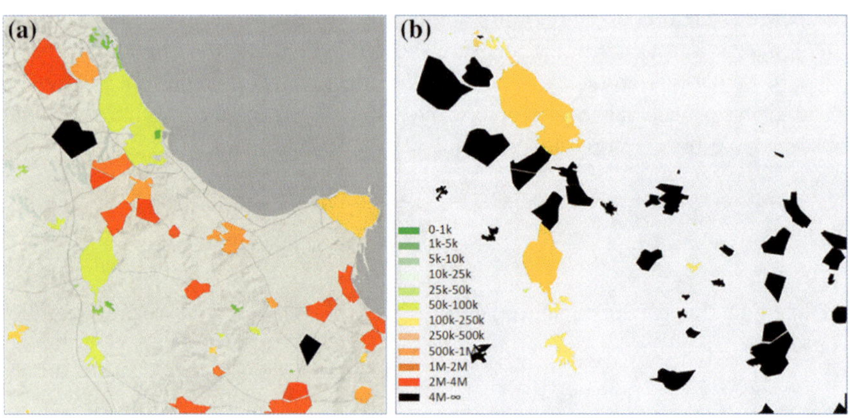

Fig. 4 **a** Reimer et al.'s (2014) scale equivalency on built-up areas in Tunisia. **b** Interpolated scale from multiple criteria LoD inference on the same dataset

Table 2 Comparison of scale inferences from both methods in all five datasets (17,402 objects)

	LoD mean	Scale from LoD (S_L)	Scale from equivalency (S_E)	Scale gap (S_L/S_E)
Mean	3.94	1:18,996,375	1:81,368	638.9
Max	5.00	1:75,375,000	1:32,017,460	11,499.7
Min	1.93	1:14,475	1:5	0.3
Median	3.94	1:411,549	1:11,149	61.1

Occasionally, SL is larger than S_E, but the mean scale gap is around 40 in most datasets (the gap is only significantly bigger in the French dataset). However, both methods are quite consistent in assessing the wide diversity of scales for built-up areas across the test datasets. When such a difference is measured, it is necessary to assess which one is closer to some form of ground truth. For this purpose, we used the French dataset where almost all built-up areas are derived from an automatic import of the Corine Land Cover European dataset, which is confirmed by the *source* tag on the objects. Corine Land Cover is produced by remote sensing for a scale of 1:100k. So, the both scale inferences should tend to 1:100k on the French dataset. S_E appears to be the more accurate scale: it slightly underestimates the scale with a mean scale of 1:43k, while S_L is much further with a mean scale of 1:2.5M.

The large bias of S_L may have several explanations. First, the multiple criteria method was not intended to infer scale but LoD, being a different and fuzzier notion (see Sect. 2). Measurements on geometry are balanced with measures on feature type, or metadata, which significantly impacts the inferred scale. Moreover, the geometry measurements were not calibrated on built-up areas but mostly on roads and buildings (Touya and Brando 2013), and they were not changed for this comparison to see if it mattered. Finally, the interpolation between LoD categories and scales is quite fuzzy and inaccurate. The ranges used for each category could be tuned and a non-linear interpolation could be used for small scales, where the scale-range is very large compared to the scale ranges of *street* and *city* categories. On the other hand, the scale equivalency also appears to be biased, as the only consideration of vertex density tends to increase scale equivalency too much. Taking granularity into account would maybe make S_L more accurate.

Although the exemplary scale equivalency was calibrated with built-up areas in authoritative datasets, it was computed on other map types of objects, as well as the multiple criteria LoD inference, in order to further explore both methods. Such evaluations can help us to know if the scale equivalency really has to be calibrated several times for several types of objects, which is the baseline assumption. Tests were carried out on rivers (tag *waterway* = river), forest (tag *landuse* = forest) that are geographically far from built-up areas and industrial areas (tag *landuse* = industrial), which is a type of object close to built-up areas (Fig. 5). Two areas with good completeness in France and Germany were selected.

In all three cases, the scale gap between S_L and S_E is less pronounced, particularly on rivers that are quite different objects, while forest and industrial areas are considered as land use areas. The gap decrease is partly due to the inference of

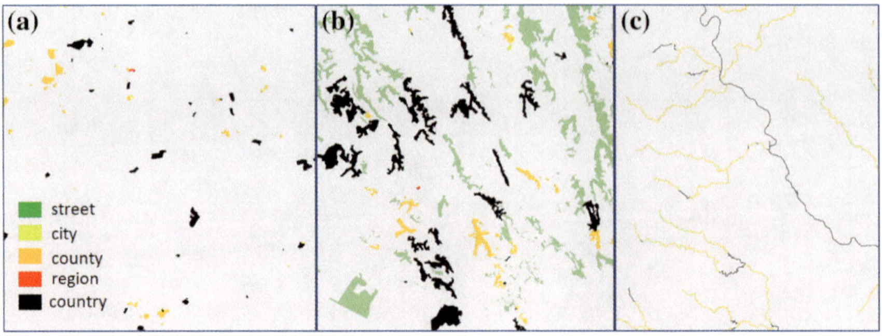

Fig. 5 LoD inference on industrial areas in **a** Germany, **b** forests in France, and **c** rivers in Germany

smaller scales by S_L compared to the inference for built-up areas. Indeed, vertex density is correlated to the shape of objects, and curved shapes like forests and rivers tend to be composed of more vertices than built-up or industrial areas with the same length.

4.2 Mixing Both Inference Methods

The exemplary scale equivalency for polygons as a function of the vertex density measure is naturally quite similar to the vertex density criterion used by Touya and Brando (2013), normalized by benchmarking on existing datasets. Hence, we tried to compute the multiple criteria scale inference replacing the vertex density criterion by the scale equivalency measure. The previous section showed that the multiple criteria method tended to yield much smaller scales compared to Reimer et al.'s (2014) scale equivalency, so adding the scale equivalency as a criterion could plausibly improve inference results.

The scale inference was computed with the vertex density criterion and with the scale equivalency instead on the same datasets of built-up areas (Table 3). The comparison was also computed on three buildings datasets (Table 4) in different

Table 3 Comparison of LoD inference with and without Reimer et al.'s (2014) scale equivalency as a criterion, on 4,858 built-up areas from the French and African datasets (tag landuse = residential)

	Mean LoD with vertex density	Mean LoD with scale equivalency	LoD difference
Ean	4.18	3.12	1.06
Max	5.00	4.20	1.72
Min	1.98	1.98	−0.03
Median	4.64	3.06	0.94

Table 4 Comparison of LoD inference with and without Reimer et al.'s (2014) scale equivalency as a criterion, on 9,153 buildings from the three chosen datasets

	Mean LoD with vertex density	Mean LoD with scale equivalency	LoD difference
Mean	2.05	1.99	0.05
Max	4.66	3.95	1.48
Min	1.64	1.61	−0.25
Median	2.02	2.00	0.03

parts of the world in order to assess the impact of the scale equivalency criterion on geographical data it was not calibrated for. The result for built-up areas shows that replacing the vertex density criterion clearly influences the LoD inference with a mean decrease of the LoD of 1 category, giving a mean scale of approximately 1:90k, which is consistent with the computation of scale equivalency on the same datasets. As most built-up areas in the tested datasets are imported from Corine Land Cover, a dataset specified for 1:100k scale, the scale inference is more accurate with Reimer et al.'s (2014) scale equivalency for these objects.

On the contrary, the LoD inference on building remains nearly the same with or without Reimer et al.'s (2014) scale equivalency, which means that the calibration of both measures for small and detailed features like buildings produce quite similar results.

5 LoD Harmonization for Large-Scale Automatic Mapping

When OSM is an input for the automatic derivation of large-scale maps (e.g. 1:10k), highly detailed objects are displayed with less detailed objects. When these objects share spatial relation (e.g. inclusion) in reality, the LoD difference may alter this spatial relation and blur the map readability (Touya 2012). In this case, the automatic inference of scale/LoD for individual objects is very useful. It helps identifying those spatial relations that are damaged because of LoD difference (Touya and Brando 2013). The spatial relations are identified by spatial relations algorithms and stored as complex objects of the dataset. Then, the LoD inference is a required input for an automatic LoD harmonization process that could restore the spatial relation (Touya and Baley 2014); here spatial relations could be made explicit as map objects in order to assure their restoration.

For instance, OSM built-up areas are often poorly detailed (see experiments above, Sect. 4.1) while buildings are quite detailed in France (imported from cadastral data). So, buildings may lie just outside built-up areas. The harmonization process extends the geometries of built-up areas to include those buildings, restoring the spatial relation of buildings included in a built-up area (Fig. 6).

Fig. 6 Automatic harmonization of a built-up area (in *red*), extended (in *grey*) to include buildings with higher LoD

6 Open Problems

6.1 Scale Inference for Point Objects

Both methods rely on some kind of point density estimation based on the series of vertices that describe objects' geometry, which is not immediately applicable to point objects (Dutton 1999; Bereuter and Weibel 2013). So, how can we infer the scale or LoD of point objects?

In the multiple criteria method, all criteria related to geometric resolution and granularity become meaningless, and the conceptual schema, semantic resolution and geometric precision criteria remain. We tested the method with the remaining criteria on several point features in OSM: trees, aerial way pylons, bicycle_rental, bus stops, towers, peaks, and power poles. These features are diverse in terms of real extent and geographical neighborhood. A fourth criterion, called version number, was added to assess the features with a crowdsourcing approach (Goodchild and Li 2012): although the Linus' law (i.e. quality increases with the number of active contributors) has not been completely proved for VGI (Haklay et al. 2010), the number of versions of a feature gives hints on the number of times a feature has been improved by OSM contributors. This criterion assesses the objects with many versions as detailed objects.

As the source tag is seldom filled, the most differentiating criterion is the *feature type* criterion. The point objects that represent a small object like a tree, bus stop or a pylon are attributed to a detailed LoD, while objects representing a large (and fuzzy) extent like peaks or bicycle rental places are given less detailed LoD. Within a feature type, the three other criteria (source, number of tags, and version number) make the objects more or less detailed: for instance, a peak captured by GPS with elevation and name tags is given a higher LoD than a peak captured from Bing imagery (Fig. 7).

These results show that the LoD inference of point objects could be improved. First, the version number criterion only uses the version number, so there is no information about the number of contributors that improved the object, and it may be a single contributor. However, the wisdom of crowd theory assumes that better decisions can be made when there are multiple independent contributors (Surowiecki 2004). Using the full history OSM database, as proposed by Barron et al. (2013), could further enhance the version number criterion with the number of different contributors involved in the object improvement. Moreover, the peaks example is the ideal case for introducing what Goodchild and Li (2012) call the geographic approach of VGI quality assurance. If we could cross the peaks with digital terrain model information, it would be easy to assess the proximity of the feature and a peak in the relief as modelled in the DTM.

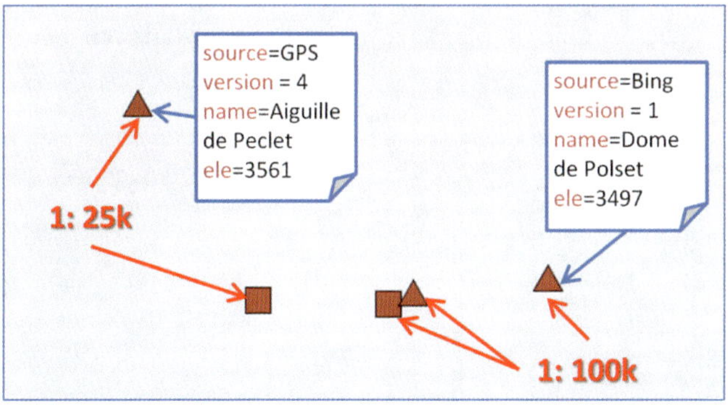

Fig. 7 Peaks (*triangle* symbols) and passes (*square* symbols) extracted from OSM in the French Alps. Scale inference (computed from the multiple criteria method) varies according to the tags on the objects

6.2 Does Feature Density Alter Scale Level?

Inferring the LoD of objects individually is sometimes insufficient, as the legibility of the object can be affected by the local density of objects. It is plausible to assume that certain dense patterns alter the LoD/scale of its composite object, potentially raising partonomic questions. Figure 8 shows the example of a cross-country ski resort in the Pyrenees. Individually, the paths are approximately inferred as 1:100k scale objects by the improved multiple criteria method. However, the density of such paths prevents any legible display at this scale (Fig. 8c). In this case, the scale inferred is not the scale at which the paths could be mapped, but the scale at which they could be individually mapped.

Regarding the multiple criteria method, two kinds of criterion could be added to integrate such density issues. First, clutter measures (Rosenholtz et al. 2007) at several scales could be made to assess at which scales the density of objects from one feature type causes clutter problems. Then, the principles from the coalescence criterion could be extended to the neighborhood of objects: as symbol size increases when scale decreases, when does the object symbol overlap with symbols from its neighbors?

6.3 The Scale of Objects with Simple Shapes

In a dataset with a homogeneous scale, simple objects like rectangular-shaped buildings are captured with simple geometries, but we know the scale is similar to all the other buildings that may have complex shapes. In OpenStreetMap, if an object has a very simple shape, it is hard to know if its scale is small, or if it is just a simple geometry that has been accurately captured. The scale equivalency measure will infer that such simple objects have a small scale, as the number of vertices to

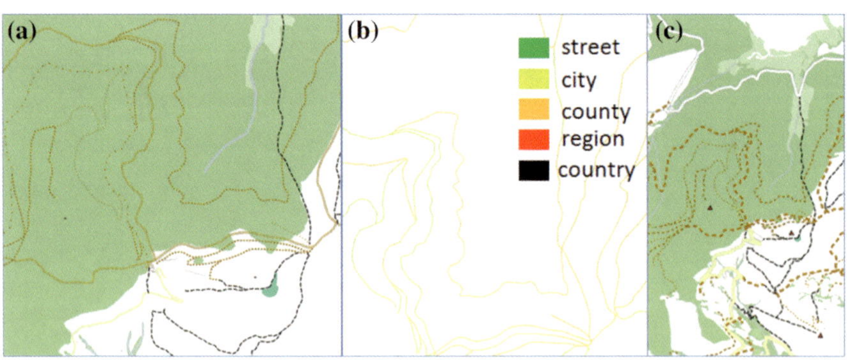

Fig. 8 **a** Multiple paths in a cross-country ski resort in the Pyrenees, **b** individual scale inference tells that paths should be mapped at 1:100k, but **c** there are too many close paths at this scale

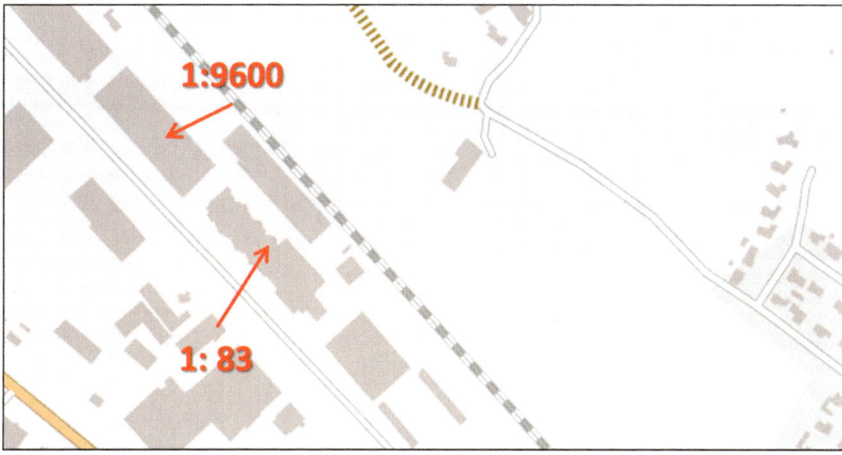

Fig. 9 Two buildings captured from the source (French cadastre) by the same contributor with quite different scale equivalency because the upper one has a much simpler shape

capture them is small (Fig. 9). Within the empirical scale equivalency approach one must come up with coping strategies for individual object types separately, as was done for city blocks (Reimer et al. 2014). LoD inference too, provides quite different results for buildings depending on the complexity of their original shape (Touya and Brando 2013).

When the application is the production of small-scale maps with a generalization process, this issue will not alter the generalization result. Indeed, if the inferred scale is small, the impact will only be a slight generalization, but if the shape is simple, it does not require as much generalization as a building with a complex shape. However, this can be an issue when the application is the production of a large-scale map, with LoD consistency checking (see Sect. 5). Then, objects with simple shapes should be identified before consistency checking, for instance, by analysis of the neighbors of the objects: if only one has a low resolution, the shape is probably simple. This analysis can be related to the geographic approach of quality of Goodchild and Li (2012).

7 Conclusions

Until the widespread availability and success of VGI and OSM specifically, scale was an easily determined metadatum. With non-government grassroots geodata becoming more encompassing by the day, users are confronted with rich but scale-wise inhomogeneous data. This is as true for geometries as it is on a conceptual level. This chapter reviewed the first two attempts at automatized scale inference for OSM and showed their relative strengths and weaknesses and ways to improve both by combining them. Both approaches currently and purposefully work on an individual object level. The open problems such as point features and feature density could plausibly be approached with meso-regions as intermediate steps, as it has been done in agent-based automated generalization. A more fundamental problem seems to be the further automation of parameterization. The current approaches make extensive use of tacit and explicit cartographic knowledge in selecting the measures and/or criteria. It is unclear at the moment how or if automation of these fundamental steps is attainable at all in the near future.

References

Barron C, Neis P, Zipf A (2013) Towards intrinsic quality analysis of OpenStreetMap datasets. In: Online proceedings of the international workshop on action and interaction in volunteered geographic information (ACTIVITY). Agile

Bereuter P, Weibel R (2013) Real-time generalization of point data in mobile and web mapping using quadtrees. Cartogr Geogr Inf Sci 40(4):271–281

Biljecki F, Ledoux H, Stoter J, Zhao J (2014) Formalisation of the level of detail in 3D city modelling. Comput Environ Urban Syst 48:1–15

Chave J (2013) The problem of pattern and scale in ecology: what have we learned in 20 years? Ecol Lett 16:4–16

Clark JH (1976) Hierarchical geometric models for visible surface algorithms. Commun ACM 19 (10):547–554

Dutton G (1999) Scale, sinuosity and point selection in digital line generalization. Cartogr Geogr Inf Sci 26(1):33–53

Feick R, Robertson C (2014) A multi-scale approach to exploring urban places in geotagged photographs. Comput Environ Urban Syst (in press)

Figueira J, Greco S, Ehrogott M (eds) (2005a) Multiple criteria decision analysis: state of the art surveys of international series in operations research and management science, vol 78. Springer, Heidelberg

Figueira J, Mousseau V, Roy B (2005b) ELECTRE methods. In: Figueira J, Greco S, Ehrogott M (eds) Multiple criteria decision analysis: state of the art surveys. Springer, Heidelberg, pp 133–162

Freitag U (1962) Der Kartenmaßstab—Betrachtungen über den Maßstabsbegriff in der Kartographie. Kartographische Nachrichten, Heft 5,12. Jahrgang, Gütersloh, pp 134–146

Girres J-F (2011) An evaluation of the impact of cartographic generalisation on length measurement computed from linear vector databases. In: Proceedings of 25th international cartographic conference (ICC'11), Paris, France. ICA

Girres J-F, Touya G (2010) Quality assessment of the french OpenStreetMap dataset. Trans GIS 14(4):435–459

Goodchild MF, Li L (2012) Assuring the quality of volunteered geographic information. Spat Stat 1:110–120

Grosso E, Perret J, Brasebin M (2012) GEOXYGENE: an interoperable platform for geographical application development. In: Innovative software development in GIS (Chapter 3), Wiley, New York, pp 67–90

Haggett P, Chorley RJ, Stoddart DR (1965) Scale standards in geographical research: a new measure of areal magnitude. Nature 205:844–847

Haklay M (2010) How good is volunteered geographical information? A comparative study of OpenStreetMap and ordnance survey datasets. Environ Plann B: Plann Des 37(4):682–703

Haklay M, Basiouka S, Antoniou V, Ather A (2010) How many volunteers does it take to map an area well? The validity of linus's law to volunteered geographic information. Cartogr J 47 (4):315–322
Keßler C, de Groot RT (2013) Trust as a proxy measure for the quality of volunteered geographic information in the case of OpenStreetMap. In: Vandenbroucke D, Bucher B, Crompvoets J (eds) Geographic information science at the heart of Europe. Lecture notes in geoinformation and cartography. Springer International Publishing, New York, pp 21–37
Kolbe TH (2009) Representing and exchanging 3D city models with CityGML. In: Lee J, Zlatanova S (eds) 3D Geo-information sciences. Lecture notes in geoinformation and cartography. Springer, Berlin, pp 15–31
Levin SA (1992) The problem of pattern and scale in ecology. Ecology 73(6):1943–1967
Mondzech J, Sester M (2011) Quality analysis of OpenStreetMap data based on application needs. Cartogr Int J Geogr Inf Geovisual 46(2):115–125
Mooney P, Corcoran P (2012) The annotation process in OpenStreetMap. Trans GIS 16(4):561–579
Reimer A, Kempf C, Rylov M, Neis P (2014) Assigning scale equivalencies to OpenStreetMap polygons. In: Proceedings of AutoCarto international symposium on automated cartography 2014 (accepted)
Rosenholtz R, Li Y, Nakano L (2007) Measuring visual clutter. J Vis 7(2):17
Sester M, Jokar Arsanjani J, Klammer R, Burghardt D, Haunert J-H (2014) Integrating and generalising volunteered geographic information. In: Burghardt D, Duchêne C, Mackaness W (eds) Abstracting geographic data in a data rich world. Springer, Berlin, pp 119–155
Skarlatidou A, Haklay M, Cheng T (2011) Trust in web GIS: the role of the trustee attributes in the design of trustworthy web GIS applications. Int J Geogr Inf Sci 25(12):1913–1930
Steinhardt U (1999) Die Theorie der geographischen Dimensionen in der Angewandten Landschaftsökologie. In: Schneider-Sliwa et al (eds) Angewandte Landschaftsökologie. Springer, Berlin
Stoter J, Burghardt D, Duchêne C, Baella B, Bakker N, Blok C, Pla M, Regnauld N, Touya G, Schmid S (2009) Methodology for evaluating automated map generalization in commercial software. Comput Environ Urban Syst 33(5):311–324
Sudgen D, Hamilton P (1971) Scale, systems and regional geography. Area 3(3):139–144
Surowiecki J (2004) The wisdom of crowds. Anchor Books
Töpfer F, Pillewizer W (1966) The principle of selection. Cartogr J 3:10–16
Töpfer F (1979) Kartographische generalisierung, 2nd edn. VEB Hermann Haack, Gotha
Touya G (2012) What is the level of detail of OpenStreetMap? In: Workshop on role of volunteered geographic information in advancing science: quality and credibility. Columbus (Ohio), USA
Touya G, Brando-Escobar C (2013) Detecting level-of-detail inconsistencies in volunteered geographic information data sets. Cartogr Int J Geogr Inf Geovisual 48(2):134–143
Touya G, Baley M (2014) Harmonizing level of details in OpenStreetMap based maps. In: Duckham M, Stewart K, Pebesma E (eds) Proceedings of GIScience 2014—Poster session, Vienna, Austria
Zielstra D, Zipf A (2010) A comparative study of proprietary geodata and volunteered geographic information for Germany. In: Proceedings of 13th Agile international conference on geographic information science. Guimaraes, Portugal

Data Retrieval for Small Spatial Regions in OpenStreetMap

Roland M. Olbricht

Abstract Using OpenStreetMap data usually means, firstly, filtering for thematic extracts. This can be done with the OpenStreetMap mirror database Overpass API (application programming interface). To do this efficiently, the database takes advantage of the assumption that data is often selected from a relatively small spatial region instead of randomly across the planet. The paper aims to investigate what design choices are required to be able to answer almost any geographic query whilst serving common use cases fast enough such that the services based on this database are fast on affordable and standard sized hardware. The usage patterns from the main instance of Overpass API on overpass-api.de are evaluated. These comprise more than 40 million requests from the years 2012 to 2013 coming from about 60 % of the global IPv4 (IP protocol version 4) space. Therefore it can be said for sure that the assumption that queries are spatially dense holds on a large share of all queries.

1 Introduction

The open data project OpenStreetMap (2014) aims at creating a geographic database of the entire world, i.e. collecting the data necessary to make any map of what is on the ground. Starting in 2004, it has now grown to more than a million registered users and is starting to outperform commercial data sources (2011) in more and more countries and on more and more topics.

Such a database is more useful and more open when it is possible to effectively extract data from it. This is the purpose of the tool Overpass API (2014). Typical use cases are extracts that are filtered by region and possibly by subject. The service is set up to be accessible by cross-origin resource sharing (CORS) (2014): CORS is

R.M. Olbricht (✉)
GIScience Research Group, Institute of Geography, Heidelberg University, 69120 Heidelberg, Germany
e-mail: roland.olbricht@gmx.de

J. Jokar Arsanjani et al. (eds.), *OpenStreetMap in GIScience*, Lecture Notes in Geoinformation and Cartography,
DOI 10.1007/978-3-319-14280-7_6

a web standard that specifies how to let the browser include content from a URL on a different domain into a website. Thus, websites can show arbitrary data overlays on top of a general-purpose background map. The development environment Overpass Turbo (2014) lowers the barrier to executing queries by hand. This paper investigates what design choices are required to achieve two divergent goals: On the one hand, the server should be able to answer almost any geographic query. On the other hand, it must serve common use cases fast enough so that services based on this database are reasonably fast on affordable standard sized hardware. The architecture of OpenStreetMap is so well modularized that this can indeed be run as a third-party-service:

The main database, *OSM* in Fig. 1, collects the edits and exports them once per week as a full database dump (2014). In addition, a replication stream is maintained that contains the changes that have happened to the database: Once per minute, an XML file with all changes is generated at (2014). All other services than editing should be catered for by servers based on this replication stream. Even the map renderer Mapnik (2014) and the main geocoder Nominatim (2014) are organized as data consumers for this replication stream. This keeps a good scalability, because the database server can be kept as simple as possible. In the same manner, Overpass API keeps an up-to-date copy of the full database.

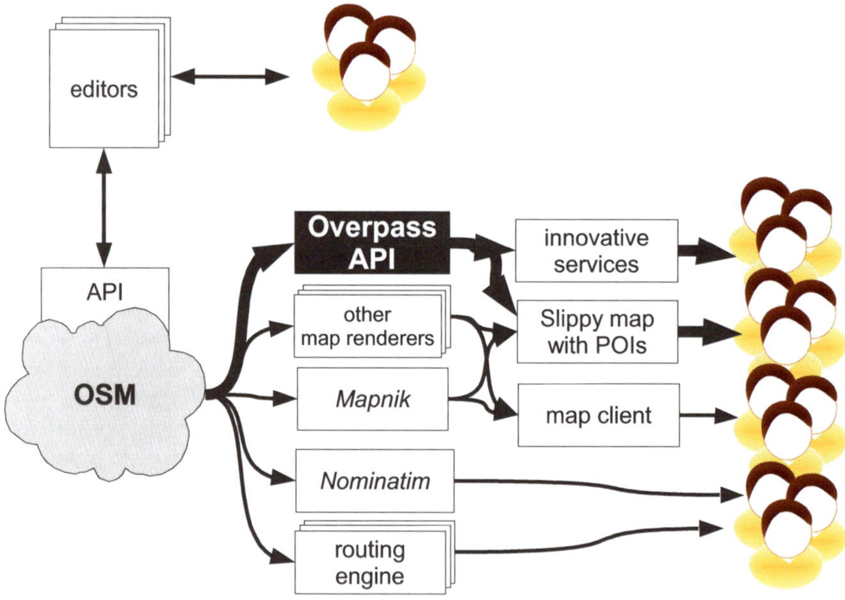

Fig. 1 Overpass API in the architecture of OpenStreetMap: The *arrows* represent the flow of data. The services in the middle column retrieve minutely updates from the main database to keep their copy of the OpenStreetMap data up to date. They deliver processed data to the services on the *right* on request of that services

Another important condition is the relatively small budget of less than 1,500 EUR per year: Running a service like Overpass API must be feasible on a single server instance, and it should require so little maintenance that this can be done during spare time.

2 Related Work

A good introductory book to OpenStreetMap is (2010). A recent overview of OpenStreetMap peer reviewed papers is given in (2012). Notable for continuous research on OpenStreetMap is the Alexander Zipf's research group: the quantitative development of OpenStreetMap is documented, e.g., in (2011; 2012).

The informal structures in OpenStreetMap have been the subject of various conference talks (2013, 2013, 2013) at the FOSSGIS as well as SotM conference series. They have shed some light on the structures: One should take into account the rendering rules for the most used maps, the rules encoded in the most used editors, and the mindset of mappers about the mapped objects, which is unfortunately hard to record.

The specific problem set here makes it necessary to have control over the data storage layout on the hard disk, i.e. to control which data items are stored close to each other on the hard disk. The two most well-known relational databases, MySQL (2014) (now forked to 2014) and PostgreSQL (2014), do not address the problem. Specifically the spatial extension PostGIS (2014) of PostgreSQL still relies on the standard database engine.

To understand which components of a hard disk contribute to performance a general introduction to model hard disks (1994) might help.

To cross-check whether custom or optimized data storage layout is offered by NoSQL databases, the overview in (2012) is used: The mentioned databases Riak, HBase, MongoDB, CouchDB, Neo4 J, and Redis all focus on scaling over multiple servers, but here effective performance is required with only a single server. Furthermore, the book suggests that for well-structured data with varying query patterns, relational databases are still the best choice.

No other work has been found to address the problem.

3 Challenges

Given the requirement to add effective query functionality on a single server instance, the following options come to mind: The main database server's software, *The Rails Port*, could be enhanced to also cover search functionality. This would have the advantage that the maintenance could be done easily by the maintainers of the main database server by caring for a fourth database server.

By contrast, a distinct software package might bring a significantly improved performance over the first approach. In that case, various options of the base system come to mind, but as has been seen in the section titled *Related Work*, a relational database, like PostgreSQL on The Rails Port, is probably already the best choice amongst ready-made database systems.

3.1 Structure of OSM Data

All real-world objects are modeled explicitly in OpenStreetMap in terms of nodes, ways and relations and more implicitly by various rules of informal consent (2010). It starts with the formal level: Only nodes carry explicitly a geometry, encoded as coordinates. Ways consist of a list of node ids, and the coordinates of these nodes are considered as the way's geometry. Relations can have node ids, way ids, and ids of other relations as members.

Type	Content	Can represent	Geometry
Node	Coordinates, tags	Single POI, support point for a way	Yes
Way	Member node ids, tags	Line string, area	Only indirectly (look up every referred node)
Relation	Member ids and type (node, way, or relation), tags	Area, abstract concepts	Double indirectly (look up every referred way, then look up for each referred way its nodes)

Ways are interpreted as either line-strings or polygons depending on their non-geometric properties, but this decision is subject not to fixed rules but part of the informal consent. There is informal consent about encoding schemes for multi-polygons and collections of geometries by using relations, e.g. to model country or lower-level administrative boundaries or coastlines.

The first aspect of the informal structures is the distribution of extent among all elements: To give some numbers, on 8 May 2014 the global OSM database consisted of more than 2.3 billion nodes. It contains about 230 million ways with the approximate sizes indicated in Fig. 2.The extent is computed as the great circle distance of the corners of the smallest enclosing bounding box. While the relative fraction of ways of large extent (i.e. more than 100 km) is small, its absolute number adds up to millions of way segments, and every such segment is in the potential result set for any spatial query. Thus if a non-trivial spatial query (e.g. intersection tests) is done then it is necessary to do sophisticated pre-filtering.

Relations expose a similar distribution pattern, but are much fewer. Furthermore, the geometry of a relation is considered as a simple union of the geometries of its node and way members. Together with the lower number of relations and in particular large relations this allows to focus filtering efforts on ways.

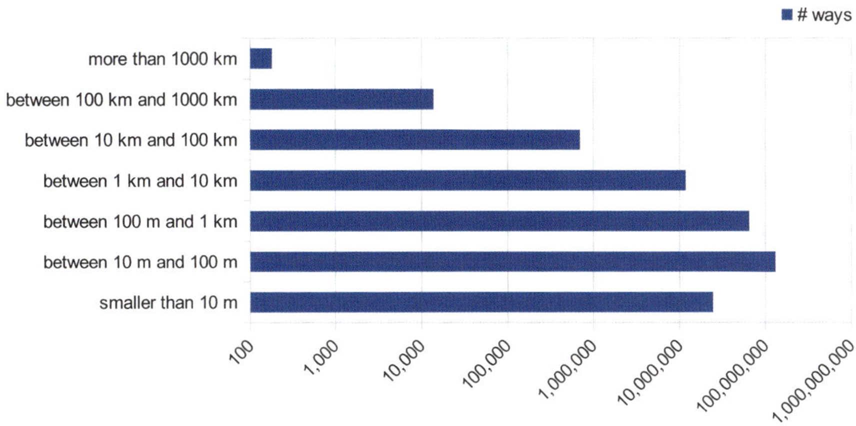

Fig. 2 Number of ways by their extent, logarithmic scale

Non-geometric properties are encoded as key-value pairs, and here again the informal structures are the essential ones: No semantic restrictions for the keys and values exist. By informal consent, an object (a node, way, or relation) carries at least one tag to characterize the type of the represented real-world object, e.g. a street or a motorway, and usually a further tag to give it a name or a similar thing. An exception are nodes that serve only as vertices of ways: these nodes normally have no tags at all. If different parts of a real-world object with a single name have different properties then they are encoded as distinct objects with different attributes but have the same value for the name tag.

Similar challenges exist for non-geometric as for geometric properties: Multiple real-world objects may have the same name. The same real-world object may have multiple names in different languages or even the same language. Furthermore, the same geometric object may have a different name for different purposes (e.g. a restaurant with a name in a historic building with a different name). Another common case is branch stores: They are often perceived as having the operating brand as their name, but other mappers put into the name tag the name of that particular store if it is known and the brand name into the "operator" or "brand" tag.

3.2 General Query Concepts

As mentioned before, it is in principle possible to get the data from the entire planet, but the dataset is too big to be useful in most use cases.

The aim of the query language, called *Overpass QL*, for Overpass API is to be sufficiently versatile but still easy to learn, understand, and remember. A secondary aim is to make predicting the runtime easier: This allows the user to understand which part of her query is resource consuming. For that purpose, queries are structured into statements.

The simplest queries need two statements: The first selects data from the database. The second prints the selected data in the specified form.

A simple example is:

```
node[amenity=restaurant](50.6,7.0,50.8,7.3);
out;
```

Here, first all elements of type node are chosen that have a tag *amenity* with value *restaurant* and are inside the bounding box north of 50.6° latitude, east of 7.0° longitude, south of 50.8° latitude, and west of 7.3° longitude. This is roughly an enclosing bounding box for the city of Bonn in Germany. Then the second statement lets Overpass API print the result in its default format.

A variant is:

```
node[amenity=restaurant](50.6,7.0,50.8,7.3);
out skel;
```

This cuts down the output to contain only the node's coordinates and not their tags.

A second variant would be:

```
node[amenity=restaurant](50.6,7.0,50.8,7.3);
out meta;
```

This enhances the output to also contain meta data (the object's version, timestamp of change, user who touched it last by her id). The result of this latter query is 6.8 times bigger than the result of the former query. This is the reason why it makes sense to have all the different degrees of detail on the output.

The selection statement can be independently adapted:

```
way[amenity=restaurant](50.6,7.0,50.8,7.3);
out;
```

This selects elements of type "way". It is useful to get a list of restaurants, but not their geographic position, because this is stored in the referred nodes.

To get their geographic position, four statements are combined:

```
(
  way[amenity=restaurant](50.6,7.0,50.8,7.3);
  node(w);
);
out;
```

After the way selecting statement, the next statement selects all nodes that have been referred to by the just selected ways. The parentheses select the union of all selected results from the statements inside the parentheses. Altogether now all ways and the nodes that are referred to in the ways are printed. This gives the complete geometry for the ways.

For relations, it is also possible to compile the geometry:

```
(
  rel[network=VRS](50.6,7.0,50.8,7.3);
  node(r)->.busstops;
  way(r);
  node(w);
);
out;
```

This is composed of six statements: Node members of a relation contribute directly to the geometry of the relation by their coordinates. Way members contribute indirectly: the coordinates for their geometry are stored in the nodes. Hence two steps are needed to collect the nodes that are members of way members.

Because collecting the geometry is an important operation, there is a special syntax to do these operations in one step:

```
(
  rel[network=VRS](50.6,7.0,50.8,7.3);
  >;
);
out;
```

In a similar manner, additional types of conditions allow the user to do more specific searches with regard to geometry or tag values. In particular, the required time of a query is just the sum of the required time of its statements and the required space of a query is the maximum of the space required by any statement.

3.3 Featured Usage Patterns

With such a versatile language, it is unlikely that the database layout could be chosen so that it is optimal for every possible query. The following use cases are focused on: The most prominent use case is to search for an object with a specific name. Specifically, this means to search globally for nodes, ways, or relations that carry a tag with a name key and the desired value. Another common scenario would be to receive all objects of a given class within a reasonable bounding box, for example the size of a suburb, a city, a county, or a country. Such a class could be all kind of streets, as opposed to buildings, railways, and other data.

It is also often useful to obtain all the data within a given bounding box. There are multiple variants of this so-called map call, mostly depending on the conventions one has on the geometry of relations. The map call on the main site, *e.g.*, considers not only the node and way members of a relation as part of their geometry but also the node and way members of the relations that are members of a relation.

To give an idea of sizes: The medium-sized city Bonn in Germany, with about 300,000 inhabitants, comprises about 700,000 nodes, about 150,000 ways, and

2500 relations. Therefore it is desirable to be able to process and return about a million objects, i.e. all data for Bonn, within less than a minute.

As explained in the previous section, more complex queries are desirable too, but these are still assembled from simpler queries.

3.4 Hardware Constraints

To answer queries of these classes within a reasonable time, it is necessary to identify the scarce hardware resource or resources. If a database does not fit into the memory, then the usual limiting candidates for database performance are CPU time or hard disk speed. Starting with the aforementioned bounding box of Bonn. The main database, being the authoritative source of the data, allows the user to fetch 50,000 objects at most at once, and retrieving such a bounding box already takes more than a minute. Some theoretical considerations have to be done first.

To estimate the influence of reading speed, the amount of data involved is estimated: Each node's id can be encoded in 8 bytes, each node's coordinates in a further 8 bytes, and its meta data (the object's version, timestamp of change, user who touched it last by her id) in a further 17 bytes. There are only 0.7 billion tags on nodes compared to 2.3 billion nodes altogether. The average tag is no longer than 10 characters for each key and value. Hence, the average amount of data per node is estimated to be 64 bytes. This makes for 700,000 objects about 45 MB. Ways and relations each add to this a list of node ids and on average more tags, but are much fewer. To cover them, the node data amount numbers are multiplied by three to get a safe upper bound. This makes about 150 MB of disk size. This is delivered by a hard disk within about 3 s.

From the CPU point of view, a million items are handled, for which any of the usual operations like sorting will take a fraction of a second: Sorting a million unsorted integers in a test program on a 2.5 GHz processor has taken on average 0.03 s.

However, reading one million lines from a MySQL table on a single and otherwise idle server has taken, even with MySQL's binary interface, about 600 s. Thus, more potential bottlenecks need to be searched for: In the setting of a relational database, it is expected that for each OSM object in the result at least one line from a table is read. Any of these line reads may require a disk seek which has on a hard disk a latency of usually 150–250 ms, or at least 50 ms (2014). Given these numbers, it is likely that reading a random million lines out of 2.3 billion lines will cause a full table scan. That would explain the observed performance delay.

On the other hand, if accepting an order of magnitude more time for disk seeks than for reading data, then 30 s or less were acceptable. Even then, the hard disk could do only 600 disk seeks within that time frame, but it is necessary to access 700,000 objects. That means that unless almost all lines are clustered, the disk latency is the most important limiting factor for the query speed. The dominance of disk latency still holds for SSDs: SSDs organize data in blocks. It takes a small but non-zero delay for the firmware to make the asked-for block available for reading. Because the block seek times are by far dominant, it is highly likely that the reduced delays still dominate the consumed time. In addition, reading blocks of data turned out to be very cache effective with the Linux kernel. Unfortunately, a reproducible and representative approach to measure the effect is unknown.

4 Design Considerations

4.1 Initial Design

The system contains two layers: The lower layer (or back end) offers multi-map storage as a file on the disk, i.e. data is organized as a dictionary of keys such that for each key one or multiple objects are stored. The upper layer organizes the OSM data as a group of various multi-maps. The rationale for implementing a dedicated back end is due to the disk latency. A standard relational database is designed to return the lines of a table independent from each other and retrieving n lines takes essentially n-times longer than retrieving one line from a table. As it is necessary to retrieve often a million or more lines, even in a standard use case like the one sketched in Sect. 3.4, this is way too slow to be effective.

To overcome this problem, Overpass API expects the key space to be totally ordered. It then keeps all values for each key and its neighbours in the same data block. If a data block overflows, then it is split in two half-filled data blocks. The size of the data blocks allows the software to keep an index on all blocks easily in memory. Furthermore, all file systems have themselves block sizes of 2 KB or bigger (2014), hence reading such a data block requires only a few or just one disk seek. Developing realistic testing conditions is difficult because disk latency depends on the disk's fragmentation, which in turn results from the operational history.

The database transactionality is also assured on this lower layer: Because there is only a single source of writes, i.e. the replication stream of the authoritative main database, it is sufficient to provide snapshot isolation. This is done by keeping multiple versions of a data block. On the upper layer, the design principle was to keep the objects in the same structure as on the main database. This means that the geometry of a way is not stored along the way but can only be reconstructed by retrieving the nodes referred to on the way. Multiple later redesigns changed that behaviour to improve performance. Relations are managed in the same way, but this has a lesser impact on the performance.

To complete the overview, the tags of an element and the meta data (timestamp, the version of the object, the user who has last changed the object) are kept in a separate key-value store. A special key value-store for each type of object allows the software to get the spatial key for an object from its id.

4.2 Adaptions

It is desired to retrieve all ways in a sufficiently small bounding box from anywhere within one second. For that purpose the database layout sketched in the previous section turned out to be unsuitable, because of the following:

The main problem is that a large-scale way will only have a rough indication of where its geometry is actually located. Retrieving all individual nodes of more than 100,000 ways for a bounding box no matter how small is not feasible.

As a first adaption, the way values are augmented from just the list of referenced nodes by a list of the spatial indices in which the individual segments are located.

That still misses the target of serving a small bounding box within a second. So the next augmentation is to store the full node geometry within the way, i.e. for each referenced node not only the id of the node but also the coordinates of the node are stored. This improves the response time close to the expected one second. A similar change for relations is not yet implemented and might be a future improvement.

5 Evaluation

5.1 Performance on Essential Queries

5.1.1 Methodology

The tests are conducted on the current main instance of Overpass API. This server has 32 GB of RAM and a solid state disk. The processor as well as the network connection are negligible: both resources have never been exhausted during the tests.

One can ask to run the tests on a clean room instance rather than a server in production. Running on a clean room instance could benefit the software from the cache much more than under production conditions. If one tries to compensate for this with any cache flushing algorithm then it is very likely that the properties of this algorithm would have a significant influence on the results. Hence, a proper setup would be to build an abstract background activity model that mimics an average of the activity from the production system. It would be desirable to have such a load profile that is both realistic and reproducible, but this would require effort far beyond this paper. And it could still get outdated by changes in the usage pattern. As OpenStreetMap is still evolving at a high rate, such changes have happened multiple times in the past. Thus, this paper takes a pragmatic approach to repeat each request several times and to take the mean value from the individual executions.

The tests are run on a sample of 109 bounding boxes that represent all city centers of cities tagged in OpenStreetMap as having a population above one million. These bounding boxes are chosen because they represent areas that are both densely populated and densely mapped in OpenStreetMap. Other cities with a population above one million might exist, but if the population is not mapped then it is also likely that the city itself is mapped worse than average in OpenStreetMap. The given coordinate is used as center of a bounding box measured in degrees. For the edge length it is taken 0.002, 0.005, 0.01, 0.02, 0.05, 0.1, and 0.2°. From this edge lengths result as approximately 200, 500 m, 1, 2, 5, 10 and 20 km.

The bounding boxes are asked for in batch mode with 5 s pause between two queries. Time is measured as it is observable for an external observer, with the system time before and after the respective call to download the data. An exact time measurement cannot be expected, because in addition to network timings, the database serves other requests in parallel.

5.1.2 Different Area Sizes

Whether Overpass API scales well was investigated first, i.e. by what order of magnitude bigger queries are slower than smaller queries.

Objective	Check if the object density is approximately constant with respect to the bounding box size
Input	The semantics of the ?map API call with varying bounding box sizes. The measured quantity is file size
Expected outputs	The file sizes should grow quadratic with the edge length of the bounding box
Actual outputs	The file sizes grow faster than linear but far slower than quadratic
Observations	The density falls but falls quite slowly with more distance to the city center. Between the observed regions the density varies by up to a factor of 5, even when excluding the lowest and highest 5 %

For that purpose how the amount of retrieved data relates to the size of the bounding box was checked first. It cannot be expected that Overpass API scales better than the file sizes, thus how the file sizes grow was investigated in Fig. 3:

How this relates to the area of the requested bounding box was cross checked in Fig. 4:

The file sizes grow slower than the size of the queried area. A second observation is that the result file sizes for equal sized bounding boxes at different locations vary by up to an order of magnitude. This phenomenon exists at any scale between 200 m and 20 km, and probably also on other, not investigated scales. This means that one cannot predict result file sizes. Therefore, one also cannot predict query processing time by just looking at the size of the bounding box.

For this reason the runtime is also investigated in absolute values (Fig. 5), in values relative to the area size (Fig. 6), and relative to the result size (Fig. 7):

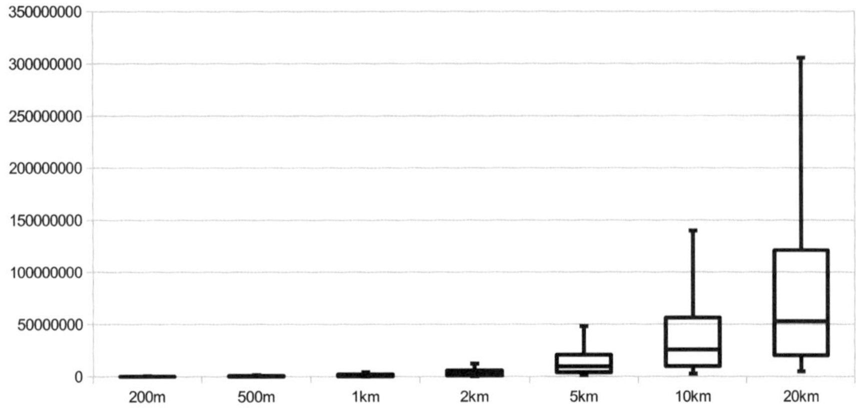

Fig. 3 Result file sizes for the sample locations by size of the bounding box. Box plot with whiskers to quantiles 5 and 95 %

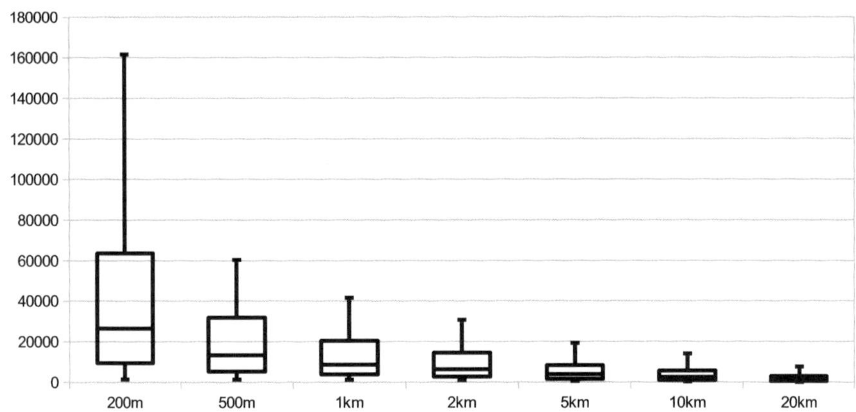

Fig. 4 Result file sizes for the sample locations divided by the size of the bounding box, in bytes per hectare. Box plot with whiskers to quantiles 5 and 95 %

Objective	Check if the runtime of a query can be predicted by straightforward indicators like the result file size or area of the bounding box
Input	The semantics of the ?map API call with varying bounding box sizes. The measured quantity is runtime
Expected outputs	The runtime of the query is proportionate to the file size of the result
Actual outputs	The runtime has a constant component of 1.5 s. The rest of the run time grows slower than the result file size, hence much slower than the area of the bounding box
Observations	The predominant components of the runtime do not scale with the bounding box. They need constant or approximately constant time

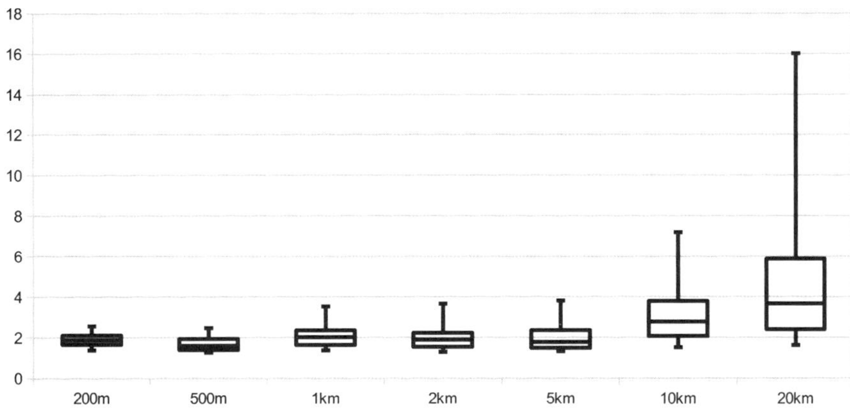

Fig. 5 Query execution time for the sample locations by size of the bounding box. Box plot with whiskers to quantiles 5 and 95 % seconds

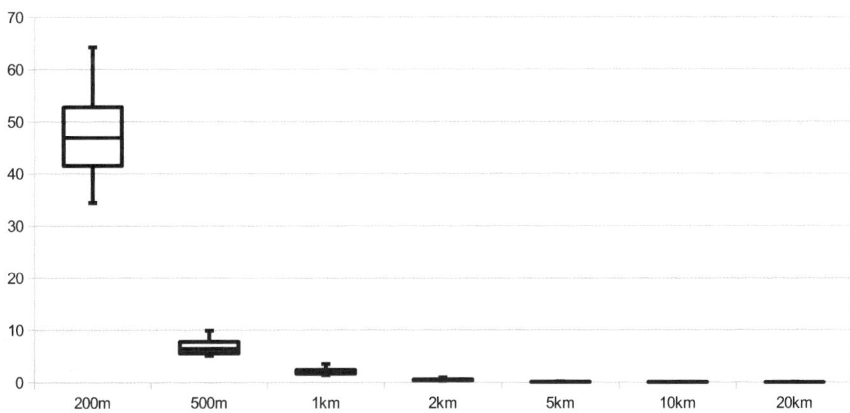

Fig. 6 Query execution time for the sample locations divided by the size of the bounding box. Box plot with whiskers to quantiles 5 and 95 % seconds per square kilometer

First of all, it is observed that each and every query takes at least 1.5 s. The following section shows that this is due to back-links to relation members of relations without geometry. Second, a significant growth in runtime is observed only for areas with an edge length beyond 5 km. Once again, the size of the queried area does not predict the runtime—the area grows much faster than the runtime. But even the result file sizes grow faster than the query runtime. The diagram in Fig. 7 has been scaled to give at least some insight into that the quotients are still decreasing on the scale of 20 km.

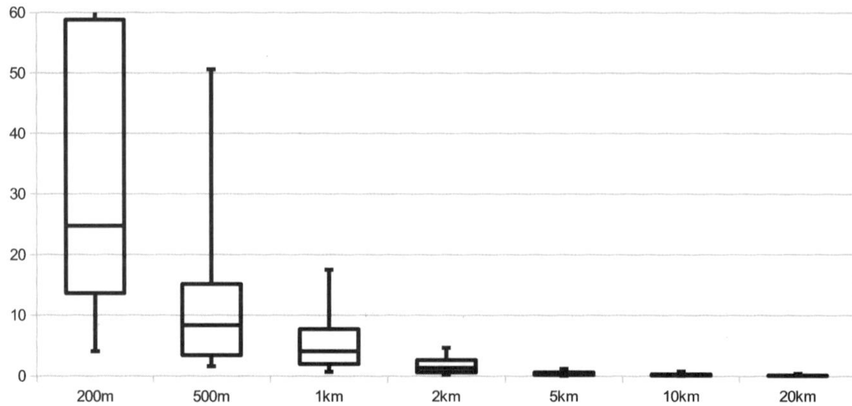

Fig. 7 Query execution time for the sample locations divided by the file size of the result. Box plot with whiskers to quantiles 5 and 95 % seconds per megabyte

An explanation for this could be that a significant amount of time is still spent on components that do not scale with size. An example is the back traversal of links from non-geometric relations: because these relations are both few and not well-structured, these back-links are resolved by scanning through all relations. This is independent from the size of the bounding box that has been asked for.

Thus, altogether it can be concluded that Overpass API indeed scales well, at least for bounding boxes up to an edge length of 20 km. For larger edge lengths the user quota mechanism of Overpass API inhibits some of the queries, because they surpass several GB for the largest areas (currently Paris). Retrieving filtered data would be possible, but it cannot be given an accurate speedup in that case.

5.1.3 Speedup of Filtering

Users of OpenStreetMap data are often interested only in a part of the data. For routing or similar purposes, it might suffice to know all the highways. Building outlines or other unrelated data can be skipped. Doing the filtering on the server anyway has the advantage of reducing the amount of transferred data. Overpass API additionally aims to make filtered queries faster than non-filtered queries for the same size of area. It is investigated in this section whether there is an effective speedup by filtering.

Objective	Check whether filtering improves query speed
Input	Various API calls that filter the output. See details below. The ratio of their runtime and the runtime of the ?map API call for the same bounding box is evaluated
Expected outputs	The queries are faster by a factor that is independent both of the size of the bounding boxes and of the location
Actual outputs	The speedups are more consistent than expected. They deviate at most by a factor of 3
Observations	Outliers are concentrated on a single bounding box size, 500 m. This could be a measurement error due to varying background load

For that purpose several types of queries are compared to each other and their relative speedup is measured.

An overview of the various filters:

Filter	Compared against	Observed speedup	Figure
Without meta data	Map API call	1.5–2	Fig. 8
Only object ids	Map API call	2–3	Fig. 9
Ignore relation back-links	All objects, without meta data	1.5–5	Fig. 10
Only nodes and ways	All objects, without meta data	2–7	Fig. 11
Only highways	All objects, without meta data	3–6	Fig. 12
Only pharmacies	All objects, without meta data	5–10	Fig. 13

For the beginning, the map query with full output is compared to the map query with output reduced to geometry and tags (Fig. 8):

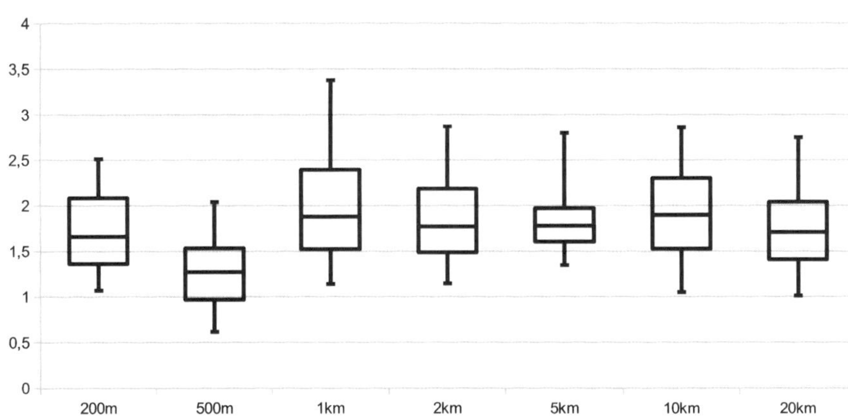

Fig. 8 Speedup for omitting meta information in the output. Box plot with whiskers to quantiles 5 and 95 %

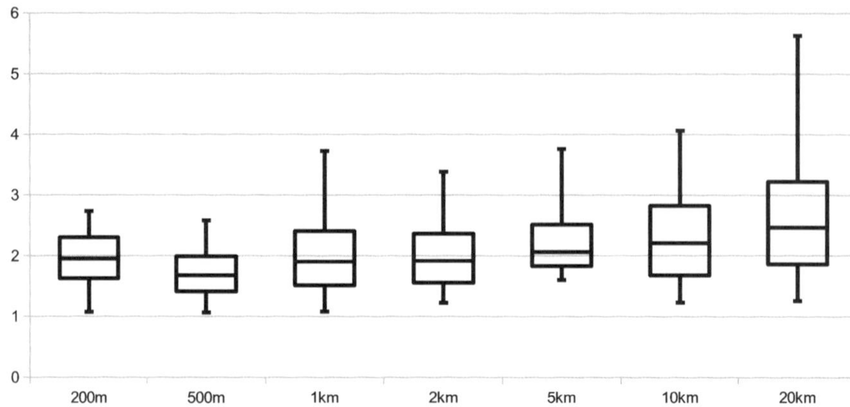

Fig. 9 Speedup for outputting only the found element ids. Box plot with whiskers to quantiles 5 and 95 %

First of all, the speedups are much more homogeneous than expected. This creates confidence that the effects of other queries, negatively by concurrence for resources, or positively by preparing cache lines, sum up to far less than an order of magnitude. A speedup between 1.5 and 2 is seen for a majority of queries on every scale. The lower speedups at 500 m have no straightforward explanation and are probably the strongest impact from the background load that is seen so far. In a next step (Fig. 9), two variants of the map query are compared: full output to the output of only the ids. In this variant, the output statement does not require any disk activity.

Again, the speedup within each group of requests by bounding box size is very homogeneous. As opposed to before, a slight increase of the speedup is seen with growing size. This could be again explained by the amount of time spent on non-geographic relations: a constant amount of time affects the speedup for shorter runtimes more than for longer runtimes.

Another observation is that there is virtually no difference between outputting tags and outputting only ids. Hence, outputting tags is sufficiently optimized. From now on a fixed output format is used for all further runs: the output format with tags and without meta data is always chosen (the object's version, timestamp of change, user who touched it last by her id), and it is also compared against the map call with this output format.

It is started with skipping the query for back-links to non-geographic relations (Fig. 10):

There is a significant speedup for small bounding boxes. The speedup then decreases with increasing bounding box sizes. The observed numbers are consistent with a model assumption that the relation back-link statement always takes between 1.5 and 2 s. For the sake of comparison the time for the map call completely

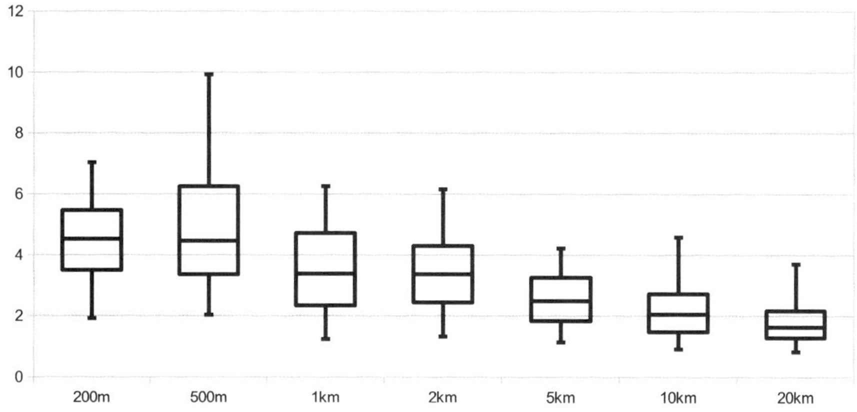

Fig. 10 Speedup of skipping relation back-links. Box plot with whiskers to quantiles 5 and 95 %

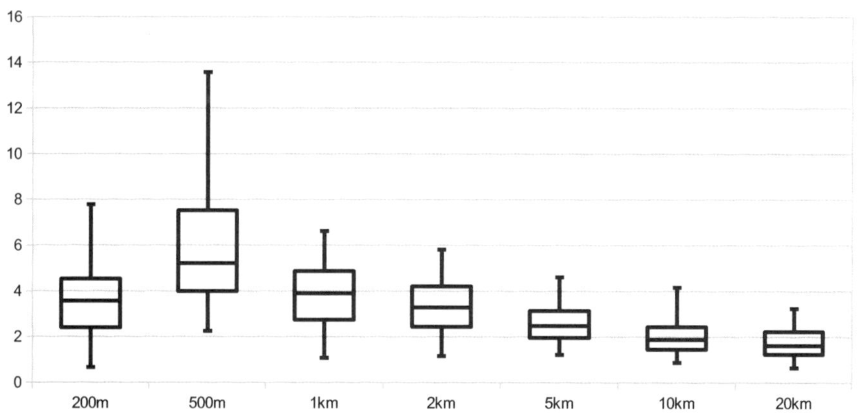

Fig. 11 Speedup of skipping all relations. Box plot with whiskers to quantiles 5 and 95 %

without relations is also measured (Fig. 11): The behaviour of the speedup is almost identical to the previous benchmark. Smaller fluctuations for 200 and 500 m could come from the same disturbance source that has been observed for the speedup measurement of omitting meta data output. This means that indeed the back-link resolution takes almost all the time of relation handling.

The last measurement is filtering for objects with specific tags only. It is started with all highways (Fig. 12). This is a relatively dense subset of all data. A consistent speedup of around 4 is observed, almost independent of the scale. Thus, it is indeed useful to filter even for such dense object categories like highways. In particular, as has been chosen the output mode without meta data, this multiplies with the speedup over meta data output. Thus there is a speedup of about 7 compared to the

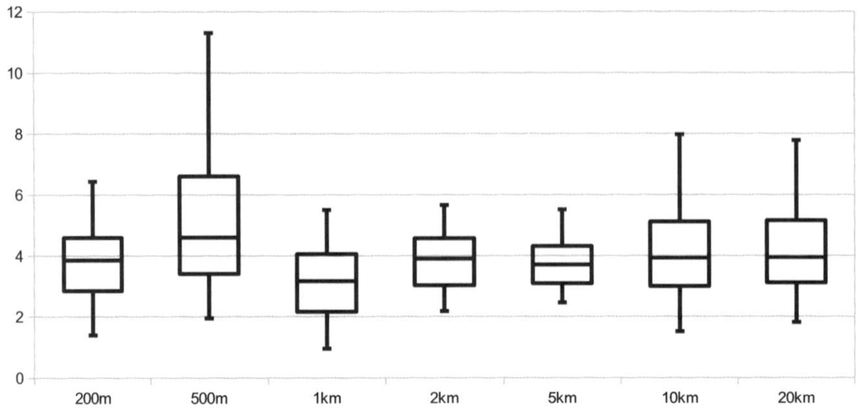

Fig. 12 Speedup of filtering for highways only. Box plot with whiskers to quantiles 5 and 95 %

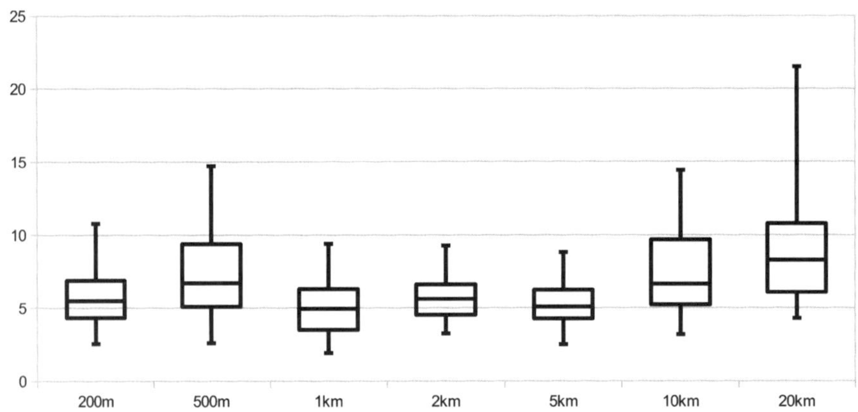

Fig. 13 Speedup of filtering for pharmacies only. Box plot with whiskers to quantiles 5 and 95 %

full map query. In the final measurement, it is filtered for pharmacies (Fig. 13). These are a sparse fraction of all data. A speedup of around 5 is seen for small bounding boxes, and this increases further for larger bounding boxes to a median of 8 for bounding boxes of 20 km edge length.

5.1.4 Comparison to the Main API

It is concluded with a comparison of the Overpass API server to the main database server. This is done to answer the hypothetical question of whether having an Overpass API server brings a higher performance than having an additional database mirror server.

Objective	Check whether Overpass API returns data faster than the main database
Input	For both servers the ?map API call. The measured quantity is runtime. It is cross-checked whether both requests return the same data
Expected outputs	Each pair of requests should return the same data. The Overpass API should be significantly faster, in particular on larger bounding boxes
Actual outputs	The requests are equal in all cases. Overpass API is faster than the main API by at least a factor of about 3
Observations	For larger bounding boxes, Overpass API is faster than the main API by a factor of up to 15

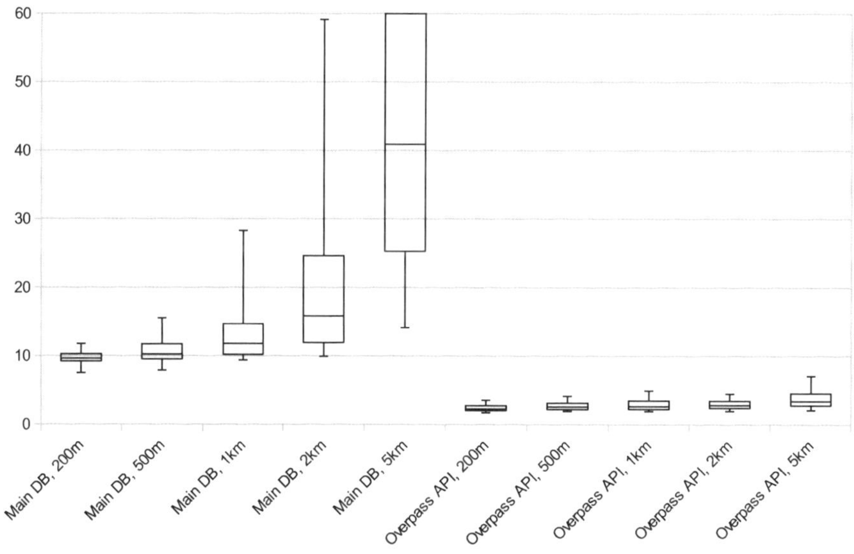

Fig. 14 Run times for the main DB and Overpass API, different edge lengths. Box plot with whiskers to quantiles 5 and 95 %

The most important observation is that the main database server has a limit of 50,000 objects per requests. Thus, Overpass API will fill the gap for all bigger requests. In that sense the following comparison (Fig. 14) is skewed because it considers only map query sizes that are still digestible for the main database server. For the sake of simplicity all requests that fail on the main database are treated as requests with infinite runtime. These are already 24 of 109 requests for the edge length of 5 km. No queries have failed on Overpass API.

Even before this limit, quite long answer times occur for the main database API. In particular, the main database API scales poorly for large bounding boxes. Thus, even when the main database API could be improved to tie on small bounding boxes then it still would be too slow on large bounding boxes to fulfill requests for more than 50,000 objects.

5.2 Performance in the Field

Beside the featured usage patterns, which requests are really important for the performance is also investigated, because they are actually used. This can still be skewed, because queries that are fast are more attractive to employ than slow queries. Data from July 2012 to September 2013 from the main Overpass instance (2014) is analyzed. There have been a total of 47,590,122 requests during that period.

To check the significance of the data, the spatial distribution is investigated: 48 % of all requests contain a bounding box in Europe, 35 % contain a bounding box in Northern America, 7 % contain a bounding box in Asia, each 1.5 % contain a bounding box in each of Southern America and Australia, each 1 % contain a bounding box in each of Africa and Antarctica. The remaining 5 % are on a latitude and longitude of zero. Although this coordinate in the sea may be considered as in Africa, it is more likely that these requests are erroneous.

No real hotspot for queries exist: The region that attracts most requests is Copenhagen, but still only 2 %. Even this might be skewed because there is a smartphone app for cyclists that is popular in Copenhagen and relies on the public instance of Overpass API.

In a next step the importance of bounding boxes is investigated. For this purpose the sizes of the requested bounding boxes are sketched by square degrees (Fig. 15):

About one third of all queries use a bounding box of a size between 10^{-4} and 10^{-2} square degrees, and a further one sixth of all queries use a larger bounding box. It has been shown in the previous section that these queries enjoy a significant speedup.

One third of all requests used a bounding box that is smaller than 10^{-4} square degrees. Queries of such a size could likely be served by both the main database and Overpass API within less than a second.

The remaining one sixth of all requests uses filtering criteria that is not available on the main database.

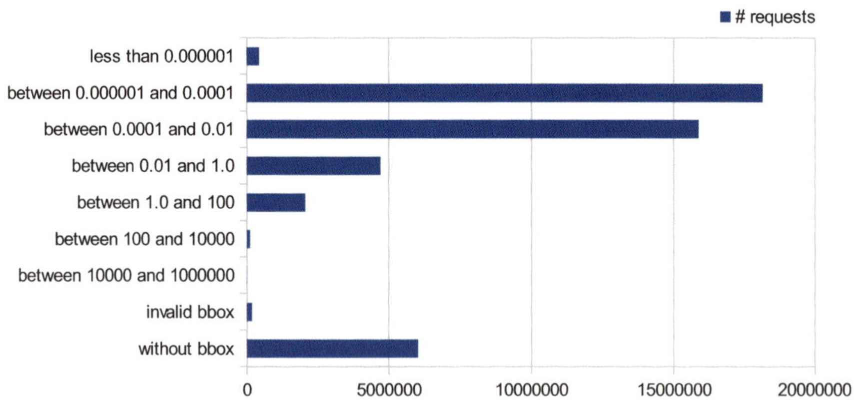

Fig. 15 Distribution of requests by the size of bounding boxes

5.2.1 Discussion and Conclusion

The geographic distribution of the requests confirms that requests come from and cover the entire world. This gives confidence that the observed patterns represent the total demand well. On the other hand, it is still possible that one or more new applications using Overpass API could alter the observed usage patterns.

A second weakness is the influence of the methodology. A request with multiple bounding boxes is counted only once, because a lot of more complex requests repeat the same bounding box multiple times. With earlier APIs, such requests would have been realized as a group of requests, hence such bounding would have been counted multiple times. A similar problem is erroneous requests. Such requests are more likely to be repeated, because the tool might misunderstand the error message or the fact of an empty result.

However, the fact of whether a certain location has been requested at all or not is not affected by these problems, and there is simply no more reliable metric. Altogether, almost two thirds of all requests are composed of requests of the expected types, and in most cases a bounding box is combined with a tag filtering criterion. The remaining requests would mostly be small enough to be run on the main database servers, but running them on the mirror service is a welcome relief (2014).

5.3 *Future Improvements*

With the general availability of SSDs it will be necessary to benchmark the remaining influence of disk latency. Although in theory no disk latency should exist on SSD, and in practice SSDs have at least a much smaller latency, the remaining gap indicates that reading blocks instead of lines may still be an advantage. A similar question arises with the availability of servers with enough memory to hold the entire database in the RAM. In such a setting, disk latency should not play a role at all. It will be essential for the further development of the software to understand which part of the hardware then becomes the bottleneck. The general one-second-delay on the map query shows that an improvement for back-links from relation members might be more important than expected.

It would also be helpful to develop a proper test environment. Testing on a production system with load always holds the risk of being inaccurate. Thus, only seen effects that are both strong enough and highly reproducible to be visible from this quite imprecise test setting are observed. Thus there are only seen effects that are both strong enough and highly reproducible to be visible from this quite imprecise test setting. A proper test setting on a dedicated system would make it possible to find more subtle effects. However, this would also require a yet-to-be-developed representative load profile to show a deterministic yet reproducible cache behaviour.

References

Mapnik. github.com/mapnik/mapnik. Retrieved 2014-05-05 04:00:00 UTC

Maria DB (2014) code.launchpad.net/maria. Retrieved 2014-05-05 04:00:00 UTC

Main database usage policy. wiki.osm.org/wiki/API_usage_policy#Heavy_Users. Retrieved 2014-05-05 04:00:00 UTC

Minute diffs. planet.osm.org/replication/minute. Retrieved 2014-05-05 04:00:00 UTC

Mooney P, Corcoran P (2012) Characteristics of heavily edited objects in openstreetmap. Future Internet 4(1):285–305

MySQL. mysql.com. Retrieved 2014-05-05 04:00:00 UTC

Neis P, Zipf A (2012) Analyzing the contributor activity of a volunteered geographic information project—the case of OpenStreetMap. ISPRS Int J Geo-Information 1(2):146–165

Neis P, Zielstra D, Zipf A (2011) The street network evolution of crowdsourced maps: OpenStreetMap in germany 2007–2011. Future Internet 4(1):1–21

Olbricht R (2014) Overpass API. github.com/drolbr/Overpass-API. Retrieved 2014-05-05 04:00:00 UTC

OpenStreetMap portal. openstreetmap.org. Retrieved 2014-05-05 04:00:00 UTC

Overpass API public instance. overpass-api.de. Retrieved 2014-05-05 04:00:00 UTC

Overpass Turbo. overpass-turbo.eu. Retrieved 2014-09-25 12:00:00 UTC

Overpass API Munin stats. overpass-api.de/munin/de/overpass-api.de/index.html#disk. Retrieved 2014-05-05 04:00:00 UTC

Planet.osm. planet.osm.org/planet. Retrieved 2014-05-05 04:00:00 UTC

PostGIS. postgis.net/source. Retrieved 2014-05-05 04:00:00 UTC

PostgreSQL. postgresql.org. Retrieved 2014-05-05 04:00:00 UTC

Quinion B, Hoffmann S et al (2014) Nominatim. github.com/twain47/Nominatim. Retrieved 2014-05-05 04:00:00 UTC

Ramm F (2013) Wer ist der Boss bei OpenStreetMap? Presented at the FOSSGIS 2013. Rapperswil, Switzerland

Ramm F, Topf J, Chilton S (2010) OpenStreetMap: using and enhancing the free map of the world. Uit, Cambridge

Ruemmler C, Wilkes J (1994) An introduction to disk drive modeling. IEEE Comput 27(3):17–29

Redmond E, Wilson JR, Carter J (2012) Seven databases in seven weeks: a guide to modern databases and the NoSQL movement. Oreilly and Associate Series. Pragmatic Bookshelf, Massachusetts

Schmidt M (2013) Frauen in OpenStreetMap. Presented at the FOSSGIS 2013. Rapperswil, Switzerland

Standard on cross origin resource sharing. www.w3.org/TR/cors/. Retrieved 2014-05-05 04:00:00 UTC

Topf J (2013) Das OpenStreetMap-Datenmodell. Presented at the FOSSGIS 2013. Rapperswil, Switzerland

Wilcox M (2014) The second extended filesystem. http://lxr.linux.no/#linux+v2.6.29/Documentation/filesystems/ext2.txt. Retrieved 2014-05-05 04:00:00 UTC

Part II
Social Context

The Impact of Society on Volunteered Geographic Information: The Case of OpenStreetMap

Afra Mashhadi, Giovanni Quattrone and Licia Capra

Abstract Volunteered Geographical Information (VGI) has been extensively studied in terms of its quality and completeness in the past. However, little attention is given to understanding what factors, beyond individuals' expertise, contribute to the success of VGI. In this chapter we ask whether society and its characteristics such as socio-economic factors have an impact on *what* part of the physical world is being digitally mapped. This question is necessary, so to understand where crowd-sourced map information can be relied upon (and crucially where not), with direct implications on the design of applications that rely on having complete and unbiased map knowledge. To answer the above questions, we study over 6 years of crowd-sourced contributions to OpenStreetMap (OSM) a successful example of the VGI paradigm. We measure the positional and thematic accuracy as well as completeness of this information and quantify the role of society on the state of this digital production. Finally we quantify the effect of social engagement as a method of intervention for improving users' participation.

Keywords VGI · Completeness · OpenStreetMap · Socio-economic factors

1 Introduction

The advent of Big Data alongside the availability of smartphones with capabilities such as positioning services (GPS) has enabled a new era of Volunteered Geographic Information (VGI). From collaborative mapping to 3D modelling of spatial objects,

A. Mashhadi (✉)
Bell Laboratories, Dublin, Republic of Ireland
e-mail: afra.mashhadi@alcatel-lucent.com

G. Quattrone · L. Capra
University College London, London, UK
e-mail: g.quattrone@cs.ucl.ac.uk

L. Capra
e-mail: l.capra@cs.ucl.ac.uk

J. Jokar Arsanjani et al. (eds.), *OpenStreetMap in GIScience*,
Lecture Notes in Geoinformation and Cartography,
DOI 10.1007/978-3-319-14280-7_7

VGI has improved the state of geographical information systems greatly over the past years by engaging participants from all the world. OpenStreetMap (OSM) alongside WikiMapia is perhaps one of the most successful examples of VGI, with over 1.7 million users (OSM Wiki http://wiki.openstreetmap.org/wiki/stats) collectively building a free, openly accessible, editable map of the world. However, these platforms have been subject to scrutiny by the research community over the years; one of the fundamental concerns is the credibility and integrity of the contributed information, as we take a task away from skilled employees and assign it to an undefined, self-selected crowd. Several studies have investigated the credibility of the volunteered content (Girres and Touya 2010; Haklay 2010; Haklay et al. 2010; Ludwig et al. 2011; Wilkinson and Huberman 2007; Zielstra and Zipf 2010) and have concluded that the content is of high quality. However, the quality of the contributed information is not the only concern that has emerged as a result of this knowledge production paradigm shift; another important issue is that of completeness of the information (Haklay et al. 2010; Koukoletsos et al. 2012). That is, how much information is contributed to the VGI systems and how would such a system grow over time. In addition to these two concerns, researchers have also been addressing the voluntary dimension of VGI by looking at what incentive models and factors motivate the contributions from individuals (Coleman et al. 2009; Harvey 2013). However, one aspect that has perhaps received less attention from the GIS research community is the impact of *society* and factors such as those of socio-economics on the contributed information. This aspect has been studied extensively in the domain of social sciences (Hargittai and Litt 2011; Schradie 2011) and has been shown to have a high impact on the content generation in Web 2.0. In the domain of GIS, as the contributions are intrinsically spatial, the potential digital production gap may contribute to some areas not being mapped. The risk of course is that if the socio-economic factors are responsible for this digital production gap, the deprived areas (e.g., less wealthy) would also remain information deprived (Graham et al. 2014). We thus aim to investigate this aspect further by studying the extent to which society's characteristics such as *socio-economic* factors determine the success of spatial crowdsourcing and VGI.

To this end, in this chapter, we study the impact of society on the quality and completeness of the contributed information in OSM. In particular we investigate over 6 years of OSM contributions to Greater London, United Kingdom, in terms of its accuracy and completeness. We first measure the quality of OSM contributions in terms of positional accuracy and thematic accuracy, paying particular attention to what properties of the editors are responsible for this quality. After presenting the accuracy of OSM in London, we then investigate the impact of society in the completeness of this contributed information. We do so by first measuring the completeness by comparing the OSM data to a proprietorial dataset. We then argue that this completeness is affected by the socio-economic factors of the society and propose a model that can capture the completeness as well as its evolution over time.

We finally end this chapter by presenting the possible interventions that could be done to improve communities' participation in OSM. We do so by measuring the impact of the social engagement of the editors on their participation, which can then in turn account for improving the sparsity of the contributed information.

2 OpenStreetMap London

OpenStreetMap is freely available to download from various repositories on the web which provide the latest snapshot of the OpenStreetMap project. We gathered the dataset of London, United Kingdom, from (Geofabrik http://www.geofabrik.de/data/download.html) which contains the history of all edits since 2006 on all spatial objects performed by all users. In OSM terminology, spatial objects can be one of three types: *nodes*, *ways*, and *relations*. Nodes are single geospatial points, defined using latitude/longitude coordinates, and they can be used to represent Points Of Interest (e.g., cafes, restaurants, hospitals, schools); ways consist of ordered sequences of nodes, and mostly represent roads (as well as streams, railway lines, and the like); finally, relations are used for grouping other objects together, based on logical (and usually local) relationships (e.g., administrative boundaries, bus routes).

We chose Greater London, United Kingdom as our subject city as we are interested in studying a metropolitan city with a diverse society which has been engaged with OSM since the very beginning. Furthermore, as we are interested in evaluating the impact of society on the contributed information, we limited our investigation to only Points Of Interest (POIs) rather than the roads. This is because the contribution to the OSM differ greatly between the two categories: road mapping is typically done by users who have high expertise in both the geography of an area and the editing tools required to digitally represent it, while POI mapping can be performed by any city dweller, with no specific cartographic skills required. The latter category is thus more representative of the broad VGI setting and the impact that ordinary citizens can have on VGI. Finally, to consider only genuine users' contributions, we have excluded contributions that most likely correspond to bulk imports. Two bulk imports were detected in the whole dataset, with tens of thousands of edits done in a single day by a single user, spread throughout Greater London (e.g., more than 20,000 post boxes spread across all Greater London appeared in OSM in only one day in 2009 from the same user). We chose to discard such data as we intend to model genuine 'bottom-up' user-generated contributions, of which massive imports are not representative of.

To evaluate the quality of the OSM POIs in London based on the dataset at hand, we need to: (1) define benchmarks against which to compare accuracy; and (2) define quality metrics for OSM objects.

2.1 Benchmarks

We considered two different commercial geographic information systems covering the same type of information (in terms of POIs) as OSM: Navteq and Yelp. Navteq (http://www.navteq.com/) is the leading global provider of maps and location data, covering not only roads but also millions of POIs of varying nature, from restaurants to hospitals and gas stations. Yelp (http://www.yelp.com/) focuses on business listings, from store-fronts (e.g., restaurants and shops) to services (e.g., doctors, hotels, and cultural venues). Being commercial services, Yelp and Navteq's primary objective is to ensure the highest level of accuracy of its data (the information contained there is factually correct and up-to-date). We then built our benchmark (or *ground truth* dataset) as the set-intersection of Navteq and Yelp data; in doing so, a POI in Navteq is considered to be the same POI in Yelp if the name is the same and the geographic distance is less than 20 m.

2.2 Metrics

In both OSM and in the ground-truth dataset, a POI is defined as a triple: $poi = \langle name, amenity, (lat,\ lon)\rangle$, where *name* is the POI's name, *amenity* is its category (e.g., cafe, restaurant), and (*lat, lon*) are the coordinates defining its geographical position. We then quantify *quality* of OSM data in terms of its positional accuracy and thematic accuracy defined based on *geographic error* and *lexicographic error*, respectively. We measure geographic error as the Euclidean distance between the OSM points and those in the ground truth dataset. The lexicographical error is computed as the Levenshtein distance between the POI names (of OSM and the ground truth dataset). This calculated Levenshtein distance captures the minimum number of single characters that are required to change the POI name as stated in OSM to the name that exists in the ground truth dataset. Finally, we consider two points to be equivalent in both datasets if their geographical error is less than 100 m and their Levenshtein distance is less than 0.35.[1]

The results based on the above benchmark and metrics, as illustrated in Fig. 1, indicate an overall *high quality* of information for OSM POIs with geographic errors almost normally distributed and their average value is less than 25 m thus revealing accurate positioning of POIs on the map with respect to the ground truth dataset. Lexicographic error is almost zero (0.13 on average), thus revealing thematic accuracy in spelling names of POIs. This overall high level of quality in OSM is an extremely positive result, encouraging further insights to measure the completeness of this information.

[1] These values were chosen after manual inspection of a number of POIs jointly present in the two datasets that we knew to be the same.

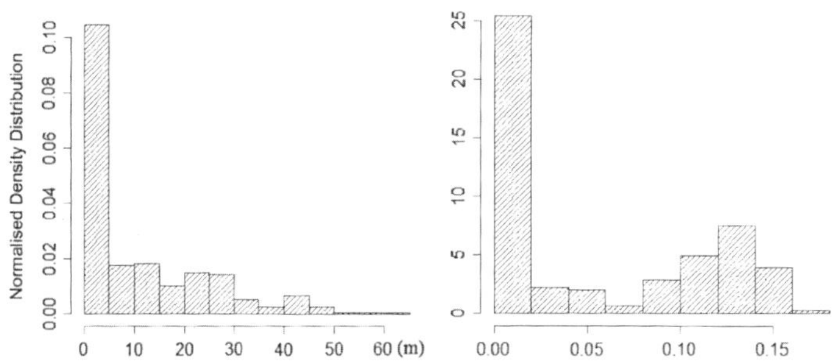

Fig. 1 Normalized density distributions of standard deviation of error for geographical error and lexical error, respectively

3 Impact of Society on Information Completeness

In the previous section we demonstrated that the OSM POIs have a high quality. Indeed, OSM's positional and thematic accuracy has shown to sometimes supersede the most reputable geographic datasets, performing especially well in urban areas (Haklay et al. 2010). However, these accuracy measures alone are not enough, and another concern is the *completeness*. In other words, *what part* of the physical world has been digitally mapped and which parts are lacking digital representation? Answering this question is necessary, in order to understand *where* crowd-sourced map information can be relied upon (and crucially where not), with direct implications on the design of applications that rely on having complete and unbiased map knowledge. To address this question, we investigate the impact of society on the completeness of information. In so doing, we first measure the completeness of information in terms of POI presence in OSM London in comparison with the described benchmark dataset.

In order to compute the completeness of POIs, we require a matching algorithm to map OSM POIs to those in the benchmark dataset. We first need to relate POIs in OSM with the same POIs in the ground-truth dataset in an automatic way. We borrowed the same matching methodology as described earlier, where two POIs are considered the same based on their lexical and geographical similarity.

Based on the above matching, we have evaluated completeness of OSM POIs for Greater London as:

$$\text{Completeness} = \frac{\#(\{\text{POIs in OSM}\} \cap \{\text{POIs in Ground} - \text{Truth}\})}{\#\{\text{POIs in Ground} - \text{Truth}\}} \quad (1)$$

with *Completeness* $\in [0, 1]$. The higher the completeness, the higher the extent to which the ground-truth POIs are also present in OSM.

3.1 A Non-uniform Completeness

This section reports on the results of our completeness analysis based on the above formulation. We first considered the area of Greater London as a whole, for which we found the completeness to be 0.35. However, this single aggregate value does not reveal much in terms of what areas of London are being digitally mapped. We thus considered the finest level of granularity for which societal information (such as population etc.) is still available. We selected *wards* representation to define the spatial granularity of our analysis. Wards are spatial boundaries defined by London Local Authorities. Currently London consists of 600 wards (London Data Store 2011). Figure 2 illustrates the choropleth map of London's POI distribution, where each tile represents a ward. As shown, completeness is non-uniformly distributed across the city. Previous studies on completeness of OSM for road networks have revealed that distance from the city center is inversely related to this completeness (Zielstra and Zipf 2010); although at a first approximation a similar pattern seems to emerge for POIs too (i.e., the further away we move from the city center, the worse the completeness), we can also identify various suburban areas with high completeness. Figure 3 further shows the histogram approximating completeness distribution at ward level. As shown, there are many wards where completeness is very low (≈0), and a few wards where completeness is quite high (≈0.6) instead.

We hypothesize that the society's characteristics influence and contribute to this non-uniform distribution of spatial information. To test our hypothesis we take a closer look at the contextual factors affecting different areas (wards) of London. We extract and consider the following society factors:

Population Using UK Census 2011 data published by the National Statistics Office (Census 2011) we have information about population at the ward level. Previous studies of OSM coverage for road networks have revealed a correlation between the number of contributors in an area and the number of OSM objects digitally mapped in that area (Girres

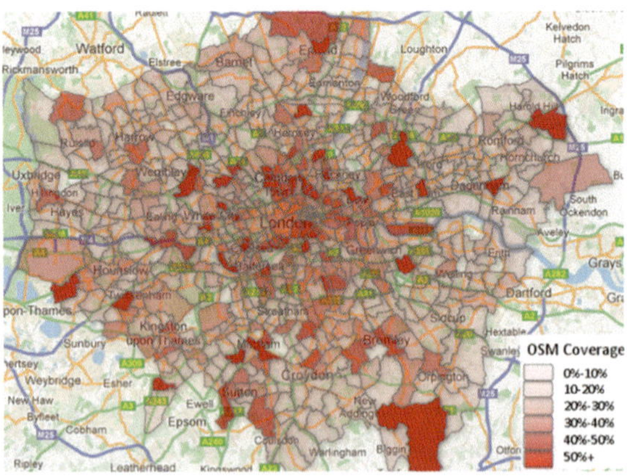

Fig. 2 Choropleth map of OSM POI's completeness for Greater London. The darker area correspond to higher completeness

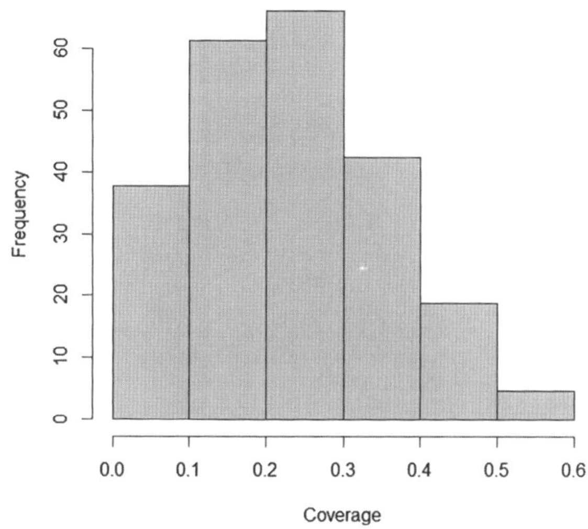

Fig. 3 Frequency distribution of POI completeness

and Touya 2010). We have thus selected population as an attribute for investigation in this study, as it can give us an expectation of contributions per area. Although higher population density does not directly translate into a higher number of contributors, we may expect more contributors per unit area to exist in denser areas. The hypothesis we thus want to test is the higher the *population density* of an area (that is, population divided by ward size), the higher the completeness

Poverty — Analyzing the relationship between poverty of an area and completeness is important, as it may reveal the impact that (lack of) technology adoption (e.g., use of Internet), as well as (lack of) available leisure time, has on it. In this regard, UK Census data contains information about the Indices of Multiple Deprivation (IMD). IMD is a set of indicators, published by the UK Office for National Statistics, measuring deprivation of small geographic areas known as Lower-layer Super Output Areas (LSOA) in England. The hypothesis under test is that poverty of an area is negatively correlated with digital mapping of its POIs

In addition to that we consider in our study the distance from where the social and economic activities happen. Previous studies on OSM have shown that road completeness decreases when moving away from the city centers (Zielstra and Zipf 2010). Similarly, we are interested in examining the effect of distance from the city center on completeness. However, metropolitan cities contain more than one center per se but include multiple urban hubs referred to as poly-centers (Brunn et al. 2003). London currently has 10 different poly-center (Roth et al. 2011), from which we consider and compute the Euclidean distance. We then used the shortest distance as our 'distance from the center' factor, and tested the hypothesis that the closer to the center, the higher the completeness.

Table 1 Pearson correlation coefficients r and p-values codes between socio-economic factors and OSM completeness at the ward level

Factor	r	p-value
Population density	0.32	***
Poverty	−0.10	*
Distance from the nearest poly-center	0.36	***

p-value significance codes: '***' 0.001 '**' 0.01 '*' 0.05 ''1

Table 1 reports the Pearson Correlation coefficients between each of the previous factors and OSM completeness as well as the p-value codes, indicating the significance level of each presented result.

The results indeed confirm that population density is positively correlated with completeness ($r = 0.32$ and p-value < 0.001). In particular, we found that an increment in population density of 50 people per hectare corresponds to a 25 % increase in completeness for the average case. Focusing on poverty, we can see that $r = -0.10$ is significantly weaker than that found for other factors such as population, suggesting that, although significant, poverty itself is only a secondary factor in explaining completeness. Turning our attention to the last factor under examination, distance to the closest poly-center, our intuition is confirmed by Table 1, which shows that distance from the closest poly-center is inversely correlated with completeness ($r = 0.36$ and p-value < 0.001). In particular, we also found that a decrement of 5 km in distance from the closest poly-center corresponds to a 28 % increase in completeness for the average case.

However, these findings raise concerns in terms of the long-term sustainability and completeness of the VGI. More specifically, is the completeness going to spontaneously grow across the city? Or are there going to be areas that will continue to be neglected? To address this question we built a set of models based on the discovered socio-economic factors that can accurately capture the *digital* growth of spatial information in VGI.

3.2 Growth of Spatial Content Production

In order to measure the growth of contributed information over time, the first step was to choose a *spatial* and *temporal* unit of analysis. In terms of the spatial unit of analysis, we have maintained the same level of granularity as before and operate at the ward level of London. In terms of the temporal unit of analysis, we tried different time units, from finer (3 months) to coarser (18 months) granularity. In the end, we chose to report the results for the smallest unit of granularity (12 months) that still afforded statistically significant results across *all areas* of Greater London.

We then needed to define a metric that reflected which areas had been digitally mapped and which had been neglected instead. To this purpose, it is worth pointing out that not all areas naturally require the same amount of OSM edits to be mapped. For example, areas containing many services and attractions will require many OSM edits to be mapped (e.g., Soho in London); however, sparse areas like parks and industrial estates will require significantly less. To capture this property, we chose as metric *OSM activity*, defined as the number of OSM edits *relative to* the number of physical POIs in each ward at that time:

$$\text{OSM activity} = \frac{\#\text{OSM edits}}{\#\text{POIs}} \quad (2)$$

#OSM edits is readily available from our OSM dataset. To estimate #POIs, that is, the actual number of POIs present in each area, we used the ground-truth dataset as before.

Figure 4 illustrates the cumulative temporal evolution of OSM activity (Eq. 2) in London from 2007 to 2012. As shown, the vast majority of areas have low cumulative activity (with only a few wards slightly above 0.5); furthermore, complex dynamics are at play, with no clear pattern emerging (e.g., no core-to-periphery spreading).

To capture the growth of OSM information, we built a regression model which takes into account the past OSM activity of each area as well as the following two features:

Community Editing — We argue that, regardless of spatial positioning of wards within a city, they attract the same OSM contributors [because they might, for example, offer related attractions/urban functions (Cranshaw et al. 2012)]. We thus incorporate the *community* feature into our model which hypothesizes that if a ward has been edited by contributors who have heavily edited other wards in the past year, the ward is likely to be edited in the future. In other words, the

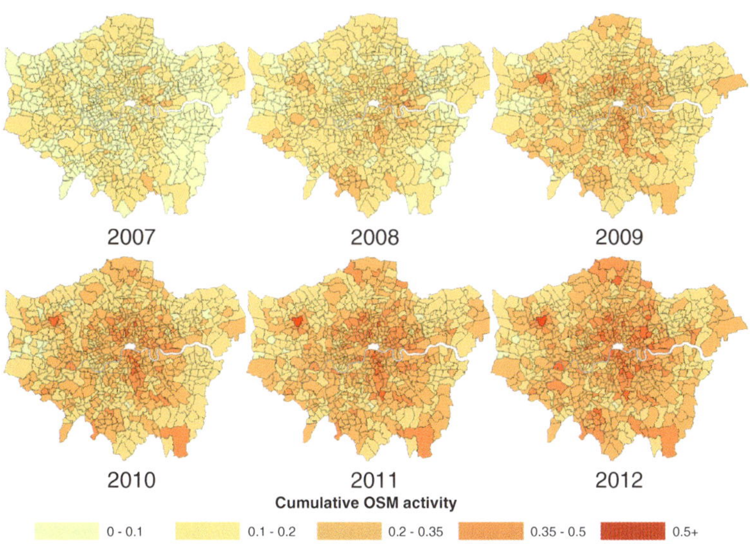

Fig. 4 Cumulative OSM activity from 2007 until 2012

Table 2 True negative rate (slow growth), true positive rate (fast growth), accuracy and sensitivity of our classification model

Predicted year	TN rate	TP rate	Accuracy	Sensitivity
2009	0.78 (+56 %)	0.75 (+50 %)	0.77 (+54 %)	0.77 (+54 %)
2010	0.81 (+62 %)	0.78 (+56 %)	0.80 (+60 %)	0.80 (+60 %)
2011	0.82 (+64 %)	0.79 (+58 %)	0.81 (+62 %)	0.81 (+62 %)
2012	0.79 (+58 %)	0.75 (+50 %)	0.77 (+54 %)	0.78 (+56 %)

Relative improvement of each model with regards to a random classifier is also reported in parentheses

activity of a ward depends on the past activity of its 'co-edited' areas, where two areas are defined as 'co-edited' if they are edited by the same shared community of editors

Society Factors — Based on our hypothesis that the society has an impact on the VGI, we incorporate population, poverty and distance from the center into our model. We hypothesize that these societal factors influence the likelihood of a ward being mapped in the future

We then built a linear regression model where the predicted outcome of OSM $act(w_i, t+1)$—the activity in a ward w_i at time $t+1$ based on the above features.[2]

To quantify the *predictive* accuracy of our model we then conducted a classification experiment, and used the discovered classification parameters to classify OSM activity for the upcoming year. For example, we used 2007/2008 to estimate the parameters, built our model, then made predictions for 2009. In this case, we divided the outcome of our models into two distinct categories: 'slow future OSM activity growth' (when $act(w_i, t+1) < 0.3$) and 'fast future OSM activity growth' (when $act(w_i, t+1) \geq 0.3$), with 0.3 being the median value of OSM activity growth for the time windows under consideration. Finally, we considered the top 75 % wards in London only, as predicting OSM activity growth of very sparse areas (e.g., parks) has little significance. Table 2 presents the results of the classification. As shown, the accuracy of our model is quite high with up to 82 % for slow growth and 79 % for fast growth.

Being able to *predict* what areas will not be digitally mapped can help to plan and execute interventions. Such interventions may span a wide spectrum: from allocating financial resources to cover neglected areas, to organizing public mapping events to direct the crowd towards specific mapping goals. Having an accurate growth model at hand implies that these limited resources (human and/or financial) can be best allocated to maximize return on investment. For example, Fig. 5 illustrates the wards of London for which our classification model forecasts slow OSM growth in 2013; various wards in the west/north-west/south-west of London are highlighted as areas at risk. Using influence maximization schemes (e.g., Singer 2012), one could decide

[2] The full details about the model can be found in Quattrone et al. (2014).

Fig. 5 OSM in 2013. Highlighted are the wards for which our model predicated a slow growth

how many resources to allocate to each of these highlighted areas, to maximize expected growth in the following year(s), both as an immediate result of investment *and* thanks to the contagion and self-reinforcement processes that should follow.

4 Impact of Social Mapping Parties

So far we have looked at the impact of society and socio-economic features on the contributed information in OSM in terms of spatial completeness and information growth. We now focus on what motivates people to contribute to OSM, paying particular attention to the social aspects of the OSM community. Social contact has been identified as a powerful motivator by many successful online communities (Kraut and Resnick 2012), Hackathons, mapathons and other similar social events are often organized, in order to bring together people with similar technical skills and interests to accomplish collaborative projects. Likewise, OSM contributors organize local social events, so called mapping parties throughout the year, to bring together the editors to *socialize*, *map*, and *engage the new comers*. In London, these events happen on a fortnightly basis (Haklay and Weber 2008) and their details (when/where it happened as well as who participated) are recorded on OSM wiki pages (http://wiki.openstreetmap.org/wiki/london/summer_2008_mapping_party_marathon/2008-05-21).

To understand whether these mapping parties are successful in encouraging participation, we address the following two research questions: (i) Do the mapping parties cause users to map more than usual *during* the collaborative event? (ii) Do the mapping parties cause users to map more than usual afterwards both in the *short* and *long term*?

To address these questions, we borrow from the field of economics (MacKinlay 1997) and quantify the direct (immediate) and indirect (subsequent) impact of a mapping party using the Abnormal Returns (AR) model. ARs are triggered by events, in our case the mapping party, and are assessed as the higher the abnormal return, the higher the impact of the event (mapping party) on the variable (user contributions in our case). In our analysis we define for each user i and time period τ after the party, we measured the *actual* returns R_i^τ as the average number of contributions per unit of time Δt made by user i during period τ. We also computed the *expected* returns E_i^δ as the average number of contributions made by the same user i per unit of time Δt during a period δ prior to the event. We then calculated the abnormal returns $AR_i^{\delta\tau}$ per unit of time Δt of each user i as:

$$AR_i^{\delta\tau} = R_i^\tau - E_i^\delta \tag{3}$$

In order to conduct impact analysis for mapping parties on users' contributions, we needed to manually construct dataset of the mapping parties in London both in terms of where it happened (the geographical area) and who took part in the event. We recorded 94 mapping parties for the period under examination. As we do not have ground truth about who took part in what event, we inferred a set of 150 'social mappers' from the list of users in the wiki who 'intended to attend' and had made an edit during the event time in the vicinity of the mapping event. Figure 6 illustrates an example of this inference for a mapping party that took place in the Isle of Dogs area of London.

Fig. 6 Map of edits made around London over 48 h during and after one of the identified mapping parties (the Isle of Dogs mapping party)

4.1 Direct Impact of Mapping Parties

The first hypothesis we tested is that users contribute more during mapping parties than outside these events. For each mapping party, and for each user who took part in it, we compute the abnormal returns as per Eq. 3, with Δt equal to one day. We further selected δ equal to 6 months prior to each party, to have enough history about users' editing behaviour, and τ equal to the 'party time' (from the day of the party up until midnight of the day after).

OSM users greatly differ in terms of the amount of contributions they make, and over what timespan (Jokar Arsanjani et al. 2013). In order to quantify the impact of mapping parties on different types of users, we have grouped them based on the number of contributions they made in the 6 months prior to each party. We do so on a log scale of 10 as in the above pre analysis, and split users into five distinct groups: *Group* 0 (just 1 edit); *Group* 1 (from 1 up to 10 edits); *Group* 2 (from 10 up to 10^2 edits); *Group* 3 (from 10^2 up to 10^3 edits); *Group* 4 (from 10^3 up to 10^4 edits). An additional group of newly joined users (*Group NA*) is considered, consisting of those who make their first edit in the system either during the mapping party or less than six months preceding it (thus not having sufficient editing history to be confidently placed in the above groups). The results for this group assess the impact of mapping parties on new comers.

Figure 7 shows average results across the 94 mapping parties that took place in London in the period under study, for each of these user groups. We use a box-and-whiskers plot, with the thick black line within each box representing the median

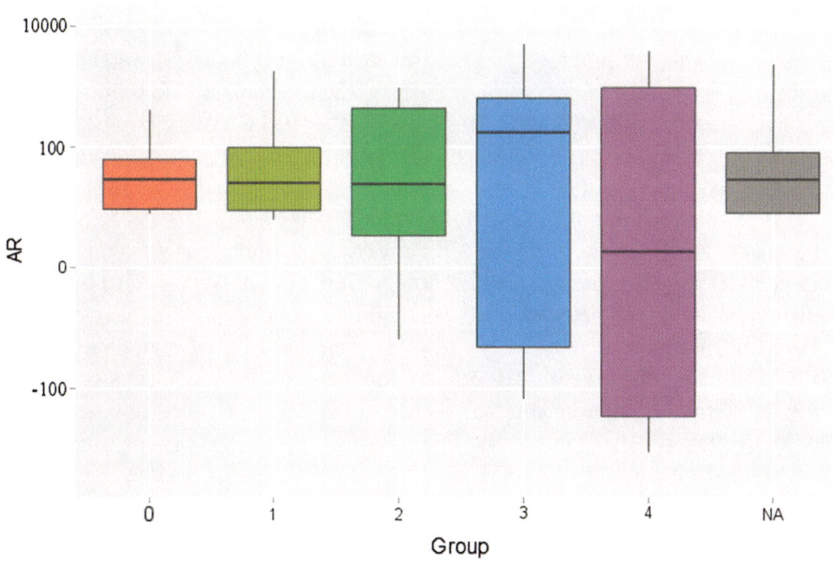

Fig. 7 Box-and-whiskers plot of abnormal returns during a party

value and the 'whiskers' of the box representing the top and bottom quartile values. Median y values above zero indicate that most users within that group exhibit a higher number of edits during the party time than before it, and vice versa (negative y values indicate reduced activity during the mapping as compared to the norm).

The results show that for Groups 0–2 (light to medium contributors) and Group NA (new comers), mapping parties have a strong positive impact in terms of contributions, with their edits being significantly more than usual. Despite more variation within it (and some negative returns too), Group 3 experienced the overall highest AR, with more than 50 % of its members (median value and above) contributing at least 100 edits *more than expected* in the observation period (i.e., party time). Perhaps surprisingly at first glance, only half of the heaviest editors (Group 4) contribute more than expected; the other half in fact perform much below par. We cross checked the names of some of these contributors against what is publicly available in OSM wikis, and found that many of these users take on organizational roles, visiting an area prior to the party, creating 'cake diagrams', and identifying 'problems' they would like the party to fix. We thus speculate that their reduced contribution during the event itself might be due to their engagement in organizational rather than editing activities (e.g., acting as demonstrators for less expert users).

4.2 Indirect Impact of Mapping Parties

The second hypothesis aims to quantify the impact that mapping parties have on users' contributions *after* they took part in an event. As before, we do so by computing AR for the 6 user categories (from light to heavy editors: Groups 0–4, and new comers: Group NA). To distinguish between the impact caused by attending a party from the impact potentially caused by external events (e.g., weather, OSM advertising), we constructed control groups for each of the six study groups. Each respective control group includes users who (i) have had a similar number of contributions as users in the corresponding study group in the $\delta = 6$ months prior to the party under examination and (ii) who did not take part in it or any other event in that time period. We then computed AR for each control group too.

To quantify both short- and long-term effects of mapping party attendance, we computed AR on four non-overlapping observation windows τ: (i) up to one week following the event, (ii) between one week and one month following the event, (iii) between one and three months following the event, and (iv) between three and six months following the event. All observations exclude the contributions made *during* the event. For an easy comparison across all plots, we chose Δt equal to one week as the unit of time to compute AR across all cases. Results for each observation window are shown in Fig. 8. Once again, we use box-and-whiskers plots, with boxes in the upper part of the plot illustrating the behaviour of the control groups, and the bottom part displaying the behaviour of the study groups (referred to as 'Target' group in plots).

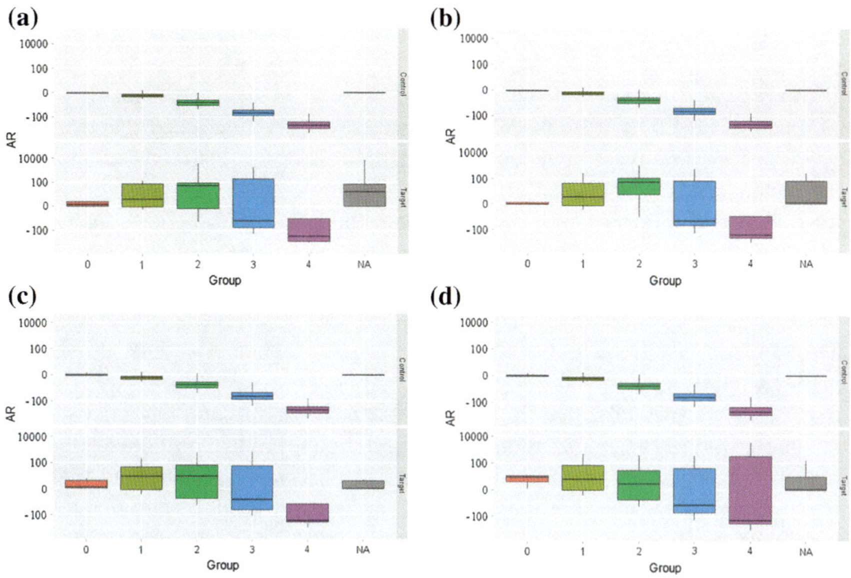

Fig. 8 Box-and-whisker plots of abnormal returns for the mapping parties. **a** 1 week following a mapping party. **b** 1 week–1 month following a mapping party. **c** 1–3 months following a mapping party. **d** 3–6 months following a mapping party

First of all, we observe a decline in contributions (negative AR) by all *control* groups across all observation windows: users who do not take part in a mapping party tend to become more and more disengaged as time passes. This loss of engagement is more pronounced for users who were previously heavily contributing to OSM (Groups 3 and 4).

Conversely, both light contributors and medium contributors (Groups 1 and 2) who attended a mapping party tend to be more engaged over time (in both 1-to-3 and 3-to-6 months—see Fig. 8c, d). As for the heavy contributors (Groups 3 and 4), we observe slightly increased engagement in the short term (Fig. 8). However, as time progresses (Fig. 7b), we observe 25 % of Group 4 participants now exhibiting positive abnormal returns, whilst the AR of its control group remains consistently low. Finally, in the longer term (3–6 months, Fig. 8d), Group 4 is indeed the only study group exhibiting significantly more engagement than what is observed in the corresponding control group. As for Newcomers (Group NA), we can observe that a strong positive AR is indeed evident in the first week following participation in a mapping event (Fig. 8). However, after the first week following the event, 50 % of the newcomers stop contributing completely, with further complete disengagement as time passes.

5 Summary and Conclusion

In this chapter, we studied the impact of society on the crowd-sourced spatial information for OSM London. We showed that the positional and thematic accuracy is high, while the completeness of this information is low and non-uniformly distributed across the city. We revealed that different societal factors, including population density, distance from the center and poverty, are correlated with the information completeness. Given the role that these factors play on the production of digital spatial content, the risk that arises is that the deprived areas might also remain *digitally deprived* on the maps. As a result they may attract even less attention from visitors and city dwellers, thus putting their economy at risk (Cain 2013; Lea 1987).

However, as we presented this low completeness is not a problem per se, if we are able to model the information growth and digital production based on societal factors. Indeed, being able to build a model that *explains* growth, and that accurately detects what areas are most likely to suffer from neglect, has enabled us to highlight these areas and so to bring them to the attention for targeted interventions. One form of these targeted interventions is the design of incentives to call the OSM community to edit specific areas by participating in OSM mapping parties. In understanding whether such an incentive model is successful, we studied OSM mapping parties and revealed that they are extremely successful in retaining users and increasing the editors' contributions both in the long and short term. However, the demonstrated results are based on the users who attended the pre-organized London mapping parties. Therefore, further research is required to understand how communities react to the directed mapping parties, and how the success of the mapping parties may vary across different culture traits around the world.

References

Jokar Arsanjani J, Barron C, Bakillah M, Helbich M (2013) Assessing the quality of openstreetmap contributors together with their contributions. In: Proceedings of the 16th AGILE conference, Leuven

Brunn SD, Williams JF, Zeigler DJ (2003) Cities of the world: world regional urban development. Rowman & Littlefield Publishers, USA

Cain S (2013) Quiet: the power of introverts in a world that can't stop talking. Random House LLC, New York

Census (2011) http://www.ons.gov.uk/ons/guide-method/census/2011/

Coleman DJ, Georgiadou Y, Labonte J et al (2009) Volunteered geographic information: the nature and motivation of produsers. Int J Spat Data Infrastruct Res 4(1):332–358

Cranshaw J, Schwartz R, Hong JI, Sadeh N (2012) The livehoods project: utilizing social media to understand the dynamics of a city. In: Proceedings of AAAI international conference on weblogs and social media (ICWSM)

Girres JF, Touya G (2010) Quality assessment of the french openstreetmap dataset. Trans GIS 14 (4):435–459

Graham M, Hogan B, Straumann RK, Medhat A (2014) Uneven geographies of user-generated information: patterns of increasing informational poverty. Ann Assoc Am Geogr, 1–19, (ahead-of-print)

Haklay M (2010) How good is volunteered geographical information? A comparative study of OpenStreetMap and ordnance survey datasets. Environ Plan B Plan Des 37(4):682–703

Haklay M, Weber P (2008) OpenStreetMap: user-generated street maps. IEEE Pervasive Comput 7(4):12–18

Haklay MM, Basiouka S, Antoniou V, Ather A (2010) How many volunteers does it take to map an area well? The validity of Linus law to volunteered geographic information. Cartographic J 47(4):315–322

Hargittai E, Litt E (2011) The tweet smell of celebrity success: explaining variation in twitter adoption among a diverse group of young adults. New Media Soc 13(5):824–842

Harvey F (2013) To volunteer or to contribute locational information? Towards truth in labeling for crowdsourced geographic information. Crowdsourcing geographic knowledge. Springer, Netherlands, pp 31–42

Koukoletsos T, Haklay M, Ellul C (2012) Assessing data completeness of vgi through an automated matching procedure for linear data. Trans GIS 16(4):477–498

Kraut R, Resnick P (2012) Building successful online communities: evidence-based social design. The MIT Press, Cambridge, pp 42–43

Lea SE (1987) The individual in the economy: a textbook of economic psychology. CUP Archive, Cambridge

Ludwig I, Voss A, Krause-Traudes M (2011) A comparison of the street networks of Navteq and OSM in Germany. Advancing Geoinf Sci Changing World 1(2):65–84

MacKinlay AC (1997) Event studies in economics and finance. J Econ Lit 35(1):13–39

Quattrone G, Mashhadi A, Quercia D, Smith-Clarke C, Capra L (2014) Modelling growth of urban crowd-sourced information. In: Proceedings of the 7th ACM international conference on web search and data mining, ACM, pp 563–572

Roth C, Kang SM, Batty M, Barthélemy M (2011) Structure of urban movements: polycentric activity and entangled hierarchical flows. PLoS ONE 6(1):01

Schradie J (2011) The digital production gap: the digital divide and web 2.0 collide. Poetics 39(2):145–168

Singer Y (2012) How to win friends and influence people, truthfully: influence maximization mechanisms for social networks. In: Proceedings of the 5th ACM international conference on web search and data mining, pp 733–742

London Data Store http://data.london.gov.uk/datastore/package/ward-profiles-2011

Wilkinson DM, Huberman BA (2007) Assessing the value of cooperation in wikipedia. First Monday 12(4)

Zielstra D, Zipf A (2010) A comparative study of proprietary geodata and volunteered geographic information for Germany. In: Proceedings of the 13th international conference on geographic information science

Social and Political Dimensions of the OpenStreetMap Project: Towards a Critical Geographical Research Agenda

Georg Glasze and Chris Perkins

Abstract Critical cartographic scholarship has demonstrated that maps (and geoinformation in general) can never be neutral or objective: maps are always embedded in specific social contexts of production and use and thus unavoidably reproduce social conventions and hierarchies. Furthermore, it has been argued that maps also (re)produce certain geographies and thus social realities. This argument shifts attention to the constitutive effects of maps and the ways in which they make the world. Within the discussion on neogeography and volunteered geographic information, it has been argued that crowd sourcing offers a radical alternative to conventional ways of map making, challenging the hegemony of official and commercial cartographies. In this view, crowd-sourced Web 2.0-mapping projects such as OpenStreetMap (OSM) might begin to offer a forum for different voices, mapping new things, enabling new ways of living. In our contribution, we frame a research agenda that draws upon critical cartography but widens the scope of analysis to the assemblages of practices, actors, technologies, and norms at work: an agenda which is inspired by the "critical GIS"-literature, to take the specific social contexts and effects of technologies into account, but which deploys a processual view of mapping. We recognize that a fundamental transition in mapping is taking

Part of this paper has been written while Georg Glasze was a visiting researcher at the Oxford Internet Institute (OII) in 2014. We want to thank Steve Chilton (London), Christian Bittner, Tim Elrick (both Erlangen), various colleagues from the OII and the three anonymous referees for their advices on aspects of this paper.

G. Glasze (✉)
Institute of Geography, Friedrich-Alexander-Universität Erlangen-Nürnberg (FAU), 91054 Erlangen, Germany
e-mail: georg.glasze@fau.de

C. Perkins
School of Environment, Education and Development (Geography), The University of Manchester, Manchester M13 9PL, UK
e-mail: chris.perkins@manchester.ac.uk

J. Jokar Arsanjani et al. (eds.), *OpenStreetMap in GIScience*,
Lecture Notes in Geoinformation and Cartography,
DOI 10.1007/978-3-319-14280-7_8

place, and that OSM may well be of central importance in this process. However, we stress that social conventions, political hegemonies, unequal economic and technical resources etc. do not fade away with crowdsourced Web 2.0 projects, but rather transform themselves and impact upon mapping practices. Together these examples suggest that research into OSM might usefully reflect more critically on the contexts in which new geographic knowledge is being assembled.

Keywords Critical cartography · GIS and society · Geoweb and society · Volunteered geographic information · Social and cultural geography · OpenStreetMap

1 Introduction

OpenStreetMap (OSM) has repeatedly been described as "free" and as a "crowd-sourced map of the world" which enables an opening and democratization of hitherto elitist practices of cartographic (re)presentations and the collection of geodata (see, for example, Chilton 2011). Other chapters in this volume offer detailed descriptions of the project, and the varying functionalities it offers, detailing the complex variety of application areas for OSM data, but for the purposes of this chapter we focus upon the extent to which OSM delivers a radical change that is significantly different from other digital mapping projects. We are concerned here with the social and political dimensions of OpenStreetMap, and with the extent to which the much-trumpeted free and open nature of the project delivers a democratized and emancipatory mapping of the world. We sketch out a critical research agenda, exploring the fixations, hierarchies, conventions and exclusions, which almost inevitably become inscribed in projects like OSM. This agenda argues that researchers might attend more to mapping modes through which OSM is practiced, by focusing on authorship, technical infrastructure, and governance. We argue that researchers might also deploy mixed and ethnographic approaches, in order to learn more about particular moments of mapping practice and illustrate this with a small case study of "mapping mosques". In discussing these aspects, we aim to reflect on the dynamic, changeable, and thus "open" status of the project.

2 Geoinformation, Cartographic (Re)presentation and Society

Geoinformation and cartographic (re-)presentations categorize, define, arrange, locate, designate, and thereby (re)produce certain conceptions of the world. They powerfully affect our thinking and acting. Critical cartographic scholarship has demonstrated that maps can never be neutral: they are always embedded in specific social contexts of production and use and thus unavoidably reproduce social conventions and hierarchies (see Harley 1989). Furthermore, it has been argued that maps also (re)produce geographies. This argument shifts attention to the ways in which maps make the world (e.g. Pickles 1992). For example, the projection used, what is shown on the map, what becomes silenced, what is emphasized, and where the map is centered on to construct a particular world-view. Critical engagement with the social contexts and implications of geoinformation and cartographic (re) presentation also has to consider technological changes and practices for the collection, organization, and use of geographic information—especially in the digital age.

2.1 Critical Cartography I: The Social Construction of Maps

From the 1980s a perspective that sees maps as socially constructed started to take form. In 1985 the Swiss geographer Raffestin proposed a "sociology of cartography" which asked why societies designed specific maps. In 1992, Denis Wood elaborated on the "power of maps" in bringing forward the argument that maps are always deployed to represent interests. The geographer and historian of cartography Brian Harley interpreted historic maps as documents, which have to be understood in particular social contexts. In his highly influential article "Deconstructing the map" (1989) he differentiates external and internal power in cartography. External power refers to the impact of social structures on the ways maps are produced: "Monarchs, ministers, state institutions, the Church, have all initiated programs of mapping for their own ends" (ibid: 12). The internal power of "cartographic processes" refers to the nexus of knowledge and power described by the French philosopher and historian Michel Foucault. Harley lays the basis for a social constructivist view of cartography. He suggested an approach inspired by text-based discourse analysis in order to analyze how maps tend to reproduce specific world views. In this view, regularities in the design of maps are seen as indices for the implicit and unstated rules of cartographic practice.

2.2 Critical Cartography II: Mapping as Socio-technical Practice

Since the 1990s a debate has developed amongst scholars of critical cartography, which focuses on the practices, conventions and techniques of map making and use and thus goes beyond former concerns with the visual design of the map. This research perspective questions how mapping practices shape our social worlds (Pickles 2004) and often draws upon ideas from science and technology studies. Dodge et al. (2009b), for example, point to the writings of Latour who took modern cartography as an example to show how specific practices and techniques were used, to produce scientific knowledge and thus authority in European centers of power. Latour (1986) argued that these practices, conventions, and techniques worked to create the preconditions for international trade, territorial expansion, and global colonization and thus new geographies, and that maps served as immutable mobiles, circulating and reifying a particular way of knowing the world.

2.3 Social Science Perspectives on the Transformation of Geoinformation and Cartography in the Age of GIS and the Geoweb

Since the 1960s analogue print-based cartography has been rapidly and comprehensively replaced by digital cartography. From the 1960s, Geographic Information Systems (GIS) were progressively developed, to capture, process, analyze, and map digital geodata. The encounter of GIS and critical social and cultural geography triggered a discussion of the social implications of the widespread use of GIS (see for example Pickles 1995; Schuurman 2000, 2009; Harvey et al. 2005; O'Sullivan 2006). This debate on GIS and society not only focused on mapped displays, but also on the practices and technologies "beyond" and "behind" these representations. Three important aspects of this debate should be highlighted. The major influence of economic and military interests in the development of GIS was an important focus for research. A second theme concerned disparities over access to production and use of geographic information arising from the complexity and cost of GI systems. Finally, the focus of GI analyses was on quantifiable and metric information with a consequent danger of a marginalization of "qualitative" interpretation. Today, research into GIS and society analyzes the socio- and politico-economic contexts of GIS, as well as the impacts of GIS on social structures and processes (Pickles 2004; O'Sullivan 2006; Harris and Harrower 2006; Pavlovskaya 2006).

With the development of the interactive internet, the so-called Web 2.0, and the rapidly growing availability of online-geodata, geoinformation and cartography are undergoing another fundamental transformation (Haklay et al. 2008; O'Reilly 2005). Global corporations with no background in geoinformation are developing new Geoweb applications (on the history of Google Earth see, for example, Dalton 2013). With the proliferation of global positioning systems in smartphones and navigational devices, the Geoweb is part of mobile and ubiquitous practices. Alongside commercial players in this field there are a growing number of "open" Geoweb-projects based on crowdsourcing, with OpenStreetMap being the most successful and prominent example. These projects involve thousands of volunteers in the creation, organization, and use of geoinformation, consequently described as "voluntary geographic information" or VGI (Goodchild 2007), leading to what has been labeled as a "neogeography" beyond the established academic field of geography (Goodchild 2009).

Research on the Geoweb in the social sciences can build on approaches developed in critical cartography and "GIS and society", but also profit from relations to the wider field of critical social and cultural geography and interdisciplinary internet studies (see Graham 2009; Caquard 2014). Sarah Elwood and her co-researchers have begun to research different aspects of collaborative and community-based mapping, to offer a critical interpretation of big data and the Geoweb, reorienting attention to the power of technical and political infrastructures in privileging certain kinds of information, moments, or affordances, and drawing attention to the exclusions that are normalized in the apparently neutral specifications of mapping projects on the Geoweb (see Elwood 2010a, b, 2011; Elwood and Leszczynski 2012). This kind of research also draws attention to the importance of the research discourse around Geoweb projects, that script a boosterist neogeographic agenda—in which VGI remains somehow separate from the powerful forces of commerce that maneuver around the technology, deploying it as part of their accumulation strategies (see Leszczynski and Elwood 2014). Technical research elides the social and political context of Geoweb projects and in so doing allows them to advance as "new", without having to think about why or how they are advancing. Other political economic research focuses on the relationship of depiction and inscription and the realpolitik of claims to space (see for example Burns 2014). Glasze (2014) suggests four main questions that might be answered in this kind of research:

(1) How are practices relating to compilation, processing, analysis, and presentation of geodata that were formerly the responsibilities of public organizations shifting to other actors? To what extent can this be interpreted as an "opening" of geoinformation or should these processes be seen more as a commodification and commercialization of geoinformation by means of a roll back of public services?
(2) What role do communities of collaborative internet activists play in this process?
(3) What are the consequences of this shift for the nature, quality, processing and presentation of geodata, and how do social (and spatial) inequalities become (re-)produced in this process?
(4) How does the growing extent of geodata enable new possibilities for (Geo-) surveillance and (Geo-)marketing? And what does this mean for questions of power, governance, resistance and privacy?

These issues can usefully be examined by focusing on OpenStreetMap, and we argue can most clearly be articulated if researchers adopt a concern with modes, moments and methods (see Dodge et al. 2009b) wrapped up in this project. By modes we mean the ways technologies, culture, and socio-economic organization come together to influence mapping practices; by moments we mean the banal taken-for-granted instances of practice on the ground; and by methods we mean how researchers might investigate such issues.

3 OpenStreetMap: Opening and Democratizing Geoinformation and Cartography?

OpenStreetMap was founded in the UK in 2004 by software developer Steve Coast. It offers a collaborative geodata project and cartography in which users capture, upload, edit, and tag tracks and points of interest, progressively building a global open and free geodatabase and map. Within the discussion around neo-geography and volunteered geographic information (VGI) it has been argued that this kind of crowd-sourcing offers a radical alternative to conventional ways of geoinformation and map making (e.g. Goodchild 2007, 2009), challenging the hegemony of official and commercial cartographies. In this view, crowd-sourced Web 2.0-mapping projects such as OSM might "begin to offer a forum for different voices, mapping new things, enabling new ways of living" (Perkins 2013).

OSM's web-based architecture facilitates many different kinds of involvement (see Ramm and Topf 2010). Users can create data, enhancing and growing many aspects of the project, and, in so doing, build a collaborative geodata project. The different OSM wikis document established practices. Different rendering styles have been developed to map the database. Code is revised and the functionality of the interface changes over time. Tools have been created by the community, to check the quality, coverage, and veracity of mapped features. Tagging standards are debated in talk lists. A community of users progressively adds to the project and meets online and in fora such as annual conferences and Mapping Parties.

From the outset, OSM offered a wiki-based capacity to share tasks. The Project offered something new to users, and the novelty lay in the notion that OSM was *open* and *free*. Its culture of participation is a central feature. Any registered user has the capacity to overwrite other people's work. Throughout the documentation about the project it is frequently repeated that practical needs of "doing" the project take precedence over more hierarchical rules governing behavior. An early impetus to establishing OSM was the desire to challenge the corporate and proprietary monopoly of national mapping agencies. In the early days of OSM the Ordnance Survey (the official cartographic authority in the UK) operated cost-recovery policies, and protected its products by aggressive policing of copyright. By way of contrast, OSM initially ran under a Creative Commons licensing regime, and from September 2012, under an Open Data Commons Open Database license, that encourages reuse of OSM data (see Chilton 2011). In contrast to commercial VGI services, such as Google Mapmaker and Navteq Map Reporter, volunteers contributing data do not hand over ownership of the data to a profit-making corporation. Over the past decade the sophistication and coverage of OpenStreetMap has grown apace. As of July 2014 there were 1,699,115 registered users, with 2,425,437,945 nodes and 242,404,181 ways in the database.

4 Mapping Modes: The Social and Political Dimensions of OpenStreetMap

The idea of a mapping mode builds on work by historian of cartography Edney, who suggested in 1993 that mapping might best be seen as an assemblage in which technologies, people, knowledge, culture and politics come together, and through which particular ways of doing mapping are enrolled. A mapping mode is thus variegated and situated in a particular time and place. It is transitory and constantly changing. At any one time, different mapping modes might coexist; there is no inevitable progression from one mode to another. The paper map survives in the digital era; the national mapping agency continues to produce maps in the face of competition from crowd-sourced alternatives; the touchscreen-based mobile interface coexists with fixed desktop screen-based displays, etc. (Dodge et al. 2009b). Here we focus on three key influences upon contemporary mapping modes: authorship, infrastructure, and governance.

4.1 Authorship: The Socio-cultural Embeddedness of OSM Practices

Authorship of OSM is collaborative. The project celebrates its open and shared ethos and tools exist to allow potentially anyone to drill down to identify who has been responsible for the creation of which parts of the database (see for example Fig. 1).

Empirical investigation of the OSM community suggests, however, that the nature of this collaboration is uneven, and that participation in OSM is, like all

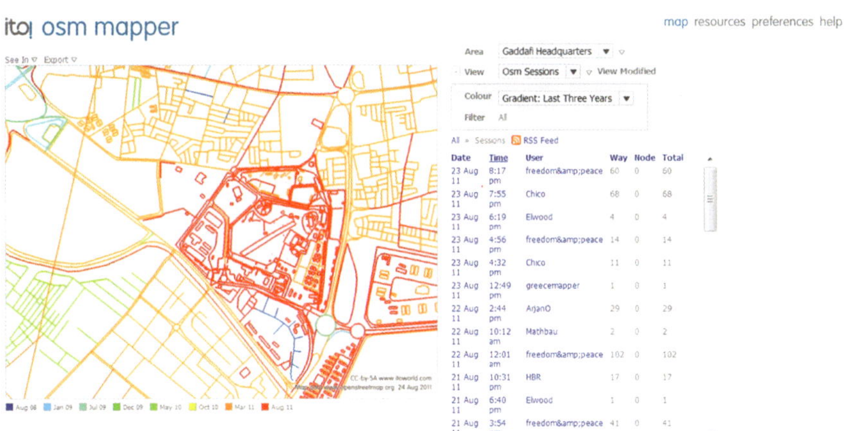

Fig. 1 The collaborative authorship of Tripoli Libya immediately after the overthrow of Colonel Gaddafi. Note mapping of the compound dates from 21st August 2011, and was mainly carried out by User IS Freedom and Peace and that almost all the immediate area has been mapped in the period since March 2011 (© ITO World, mapping data from © OpenStreetMap contributors, CC-BY-SA)

crowd-sourced projects, very unequal. A small and elite group end up taking most of the important decisions, which effectively determine project impetus and directions. The vast majority of OSM users do not register for the project. Of those who do register the majority does not stay with the project for long or contribute much to the database (Neis and Zipf 2012).

The overwhelming majority of users are male. Stephens (2013) compared gender participation in VGI projects and concluded that "Women are less aware of OSM… than their male counterparts, and those who are aware of OSM are significantly less likely to contribute spatial data. As a result of low female participation, the features and attributes on OSM reflect a male view of the landscape." She highlights the gendered nature of the tagging process that has allowed men to exercise their democratic rights to vote down a detailed classification of amenities that do not meet their immediate needs, such as childcare facilities, whilst supporting the inclusion of tags relating to stereotypically male sexualized spaces such as brothels (see also Steinmann et al. 2013).

Other inequalities are charted in empirical studies of participation such as Neis and Zipf (2012) and Budhathoki (2010). These reveal that most OSM users are wealthy and educated. Most come from the northern hemisphere (see Fig. 2).

Also within urban areas there are significant disparities of geodata-density. For example in the case of Jerusalem, Bittner (2014) shows that the data density is much higher in the neighborhoods mostly populated by secular Jews, compared to the quarters predominantly inhabited by Orthodox Jews and Palestinians (see Fig. 3).

The world mapped by the OSM community reflects its interests. Urban and wealthier areas tend to be more densely mapped (see Haklay 2010) Areas of rapid change or under crisis get mapped (see Bittner et al. 2013; Zook et al. 2010; Burns 2014) on emergency/crisis mapping.

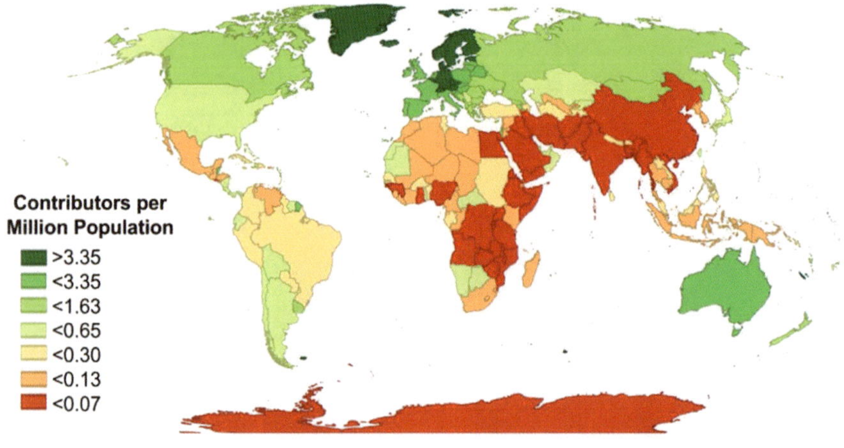

Fig. 2 Distribution of active OSM contributors per day and per population (1 August–31 October 2013) (*Source* Neils and Zielstra 2014)

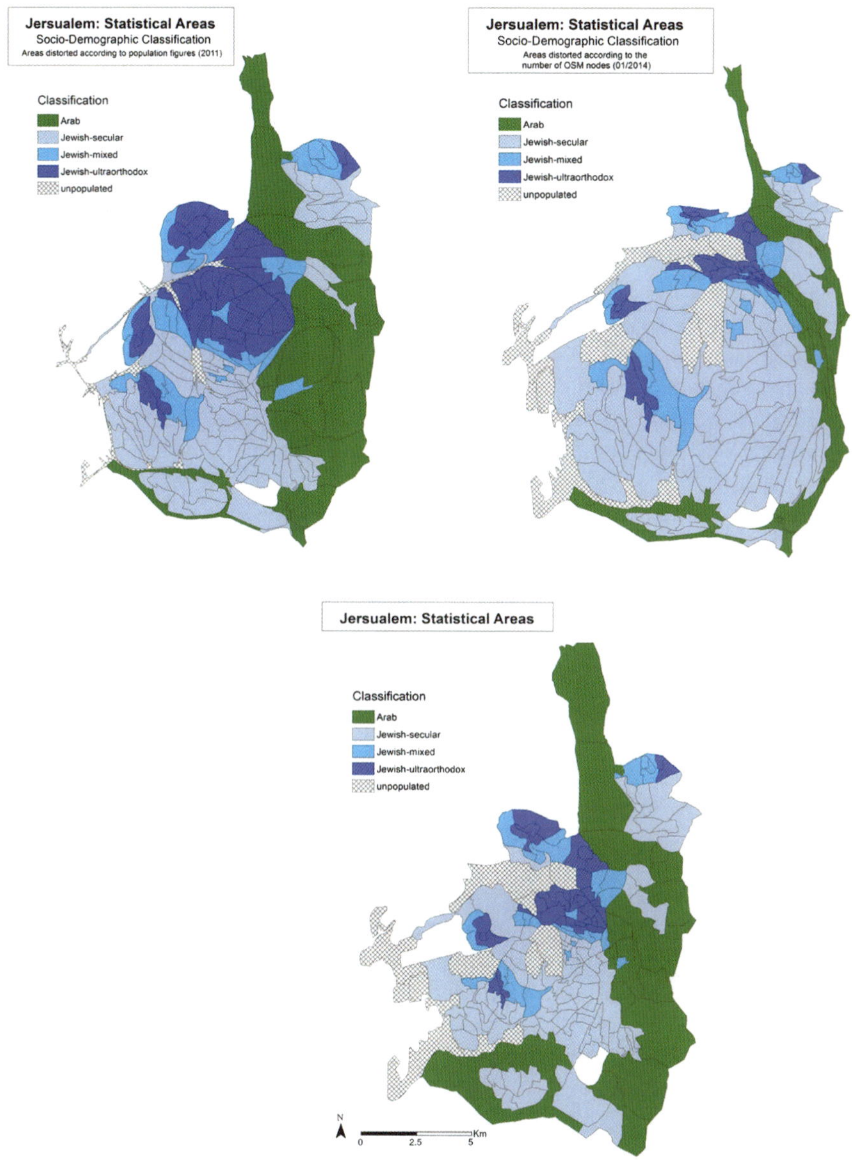

Fig. 3 Data density of OSM within different quarters of Jerusalem compared to demographic data (*Source* Bittner 2014)

Many of the mappers that stay with the project have specific technical skills. The majority of participants drift out of OSM instead of continuing to map. So, instead of becoming a genuine peoples' map it has been argued that the project represents a new kind of expert knowledge (Perkins 2013).

4.2 Algorithms and Other Blackboxes: Unpacking the Technical Infrastructure of OSM

The infrastructure of OSM is of central influence for the performance and development of OSM. Dodge et al. (2009b) use the term to highlight the role played by underlying social foundations: the often unseen and taken for granted structures through which work is done. They observe that "critical studies of infrastructures are made more difficult because of the ways in which institutions deliberately structure them as 'black-boxed' systems to keep people from easily observing (and questioning) their design and operational logic" (ibid: 227).

People interacting with OSM infrastructures mostly do so via interfaces. "Interfaces en-frame and exclude, working as mediating windows onto the world" (Dodge et al. 2009b, p. 222) They deliver different mapping functionalities. These screen spaces usually hide the apparatuses and processes through which online navigation takes place. Their layered potential confers a navigational logic that is usually unquestioned by users (Verhoeff 2012). The default OSM interface strongly impacts on affordances.

The operation of OSM depends upon the operation of algorithms and code that come together to make the map and its interfaces possible. Algorithms are hidden and often inaccessible in geoinformation and mapping systems. Although OSM makes its API freely available and the OSM wikis help to access algorithms as well as codes and, last but not least, the codes underpinning OSM are there to be shared or changed by community members, the required technical expertise limits the ability to change and redirect the algorithms and codes of OSM to a small but influential group of people [see the discussion on levels of hacking in Haklay (2013)]. Here we highlight three examples of code and draw out some of the ways in which they impact mapping practice.

Firstly, the project rests upon editing software, which allows users to amend or extend mapping coverage. Editing software suggests classifications of the world to users, implicitly encouraging "things" that might be included into the OSM database or excluded. Its form and configuration arguably influences whether a user actually changes the database and channels day-to-day mapping practices (Weber and Jones 2011).

Secondly, rendering software allows features tagged in the database to be symbolized. It structures the world, leaving many tags un-rendered, and through a visual display enables or disables different uses and evokes different feelings for the map. Chilton (2011) documents the development of the default and widely praised Mapnik style, showing how a meeting between a single coder, two cartographers, and the project founder led to a style that has impacted beyond OSM and which incorporated subsequent community enhancement of the data.

Thirdly, software to check the "quality" of the database has also proliferated as the project matures (see the Wiki-page: http://wiki.openstreetmap.org/wiki/Quality_Assurance). This tends to focus attention upon coverage and standardization, by highlighting inconsistencies and directing attention to "faults" that might be rectified in the database and thus risks blocking from view more fundamental questions.

4.3 Governance of OSM: Doocracy—Meritocracy—Technocracy—Bureaucracy

It is difficult to specify who actually 'runs' OSM. Ramm (2013) examined the question of "who the boss of the project actually is", and rejected any simple answer. However, a rich documentation of infrastructure exists on the OSM wiki, and is also described by Eckert (2010). The project employs no staff. It is answerable to the OSM Foundation, which currently has 480 members: anyone can pay to join this group, which is "dedicated to encouraging the growth, development and distribution of free geospatial data and to providing geospatial data for anybody to use and share" (OpenStreetMap Foundation 2013). Foundation members elect a Board that currently includes six members. There are also eight Working Groups focusing upon: communications, data, licensing, operations, local chapters, engineering, the State of the Map Conference, and strategy. A management team implements day-to-day decisions.

In addition, OSMappers come together in various State of the Map conferences and in Mapping Parties. Their ideas for project trajectories are played out online in blogs and user diaries and in the project wiki, and ideas are debated in numerous discussion lists. Spinoff consultancies progress the project whilst also deriving profit from the crowd.

The implicit ethos of OSM is frequently described as open, democratic, and anti-establishment. In practice, however, new mappers are encouraged to follow established ways of doing the project. Ways of doing OSM impact significantly on progress, and whilst the culture of OSM delivers what has been described as a 'do-ocracy' (see for, example, Perkins 2013), in practice the project works as a mixture of a do-ocracy, meritocracy, technocracy, and bureaucracy (see Fig. 4). The governance is meritocratic in the sense that voluntary work is rewarded by community esteem, or by external financial reward. It is technocratic in the sense that technical coding skills are most valued. These skills fix and "blackbox" classifications and practices in editing and rendering software and strongly influence the development of the project. The bureaucratic aspect of governing is less significant in OSM—compared, for example, with Wikipedia (see Ramm 2012 and several contributions in Lovink and Tkacz 2011). The OSM Foundation tends to enable rather than steer and the OSM community itself has few formalized organizational structures; there is no official and formalized hierarchy of users as, for example, in Wikipedia. There are some mechanisms for building a consensual view, with procedures for voting about the creation of new tags, for example. In comparison to Wikipedia, however, these mechanisms are much less formalized and used.

As in all collaborative projects "edit wars" can take place. For example, in contentious areas such as Cyprus, Jerusalem, or Crimea different place names and borderlines have been recorded, and overwritten.[1] It is, however, interesting that in

[1] http://wiki.openstreetmap.org/wiki/Data_working_group/Disputes (23.07.2014).

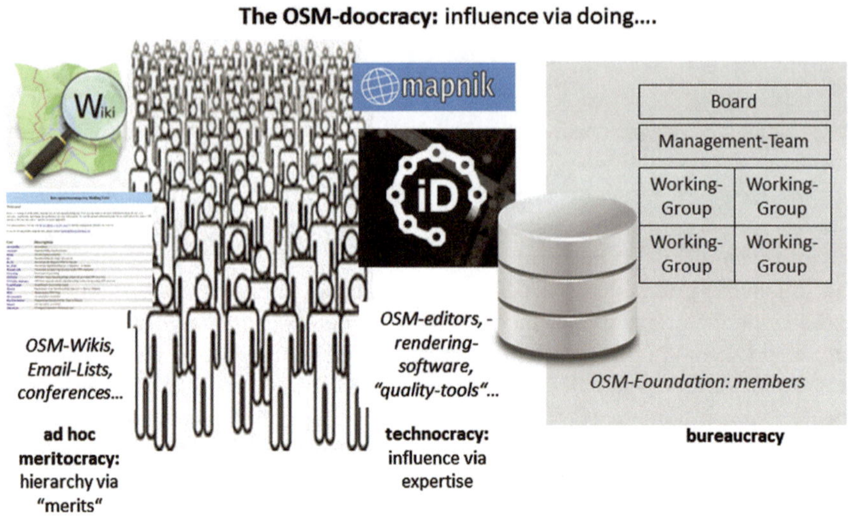

Fig. 4 Governance of OSM [based on an idea of Ramm (2013), supplemented and changed by Glasze/Perkins]

contrast to text-based open projects, such as Wikipedia, there seems to be much less interaction between different editors (Mooney and Corcoran 2013) and much less controversy over the status of objects (Wroclawski 2014). This may be attributed to the ethos of an "on the ground truth" with the basic idea of OSM representing an objectively verifiable, and knowable world.

So falsifying is strongly discouraged. Vandalism, whether for artistic or commercial purposes, is carefully policed (see Ballatore 2014).

5 Methods: Analyzing OSM

A shift towards a research focusing on social and political dimension as highlighted in our introduction has clear methodological implications. Dodge et al. (2009a) suggested that approaches drawn from Actor-Network Theory, Science and Technology Studies (STS), ethno-methodology, and non-progressive genealogy might usefully be adopted to advance our explanations of these social and political dimensions of geoinformation and cartography.

5.1 Data-Driven Research on OSM

Almost all the research on OSM reported in the meta study carried out by Neis and Zielstra (2014) deploys data-driven tools to answer practical questions about OSM.

The remit and format of the project delivers data to the research community in a much more transparent fashion than in other proprietary databases. "The crowd" leaves traces behind, that reveal things about mapping, in ways that are hidden in projects such as Google.

Neis and Zielstra (2014) highlight research into OSM that largely relies upon archived tracks and traces. This research predominantly adopts progressive scientific ways of knowing the world. Data quality analysis inevitably looms large in this field. Road networks have received significant attention and there has also been a focus on the quality of different points of interests (POIs) in the database. This kind of research inevitably compares OSM to other proprietary databases. Recent attention has also begun to focus on questions of trust and vandalism, but again largely as practical measures to investigate quality.

A second trend has been an increase in the amount of research investigating contributors, in terms of temporal trends, areal distribution, and gender balance. Methodologies deployed to chart differences depend upon large-scale generalization from big data sets, instead of detailed processual investigation of individual and qualitative data. Haklay's (2010) influential investigation of the social composition of the database is typical and foundational here. Neis and Zielstra (2014) also designate a final category of research, focusing upon other work, that does not fit into quality evaluation or participation studies, and highlight work on routing packages, 3D mapping, and application areas relating to access mapping and disaster management.

The implication from this meta-study is that the shift towards a crowd-sourced model has not so far encouraged the kinds of methodological shifts signaled by Dodge et al. (2009a). However, a careful analysis of the published literature reveals work that is beginning to approach OSM in different ways, and focusing in particular on ethnographic work on mapping practices, and on the application of multiple methods to case evidence.

5.2 Ethnography and Auto-ethnography

There has also been an increasing interest in using anthropological approaches to mapping, and in phenomenological ways of understanding mapping practice. Long-established ethno-methodological tools have begun to be applied to people deploying OSM in real-world contexts, to code up apparently banal day-to-day mapping. This kind of focus on everyday politics with a small 'p' underpins for example Hind's (forthcoming) work on OSM and protest mapping. Other ethnographic work has been carried out in spaces where OSM has been deployed, and explicitly stresses the performativity of mapping, instead of any inherent meaning [see, for example, Gerlach (2014) on everyday mapping practices]. Kitchin and Dodge (2013) draw upon these kinds of ideas in their analysis of OSM as emergent processual knowledge. Other examples of research also focus on the contexts in which the map is situated. For example, Lin (2011) attended State of the Map

Conferences in order to understand and explain how open source communities function. Perkins (2013) reports on various spaces where OSM is deployed, highlighting the differences that emerge according to sociality, and Perkins and Dodge (2008) report an ethnography of an early Mapping Party.

5.3 The Need for Mixed Methods Approaches

Different insights flow from direct participation in an event, to those that can be inferred from quantitative analysis. For example, Hristova et al. (2013) also focus on mapping parties, but deploy data sourced from the OSM web site to explore the effectiveness of the party as a device for encouraging participation. By way of contrast, Budhathoki and Haythornthwaite (2013) rely upon questionnaires in their analysis of motivations of individual OSM participants, but neither of these sources provides case evidence about cultural practice.

In the first decade of the new millennium, scholars concerned with critical approaches to GIS and the Geoweb increasingly came to realize that a mixture of qualitative and quantitative evidence can be important and can document general patterns as well as individual processes (see Kwan and Schwanen 2009). Projects like OSM offer huge potential for such mixed methods approaches (see Elwood 2010b; Elwood et al. 2013; DeLyser and Sui 2012; Crampton et al. 2013). In OSM all edits in the database and the wiki, and all discussion in the email-lists are recorded and can be traced back to individual participants.[2] This enables analysis of mapping practices and collaboration within the community (see, for example, Kremer and Stein 2014; Elrick 2014). OSM databases make it easy to combine quantitative approaches with qualitative interviews (see, for example, Bittner 2014). Individual tracks can be documented and the history of the unfolding map can be unpacked. Big data can actually greatly facilitate critical multi-method approaches to the project.

6 Maps and Mosques: A Case Study on the Transformation of Techniques, Practices, and Conventions Within OSM

The tension between openness and fixation revealed in OSM practices will be exemplified by a short case study on the depiction and non-depiction of mosques in OSM.

[2] See for example the "How Did You Contribute to OpenStreetMap tool" available at http://hdyc.neis-one.org/ deploys charting and tabulation and mapping to document individual user name participation in the project, and the user diaries attached to the site.

6.1 Concealed Mosques in State-Based Cartography

In the late 1980s, Brian Harley charted what he described as silences, highlighting many of the social reasons why maps omit, simplify, and homogenies landscapes (Harley 1988). The social context of map making establishes accepted ways of fixing what is included or left out (Harley 1989). A striking example of such impacts is how maps choose to depict and include (or not include) places of worships (Glasze 2009). A quick overview of the topographic maps currently produced by state-run cartographic organizations in France, Germany, and the United Kingdom reveals, for example, that none of the topographic map-styles offers a symbol for mosques. However, all of these topographic maps include symbols for "places of worship", and in the whole of Western Europe the iconography of a Christian tradition is deployed for these sites (Kent and Vujakovic 2009). The religious tradition in Western Europe normalizes current cultural diversity, and mapping styles respond only very slowly to social and cultural change, leading to an effective cartographic concealment of mosques (see the example of a cartographically concealed purpose-built mosque in Mannheim, Germany; Fig. 5).

Fig. 5 The Yavuz Sultan Selim Mosque and the Liebfrauenkirche in Mannheim (Germany) as (not) shown in the official topographic map, in Google Streetview and in different OSM rendering styles (*Sources* Landesamt für Vermessung Baden Württemberg; Google, OSMcontributors/Geofabrik)

The interesting question is whether this silencing of Muslim sites of worship continues in the crowdsourced world of VGI—and whether the possibilities of OpenStreetMap enable new mapping moments.

6.2 Newly Open but Fixed Practices in OSM

OSM's database structure enables some practices, and limits others. Open questions and conflicts are discussed within the community; results of such discussions are codified in the OSM Wiki. With the success of the project more and more applications interpret and use OSM data (e.g. software for rendering and for routing), other applications try to facilitate and analyze mapping practices (e.g. software for editing and analyzing OSM data). As an inevitable consequence, OSM mapping practices become conventionalized and fixed.

In order to understand these processes and moments we take the example of "mapping Mosques" and highlight four themes that contribute to openness or fixation.

6.2.1 Data Structure and the Wiki

The OSM Wiki suggests tagging "places of worship" in the database as nodes with the amenity value "place of worship", and to further differentiate "religion" and "denomination". The respective Wiki-page was set up as early as in 2007 by one of the key figures in the OSM community (Fig. 6).

Since 2008, there have been distinct Wiki pages to explain "religion" and "denomination", which recommend increasingly detailed categorization of religions and religious denominations. However, these lists are not the only way to classify religious affiliations.[3] There have been lively discussions in the OSM community on the attribution, acceptance, and integration of different categories.[4] While the majority of the tag values follow the categorization suggested in the Wiki, the OSM database (still) includes other tags. The wiki offers guidance, but ethnographic work suggests practice by individual mappers does not always follow these procedures (Perkins 2014).

[3] As an example the Wiki suggests to classify "druse" as a denomination of "religion = muslim"—a classification which is contested for example by many Druze living in Israel who see themselves not as Muslims but as a proper religious group.

[4] See for example the broad discussion on places of worship in OSM triggered by the debate on the Pastafarians (https://lists.openstreetmap.org/pipermail/talk/2010-January/046620.html; 10.07.2014).

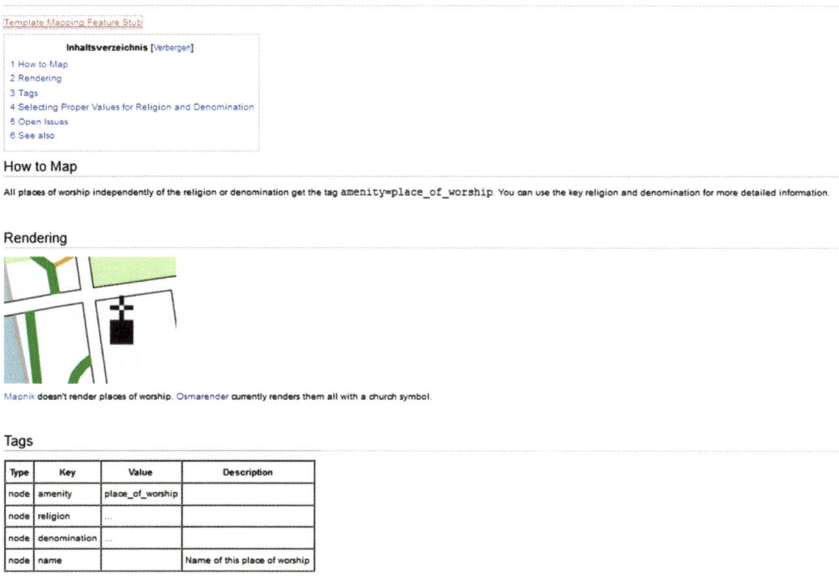

Template Mapping Feature Stub

Inhaltsverzeichnis [Verbergen]

1 How to Map
2 Rendering
3 Tags
4 Selecting Proper Values for Religion and Denomination
5 Open Issues
6 See also

How to Map

All places of worship independently of the religion or denomination get the tag amenity=place_of_worship. You can use the key religion and denomination for more detailed information.

Rendering

Mapnik doesn't render places of worship. Osmarender currently renders them all with a church symbol.

Tags

Type	Key	Value	Description
node	amenity	place_of_worship	
node	religion	...	
node	denomination	...	
node	name		Name of this place of worship

Fig. 6 First version of the OSM Wiki on places of worship in 2007 (*Source* OSM Wiki; 20.07.2014)

6.2.2 Rendering and Editing Software

The OSM database becomes "visible" in cartographic presentations through rendering software. There have been discussions in the OSM community since 2007 regarding appropriate rendering of places of worship and the respective symbology. Until 2007/2008 the most important renderers (OSMarender and Mapnik) translated all places of worship as a cross, which triggered several critical statements in the OSM discussion lists—especially with regard to the rendering of mosques with a cross.[5] As a direct reaction to this discussion one of the central actors of the British OSM community added specific symbols for religion = muslim, = jewish and = sikh to the current default renderer Mapnik in 2008. New symbols have been suggested since, for example for Buddhist or Hindu places of worship, but these have not been integrated into the rendering software (see Fig. 7).

Most OSM mappers do not deal directly with the database, but use editing software. The classifications offered by these tools are often not completely in accordance with the categorizations in the Wiki, and largely structure mapping practices, which gives developers of successful editors enormous influence (see the classifications proposed by the new ID editor, Fig. 8).

[5] See for example: http://gis.19327.n5.nabble.com/Rendering-places-of-worship-in-Mapnik-td5379077.html (10.07.2010).

Religion	proposed rendering	current rendering
religion=animist		
religion=bahai (Baha'i)		
religion=buddhist		
religion=christian		
religion=hindu		
religion=IglesiaNiCristo		
religion=jain		
religion=jewish		
religion=multifaith		
religion=muslim		
religion=pagan		
religion=pastafarian		
religion=scientologist		
religion=shinto		
religion=sikh		
religion=spiritualist		
religion=taoist		
religion=unitarian		
religion=yazidi		
religion=zoroastrian		
default symbol		

Fig. 7 OSM Wiki on key-religion (http://wiki.openstreetmap.org/wiki/Key:religion; 16.07.2014)

6.2.3 Community Practices

A case study of places of worship within OSM for the federal state of Bavaria reveals that OSM contains several non-Christian places of worship (see Fig. 9)—in contrast to the official governmental geodatabase, which lists only Christian places of worship.

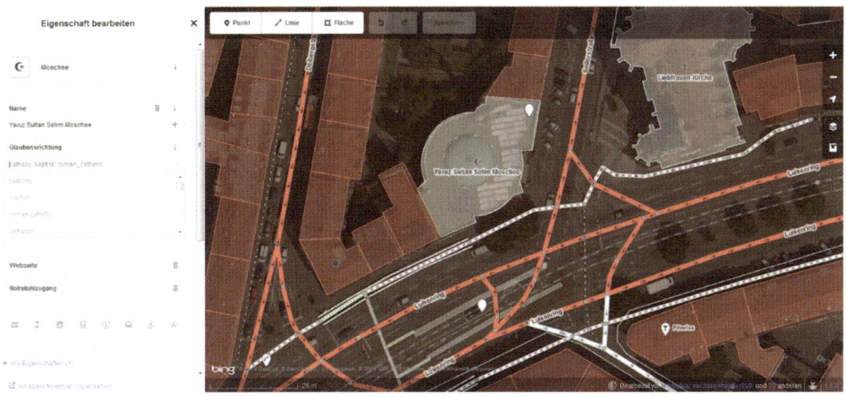

Fig. 8 Online editing of OSM

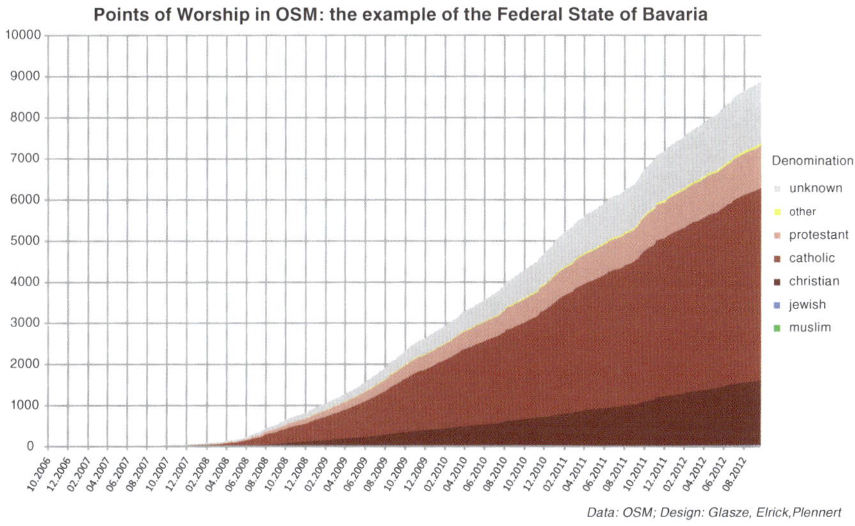

Fig. 9 Places of worship in Bavaria in the OSM database until 12/2012

A more detailed look at OSM data in Nuremberg reveals that by the end of 2012 the OSM database contained almost as many Christian places of worship as the governmental geodatabase[6], as well as two mosques. However, the German map mashup "Moscheesuche" (an application intended to help practicing Muslims to find mosques) listed 10 mosques in Nuremberg. There appears to be systematic

[6] The governmental database contains 124 places of worship—all Christian. OSM contains 119, the biggest part with 106 being qualified as Christian (mostly protestant and catholic), 8 unknown, 2 other, 2 Muslim, and 1 Jewish.

under-representation of mosques in the database: smaller and non-purpose-built mosques are often missing. The classification systems in the Wiki, the rendering and editing software each support inclusion of Muslim sites, so the difference stems from everyday mapping practices and cultural preferences of OSM mappers.

Mosques can be contentious features in Western European urban life (e.g. Schmitt 2004). They are frequently opposed by Islamophobic and right-wing groups who exploit “Not in My Backyard”-like opposition to projects. Many of the mosques that appear in the “Moschee Suche” are visually less prominent than huge recently constructed, purpose-built central mosques. They may share functions with other elements of Islamic life that fit uneasily into OSM classifications, such as Cultural Centers or Madrassars. They are also often transitory, occupying temporary spaces converted from buildings with a previously secular function and have been designated “backyard mosques”. These sites play an important role in the life of the faith community, but may be less significant to mappers who tend to tag prominent POIs or follow up the detail of something that is of direct relevance to their interests. We might further speculate that in Western Europe few members of the Islamic faith community are actively involved with OSM.

6.2.4 New Openings and Fixations

Our example shows that OSM offers openings and fixations. There are certainly new voices being articulated in the project, with evidence of open and transparent discussions, and rapid and self-evident change in the urban fabric gets mapped by the grass roots OSM community. Places of worship are separated out in feature classifications and symbolic rendering attached to buildings offers a more timely and appropriate depiction of these sites than that still delivered by official state mapping.

However, the case also reveals newly fixed codification in the wiki, and standardization in editing and rendering software, as well as cultural biases of the OSM community. The “on the ground” mapping rule[7] tends to emphasize physical structures, and under-represents practices of faith communities using mosques. This leads to a reproduction of traditional cartographic patterns—favoring concrete and other physical structures over use and meaning. As a consequence, less prominent “backyard mosques” quite often still wait to be integrated into the OSM database.[8] Last but not least, mapping practices are heavily influenced by personal preferences, knowledge, and habits—leading in the case study on Nuremberg to an under-representation of mosques in OSM, compared to Christian places of worship.

[7] See: http://wiki.openstreetmap.org/wiki/Map_Features (20.07.2014).

[8] The tagging structure with its separation of use (e.g. amenity = place of worship > religion = *) and building = * in principle enables the separation of use and physical structure and thus is more sophisticated than many tagging schemes in state-based topographic cartography.

7 Conclusions

This chapter offers a critical angle supplementing other work in this volume. It argues that mapping is a socio-technical practice—a socio-technical practice which is embedded in specific and often unequal socio-spatial structures, and which runs the risk of reproducing old and producing new inequalities. These social and also political dimensions of mapping need to be studied in novel ways. We have suggested a research agenda that addresses these concerns by focusing on aspects of the mapping modes through which OSM is practiced, highlighting the importance of authorship, technical infrastructure, and governance. Methods for analyzing these modes have so far largely relied upon quantitative analysis of data relating to the project. We suggest that research might profitably deploy more mixed approaches to data, incorporating case evidence into analyses, and also placing a greater emphasis on ethnographic studies of mapping practice. We illustrate the potential of this agenda with a limited case study of the mapping of mosques and suggest that this broadening of research interests might help OpenStreetMap to deliver the promise offered in its free and open ethos.

References

Ballatore A (2014) Defacing the map: cartographic vandalism in the digital commons. Cartographic J 5:1–24

Bittner C (2014) Reproduktion sozialräumlicher Differenzierungen in OpenStreetMap: das Beispiel Jerusalems. Kartographische Nachrichten 64(3):136–144

Bittner C, Glasze G, Turk C (2013) Tracing contingencies: analyzing the political in assemblages of web 2.0 cartographies. GeoJournal 78:935–948. doi:10.1007/s10708-013-9488-8

Budhathoki N (2010) Participant's motivations to contribute geographic information in an online community. Dissertation, University of Illinois at Urbana-Champaign

Budhathoki NR, Haythornthwaite C (2013) Motivation for open collaboration crowd and community models and the case of OpenStreetMap. Am Behav Sci 57(5):548–575

Burns R (2014) Moments of closure in the knowledge politics of digital humanitarianism. Geoforum 53:51–62. doi:10.1016/j.geoforum.2014.02.002

Caquard S (2014) Cartography II: collective cartographies in the social media era. Prog Hum Geogr 38(1):141–150. doi:10.1177/0309132513514005

Chilton S (2011) OS and OpenStreetMap. Sheetlines 91:20–27

Crampton JW, Graham M, Poorthuis A, Shelton T, Stephens M, Wilson MW, Zook M (2013) Beyond the geotag: situating 'big data' and leveraging the potential of the geoweb. Cartography Geogr Inf Sci 40(2):130–139

Dalton CM (2013) Sovereigns, Spooks, and Hackers: an early history of google geo services and map mashups. Cartographica Int J Geogr Inf Geovisualization 48(4):261–274. doi:10.3138/carto.48.4.1621

DeLyser D, Sui D (2012) Crossing the qualitative-quantitative chasm I: hybrid geographies, the spatial turn, and volunteered geographic information (VGI). Prog Hum Geogr 36(1):111–124

Dodge M, Kitchin R (2013) Crowdsourced cartography: mapping experience and knowledge. Environ Plann A 45(1):19–36. doi:10.1068/a44484

Dodge M, Kitchin R, Perkins C (eds) (2009a) Rethinking maps. New frontiers in cartographic theory. Routledge, London

Dodge M, Perkins C, Kitchin R (2009b) Mapping modes, methods and moments: a manifesto for map studies. In: Dodge M, Kitchin R, Perkins C (eds) Rethinking maps. New frontiers in cartographic theory. Routledge, London, pp 311–341

Eckert J (2010) Tropes 2.0: mobilization in OpenStreet-Map. Unpublished Master thesis, University of Washington

Edney MH (1993) Cartography without progress: reinterpreting the nature and historical development of mapmaking. Cartographica Int J Geogr Inf Geovisualization 30(2):54–68. doi:10.3138/D13V-8318-8632-18K6

Elrick T (2014) Sozialwissenschaftliche tag-Analyse mit OpenStreetMap-Daten am Beispiel religiöser Andachtsstätten in Deutschland. Kartographische Nachrichten 64(3):152–156

Elwood S (2010a) Geographic information science: emerging research on the societal implications of the geospatial web. Prog Hum Geogr 34(3):349–357

Elwood S (2010b) Mixed methods: thinking, doing, and asking in multiple ways. In: DeLyser D (ed) The SAGE handbook of qualitative geography. ausgewählte Kapitel, Sage, Los Angeles pp 94–113

Elwood S (2011) Geographic information science: visualization, visual methods, and the geoweb. Prog Hum Geogr 35(3):401–408. doi:10.1177/0309132510374250

Elwood S, Leszczynski A (2012) New spatial media, new knowledge politics. Trans Inst British Geogr. doi:10.1111/j.1475-5661.2012.00543.x

Elwood S, Goodchild MF, Sui D (2013) Prospects for VGI research and the emerging fourth paradigm. In: Sui D, Elwood S, Goodchild MF (eds) Crowdsourcing geographic knowledge. Volunteered geographic information (VGI) in theory and practice. Springer, New York, pp 361–375

Gerlach J (2014) Lines, contours and legends: coordinates for vernacular mapping. Prog Hum Geogr 38(1):22–39

Glasze G (2009) Kritische Kartographie. Geographische Zeitschrift 97(4):181–191

Glasze G (2014) Sozialwissenschaftliche Kartographie-. GIS- und Geoweb-Forschung. Kartographische Nachrichten 64(3):123–129

Goodchild MF (2007) Citizens as sensors: the world of volunteered geography. GeoJournal 69:211–221

Goodchild M (2009) First law of geography. In: Kitchin R, Thrift N (eds) International encyclopedia of human geography. Elsevier, Oxford

Graham M (2009) Neogeography and the palimpsests of place: web 2.0 and the construction of a virtual earth. Tijdschrift voor Economische en Sociale Geografie, pp 1–15 (online)

Haklay M (2010) How good is volunteered geographical information? a comparative study of OpenStreetMap and ordnance survey datasets. Env Plann B 37(4):682–703

Haklay M (2013) Neogeography and the delusion of democratisation. Env Plann A 45(1):55–69. doi:10.1068/a45184

Haklay M, Singleton A, Parker C (2008) Web mapping 2.0: the neogeography of the GeoWeb. Geogr Compass 2(6):2011–2039

Harley JB (1988) Maps, knowledge and power. In: Cosgrove D, Daniels S (eds) The iconography of landscape: essays on the symbolic representation, design and use of past environments, vol 9. Cambridge University Press, Cambridge, pp 277–312

Harley JB (1989) Deconstructing the map. Cartographica 26(2):1–20

Harris LM, Harrower M (2006) Introduction. Critical interventions and lingering concerns: critical cartography/GISci, social theory, and alternative possible futures. ACME Int E-Journal Crit Geographies 4(1):1–10

Harvey F, Kwan M, Pavlovskaya M (2005) Introduction: critical GIS. Cartographica 40(4):1–3

Hristova D, Quattrone G, Mashhadi A, Capra L (2013) The life of the party: impact of social mapping on OpenStreetMap. In: Proceedings of the AAAI international conference on weblogs and social media

Kent AJ, Vujakovic P (2009) Stylistic diversity in European State 1:50,000 topographic maps. Geogr J 46(3):179–213

Kremer D, Stein K (2014) Ein Analyseansatz für Nutzerverhalten auf Basis von OSM-Daten. User analysis methods for OSM. Kartographische Nachrichten 64(3):144–152

Kwan M, Schwanen T (2009) Quantitative revolution 2: the critical (Re)turn. Prof Geogr 61 (3):284–291

Latour B (1986) The powers of association. In: Law J (ed) Power, action and belief. A new sociology of knowledge?. Routledge and Kegan Paul, Boston, pp 264–280

Leszczynski A, Elwood S (2014) Feminist geographies of new spatial media. Can Geogr, pp 1–17. doi: 10.1111/cag.12093

Lin YW (2011) A qualitative enquiry into OpenStreetMap making. New Rev Hypermedia Multimedia 17(1):53–71

Lovink G, Tkacz N (eds) (2011) A wikipedia reader critical point of view INC reader, vol 7. Institute of Network Cultures, Amsterdam

Mooney P, Corcoran P (2013) Analysis of interaction and co-editing patterns amongst OpenStreetMap contributors. Trans GIS:n/a. doi:10.1111/tgis.12051

Neis P, Zielstra D (2014) Recent developments and future trends in volunteered geographic information research: the case of OpenStreetMap. Future Internet 6(1):76–106. doi:10.3390/fi6010076

Neis P, Zipf A (2012) Analyzing the contributor activity of a volunteered geographic information project—the case of OpenStreetMap. IJGI 1(2):146–165. doi:10.3390/ijgi1020146

O'Reilly T (2005) What is web 2.0. http://www.oreillynet.com/pub/a/oreilly/tim/news/2005/09/30/what-is-web-20.html

OpenStreetMap Foundation (2013) OpenStreetMap foundation 2011. http://www.osmfoundation.org/wiki/Main_Page

O'Sullivan D (2006) Geographical information science: critical GIS. Prog Hum Geogr 30(6):783–791

Pavlovskaya M (2006) Theorizing with GIS: a tool for critical geographies? Env Plann A 38 (11):2003–2020

Perkins C (2013) Plotting practices and politics: (im)mutable narratives in OpenStreetMap. Trans Inst Br Geogr:n/a. doi:10.1111/tran.12022

Perkins C (2014) Plotting practices and politics: (im)mutable narratives in OpenStreetMap. Trans Inst Br Geogr 39(2):304–317. doi:10.1111/tran.12022

Perkins C, Dodge M (2008) The potential of user-generated cartography: a case study of the OpenStreetMap project and mapchester mapping party. NW Geogr 8(1):19–32

Pickles J (1992) Texts, hermeneutics and propaganda maps. In: Barnes TJ, Duncan J (eds) Writing worlds discourse. Text and metaphor in the representation of landscape. Routledge, London, pp 193–230

Pickles J (1995) Ground truth: the social implications of geographic information systems, mappings, vol 1. Guilford, New York

Pickles J (2004) A history of spaces: cartographic reason, mapping, and the geo-coded world. Routledge, London

Ramm F (2012) What we can learn from wikipedia. http://osm.gryph.de/2012/04/learn-from-wikipedia/#more-95

Ramm F (2013) Wer ist der Boss bei OpenStreetMap? http://wiki.openstreetmap.org/wiki/FOSSGIS_2013/Videomitschnitte

Ramm F, Topf J (2010) Open Street Map. Die freie Weltkarte nutzen und mitgestalten, 3rd edn. Lehmanns Media, Berlin

Schmitt T (2004) Religion, Raum und Konflikt. Lokale Konflikte um Moscheen in Deutschland. Das Beispiel Duisburg. Berichte zur deutschen Landeskunde 78(2):193–212

Schuurman N (2000) Trouble in the heartland: GIS and its critics in the 1990s. Prog Hum Geogr 24(4):569–590

Schuurman N (2009) Critical GIS. In: Kitchin R, Thrift N (eds) International encyclopedia of humang geography, vol 2. Elsevier, Oxford, pp 363–368

Steinmann R, Häusler E, Klettner S, Schmidt M, Lin Y (2013) Gender dimensions in UGC and VGI—a desk-based study. AGIT 2013, Salzburg

Stephens M (2013) Gender and the GeoWeb: divisions in the production of user-generated cartographic information. GeoJournal 78(6):981–996. doi:10.1007/s10708-013-9492-z

Verhoeff N (2012) Mobile screens. The visual regime of navigation. Amsterdam University Press, Amsterdam

Weber P, Jones CE (2011) Usability of editors: what to improve. In: Schmidt M, Gartner G (eds) Proceedings of the 1st European state of the map conference, Wien, pp 14–33

Wroclawski (2014) Edit wars in OpenStreetMap. http://blog.emacsen.net/blog/2014/01/17/edit-wars-in-openstreetmap/ 23 July 2014

Zook MA, Graham M, Shelton T, Gorman S (2010) Volunteered geographic information and crowdsourcing disaster relief: a case study of the Haitian Earthquake. World Med Health Policy 2(2):7. doi:10.2202/1948-4682.1069

Spatial Collaboration Networks of OpenStreetMap

Klaus Stein, Dominik Kremer and Christoph Schlieder

Abstract The interaction of the editing operations from different OpenStreetMap (OSM) contributors provides valuable information on collaboration patterns. This chapter describes a new type of spatial collaboration network which can be extracted from OSM edit history data. Drawing from current literature in the field of Social Network Analysis different concepts of collaboration are discussed. It is shown how to apply the measurement of interlocking responses known from research on non-spatial collaboration in wikis to collaboration in OSM. The advantages of the approach are demonstrated by an analysis of collaboration on OSM sample data.

Keywords OSM · User contribution · User collaboration · Social network analysis · Interlocking

1 Introduction

Many aspects of Volunteered Geographic Information (VGI)—starting with its feasibility and success—are surprising to the naïve observer. Research on OpenStreetMap (OSM) contribution has explained some of the puzzles by revealing, for instance, the role played by highly active contributors or users (e.g. Neis and Zipf 2012). Only recently, research has moved beyond the study of contribution and has started to analyze collaboration, that is, the interaction between individual contributions. Such interactions establish implicit social ties, which is why they lend

K. Stein (✉) · D. Kremer · C. Schlieder
Chair of Computing in the Cultural Sciences, University of Bamberg, Bamberg, Germany
e-mail: klaus.stein@uni-bamberg.de

D. Kremer
e-mail: dominik.kremer@uni-bamberg.de

C. Schlieder
e-mail: christoph.schlieder@uni-bamberg.de

J. Jokar Arsanjani et al. (eds.), *OpenStreetMap in GIScience*,
Lecture Notes in Geoinformation and Cartography,
DOI 10.1007/978-3-319-14280-7_9

themselves to modeling within the methodological framework of Social Network Analysis (SNA). Mooney and Corcoran (2012a, 2013a) have looked at the social networks arising from a specific form of collaboration, namely, co-editing interactions of OSM users. By analyzing the number of direct responses to other users, they found that highly active users do not work entirely on their own but interact with low-activity users.

This analysis did not take into account different types and intensities of interaction, however. From non-spatial social network analysis research it is known that not just the breadth but also the depth of collaboration conveys important information about interaction patterns between users (Stein and Blaschke 2010). This chapter extends the analysis of collaboration depth to the spatial domain and describes a data analysis method which permits a detailed description of user collaboration on OSM data. The method is based on the idea of measuring the interlocking of user responses, an approach known for having provided interesting results for the analysis of collaborative text creation in wikis.

The chapter is structured as follows: In Sect. 2 we provide a review of the social network literature on modelling collaboration and summarize the findings on collaboration in OSM from the VGI literature. Section 3 argues that in VGI, collaboration is not necessarily limited to edits of the same OSM object. The situation is more complex: users editing different OSM objects in the same geographic neighborhood may very well influence the work of each other whereas users editing OSM objects referred to by other OSM objects (like different ways of one large OSM relation) may not even take notice of each other. Based on the concept of interlocking response, Sect. 4 introduces a new kind of spatial collaboration network, the interlocking response network. It permits to analyze patterns of alternate edits of pairs of users and describes different aspects of collaboration, especially interaction breadth and depth. Finally, Sect. 5 uses OSM data sets to compare the interlocking response networks with the co-edit networks on which previous studies were based. A computation of collaboration measures for both types of networks shows that interlocking networks are able to reveal user interaction patterns missed by the analysis of co-edit networks.

2 Non-spatial and Spatial Forms of Collaboration

SNA has been successfully applied to a variety of collaboration processes (Newman 2001), whether in traditional media such as printed scientific journals with co-authored articles or in digital media such as wikis, where a detailed interaction history is accessible to computational analysis tools. Different measures of centrality on weighted graphs allow identifying important stakeholders (e.g. Freeman et al. 1991; Opsahl and Panzarasa 2009). Provided that a history or a timeline of the interaction process is available in sufficient resolution, the emergence of sub-networks and their changes can be analyzed (Gibson 2005). In non-spatial SNA, weighted graphs derived from the interaction history of a social network are used to measure the depth or intensity of collaboration of two contributors (Stein and Blaschke 2010) as well as how it changes over time.

Location based social networks (LBSN) account for the fact that social ties are established between actors situated in space and time (Zheng 2011; Zheng and Xie 2011). Information about user location has been used in various ways, among others, for identifying frequently visited places (e.g. Girardin et al. 2009). Some approaches correlate visit preferences for places with different user properties (e.g. Stefanidis et al. 2013), while others try to predict friendship (or at least acquaintance) from co-presence, i.e. users having been at the same place at the same time (Crandall et al. 2010; Cranshaw et al. 2010; Xiao et al. 2014). Whereas place in LBSN is mostly modelled as an attribute of the users, it has a more fundamental importance in OSM, since it constitutes the main purpose and product of collaboration.

Research has shown that the contribution processes to OSM are complex and far from egalitarian (Haklay 2013). Users are predominantly male (Schmidt and Klettner 2013), well-educated and mostly European (Budhathoki and Haythornthwaite 2013). These and similar findings motivate taking a closer look at how intensely different types of contributors interact. Rehrl et al. (2013) suggest a conceptual model of contribution to the OSM database which allows filtering for different types of actions. Statistical analysis reveals that only a small share of users creates and edits most of OSM objects (Mooney and Corcoran 2012a, c). Neis and Zipf (2012) identified senior, junior and nonrecurring mappers. Applying temporal analysis to user edits, Neis et al. (2013) distinguish local, vicinity and external mappers.

Recently, concepts from SNA have been applied to describe the contribution behavior in OSM. Instead of comparing OSM data to some kind of ground truth (e.g. taken from Ordnance Survey, cf. Hecht et al. 2013) research by Keßler and Groot (2013) uses the contribution behavior of individual users from the OSM history to compute trust as a proxy variable for the purpose of assessing data quality. Mooney and Corcoran (2012b, d) introduce SNA to model the interactions of users contributing to the OSM database. They identified important stakeholders of the contribution process by applying different centrality measures to social networks gained from direct responses of contributions (Mooney and Corcoran 2013c, a). Kremer and Stein (2014) point out that interlocking, a measure used to describe the depth of interaction on wikis (Stein and Blaschke 2010), can be applied to such networks as well.

3 Spatial Collaboration Network Analysis

Haythornthwaite (2012) distinguishes between lightweight versus heavyweight collaboration. Lightweight collaboration ("Crowd") involves contributors working almost independently to collect well-defined, small pieces of information, whereas heavyweight collaboration demands for deeper interaction between contributors to reach a certain purpose ("Community"). Both kinds of interaction are visible in the contribution behavior of OSM since senior and local mappers show a high degree of dedication to their work, whereas non-recurring, external mappers often act as a crowd.

Collaboration in OSM is of a different kind than collaboration on a scientific journal paper. The co-authors of a publication are in most cases personally acquainted with each other and in all cases, they are aware of working together. Neither must be the case in the anonymous and complex collaboration processes of digital media such as wikis or VGI. OSM users often do not know explicitly who they are working together with. They are collaborating in the sense that their work affects the same dataset and that many edit operations directly relate or respond to edits of other users. In addition, the OSM database is intended to be a representation of the geofeatures representing some ground truth (the "real" topography of the world).

Relating to previous work is not always an act of affirmation as the user's intention might consist in explicitly removing contributions made by others. Some of the negating edit operations are widely agreed upon, for instance, the removal of spoilers or bad data, while others of them reflect competing standards or opinions (cf. Kittur et al 2007 about edit wars in wikis). As other SNA of OSM collaboration, we do not distinguish between affirming and negating edit operations. Different types of edits could, however, be studied by scanning for revert patterns in the edit history (cf. Keßler and Groot 2013).

OSM defines three main object types: nodes, ways, and relations. For any OSM object an edit history, i.e. a list of revisions with user and timestamp, is stored. Formally, an OSM collaboration network is a graph $G(V, E)$ in which the users are modelled by the set of vertices V. Two users are connected by an edge from E, if they collaborate according to some formal criterion to be established. In the broadest sense of collaboration, all OSM users work together as they share the common goal of improving the OSM database. More specific concepts of collaboration arise from spatial or thematic constraints such as working on the same topic, working on the same OSM object or working in the same area.

While modelling collaboration on OSM nodes is straightforward, the case of OSM ways requires more attention. A OSM way is an object (with tags) referencing a set of nodes. As a consequence, the most prominent feature of a way—its shape—is not defined by the way itself but by the geolocations of the nodes which it references and which will not appear in the version history of the way. From a database point of view, after an edit operation which relocates a node, the version number of the way itself is not increased. Nevertheless, it makes sense to assume that two users who change the shape of one way are collaborating as the partonomic principle applies to collaboration on ways: as ways rely on nodes constituting their shape, edits of the node location will affect the shape of the way (Fig. 1).

On the other hand when a user changes the tags of one of these nodes, the shape of the way is not changed. Therefore tag-changing edits on different nodes referenced from the same way are not a sign of working together. A similar type of partonomic principle does not apply to relations, since relations can recursively reference arbitrary combinations of other relations, ways or nodes.

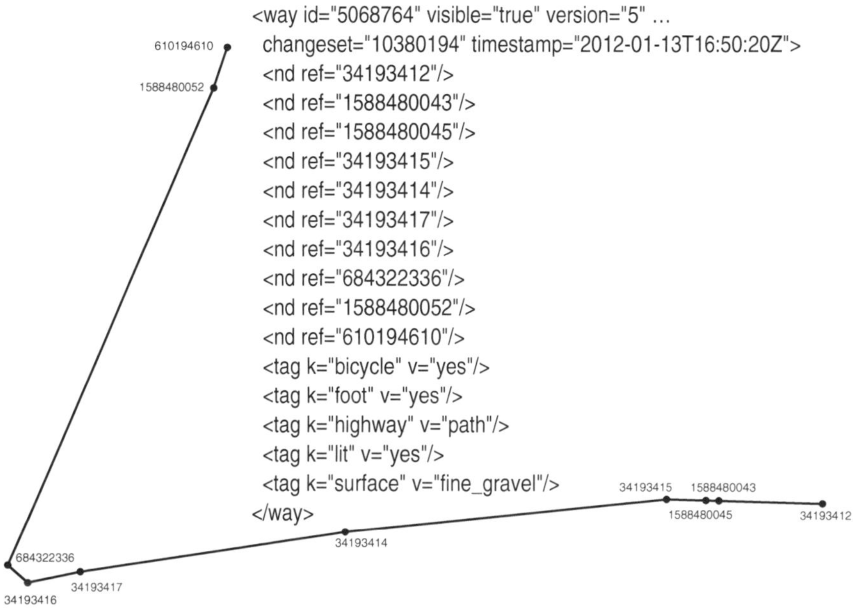

Fig. 1 OSM way object with node references

Because of the recursive nesting, in many cases it is almost impossible for a user to understand which OSM objects she or he is contributing to. Users editing different ways which are referred to by the same relation are often not aware that the ways are connected by the relation on a higher level of abstraction. In addition, OSM relations can grow huge, e.g. the Russian border (id = 309066). It is questionable whether a user moving the position of a node at the Pacific coastline is actually sharing the objective of modifying the same shape with someone editing the Finnish-Russian border. One could discuss to apply recursion to small relations (e.g. with a certain maximum diameter), for this paper we decided to limit the recursive approach to ways.[1]

Identifying the nodes of a collaboration network is straightforward, whereas defining edges depends on the collaboration model chosen (cf. Rhodes and Keefe 2007). The simplest approach consists in connecting users by a collaboration edge if they work on the same OSM object. More complex concepts of collaboration take spatial neighborhood into account, assuming that a user editing in a well-mapped area is aware of the data modified by other contributors. Keßler and Groot (2013), for example, regard it as an act of confirmation, if an edit leaves nearby (<50 m) features untouched.

[1] Körner identified and solved the versioning problem for ways in a similar way (http://mazdermind.de/Slides/Workshop-Erlangen-2013/Slides.pdf, 2014-05-23).

The collaboration concept closest to our approach is that of Mooney and Corcoran (2012b; 2013a, c). It adds a collaboration edge, if user A directly responds to an edit of B on a OSM way. The resulting graph is a direct response co-authorship graph.[2] Mooney and Corcoran (2013a) suggest to consider only successive edits falling within a small time span for collaboration edges. We argue that by looking additionally to interaction patterns, i.e. the number of alternating edits between pairs of users, we can measure collaboration intensity.

4 Measuring Collaboration Depth by Interlocking

The collaboration intensity (edge weight) of two users can be computed by counting the number of co-edits, that is, the number of geoobjects they co-authored. Similar to collaboration on wikis, repositories and other databases providing a revision history, OSM permits to look in the interaction process itself. Similarities identified between OSM and Wikipedia (Mooney and Corcoran 2013b) allow transferring methods applicable to wikis to OSM.

Consider the following example. To an OSM node created by user A, a second user B added some tags, a third user C corrected some spelling mistakes and slightly shifted the position of the node. Then B added more tags and finally D changed the node position again. All four users A, B, C and D worked on the node. Note that the collaboration may be implicit: we do not know whether A noticed the work of B, C and D, and whether D did see the result of the edit operation of A since D only works on the node position set by C.

Formally, collaboration is described as a directed edge in the collaboration graph. Two modelling options are available. A directed edge either points from the user to the user of the previous edit (direct response graph) or it points to all users editing before (group response graph) as the user adds to the collaborative work of its predecessors as the current state of the OSM object is the result of collaborative work done before. The following analysis uses the interlocking collaboration measure defined in Stein and Blaschke (2010).[3] The basic idea of interlocking is to count how often two users alternate in editing an OSM object. Figure 2 shows three OSM nodes with their node histories.

Node 1 and 3 are edited 7 times by A and 5 times by B while node 2 is edited once by A and by B. Nevertheless, there is considerable interaction between A and B on node 1, while on node 2 and 3, B only once responds to an edit of A, and A may not even have noticed B 's edits at all. As the example illustrates, the counting of the edits does not provide an adequate measure of collaboration intensity.

[2] As they refer to ways as polygons and polylines in their approach it is not obvious to us how they deal with the problem of way shape changes through node location shifts.

[3] and originally applied to revision histories of wiki pages.

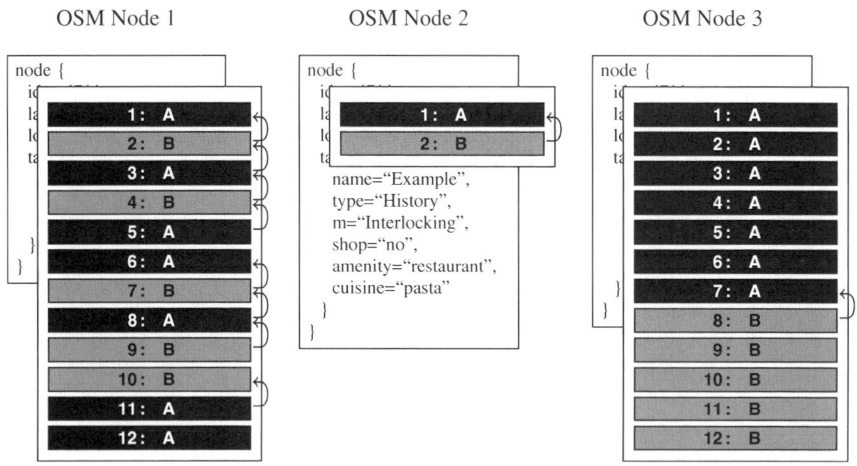

Fig. 2 Node histories with user edits

A much better characterization of collaboration is given by counting how often two users respond to one another. This is because a long revision history does not imply deep interaction.

Generally, more than two users appear in the edit history of an OSM object. In principle, only direct responses could be counted in such cases (Fig. 3a). While some edits may be direct reactions to other edits, mostly a user responds to the current state of an OSM object, i.e. its spatial configuration and its tags, which are a

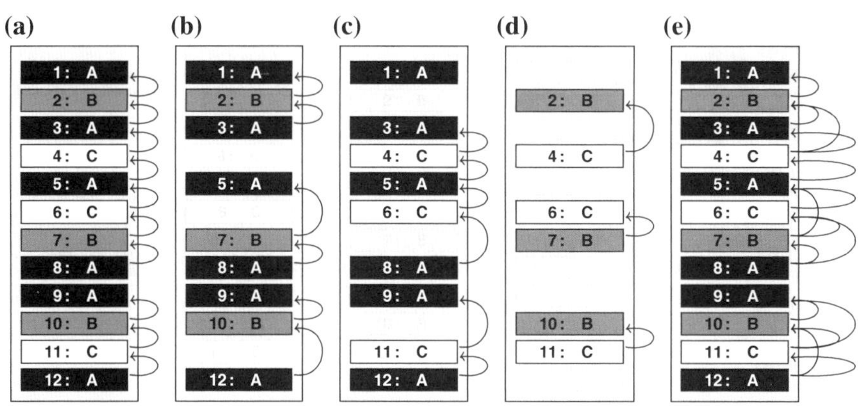

(a)

→	A	B	C
A		2	1
B	2		1
C	2	1	

(b) $A \rightarrow B = 3$, $B \rightarrow A = 3$

(c) $A \rightarrow C = 3$, $C \rightarrow A = 3$

(d) $B \rightarrow C = 1$, $C \rightarrow B = 2$

(e)

→	A	B	C
A		3	3
B	3		1
C	3	2	

Fig. 3 Interaction between multiple users

result of the collaboration of the users who edited before. Interlocking therefore counts the interactions for each pair of users independently (Fig. 3b–d) to compute the interlocking values for these users (Fig. 3e). The interlocking values describe the collaboration intensity between users editing the same OSM object. To obtain a user collaboration network for all users contributing to a certain region, the interlocking values are determined for each pair of users and across all OSM objects in the region.

Another example illustrates the consequences of measuring collaboration by interlocking. Figure 4 shows the edit histories for nine OSM objects. A and B only interact on one OSM object, but they interact with high intensity. A and C interact on a large number of OSM objects, but only once on each. This allows distinguishing different types of mappers. A and B are mappers who somehow care for this one node, either improving it step by step or perhaps fighting some kind of edit war. C on the other hand steps into the region, edits a number of nodes, edits other nodes and leaves again without coming back. This pattern differs from a mass import which normally does not change existing objects but only adds new ones and therefore would always have no outgoing edges at all and incoming edges with at most weight 1. This distinction allows us refining the typology presented by Neis and Zipf (2012).

The simplest approach to obtain a weighted edge for a pair of users is to sum up across all OSM objects:

$$\mathrm{il}^{+}(\mathrm{A} \leftarrow \mathrm{B}) := \sum_{p \in P} \mathrm{il}_p(\mathrm{A} \leftarrow \mathrm{B}). \tag{1}$$

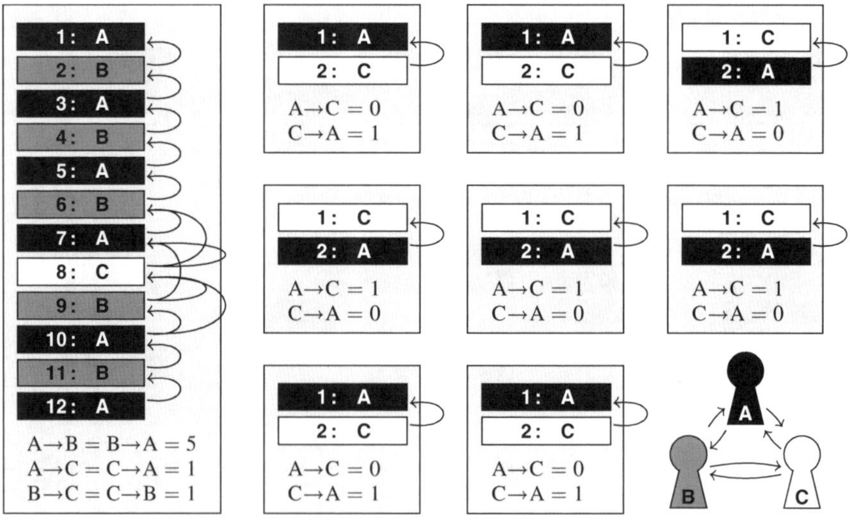

Fig. 4 Interlocking communication events across OSM objects

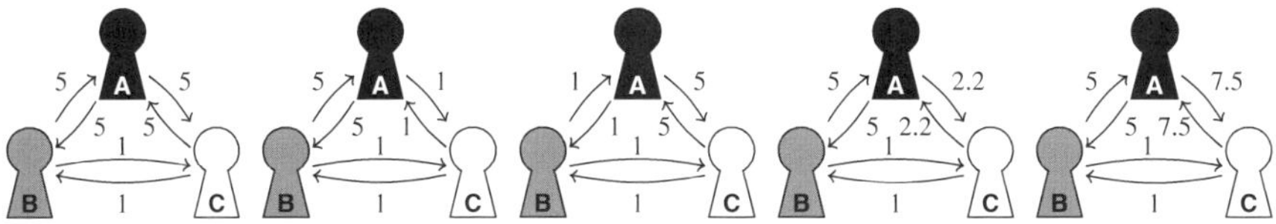

→	A	B	C
A		5	5
B	5		1
C	5	1	

il^{+}

→	A	B	C
A		5	1
B	5		1
C	1	1	

$\mathrm{il}^{\top}$

→	A	B	C
A		1	5
B	1		1
C	5	1	

$\mathrm{il}^{\#}$

→	A	B	C
A		5	2.2
B	5		1
C	2.2	1	

$\mathrm{il}^{\lceil 2}$

→	A	B	C
A		5	7.5
B	5		1
C	7.5	1	

$\mathrm{il}^{\lceil 0.8}$

Fig. 5 Edge weights between users for different interlocking measures

Unfortunately, this does not permit to distinguish the different user types described above. Two complementary measures are needed, the breadth (2) and depth (3) of collaboration.

$$\mathrm{il}^{\#}(\mathrm{A} \leftarrow \mathrm{B}) := \left|\left\{p : \mathrm{il}_p(\mathrm{A} \leftarrow \mathrm{B}) > 0\right\}\right| \tag{2}$$

$$\mathrm{il}^{\mathrm{T}}(\mathrm{A} \leftarrow \mathrm{B}) := \max_{p \in P}\left\{\mathrm{il}_p(\mathrm{A} \leftarrow \mathrm{B})\right\} \tag{3}$$

Both of these measures are extreme cases as they totally disregard either the collaborative depth between two users or the number of OSM objects they collaborated on.

We are most interested in an edge weight that combines both, the width and depth of collaboration. Interlocking provides an adjustable measure based upon interlocking communication events

$$\mathrm{il}^{\lceil k}(\mathrm{A} \leftarrow \mathrm{B}) := \sqrt[k]{\sum_{p \in P} \left(\mathrm{il}_p(\mathrm{A} \leftarrow \mathrm{B})\right)^k}, \tag{4}$$

where the factor k allows to adjust it either to the breadth or depth of collaboration. $k = 1$ gives the sum across all objects, $k < 1$ puts the emphasis on the width, and $k > 1$ on the depth of collaboration (see Fig. 5). Note that all of these networks are structurally equal (same nodes, same links) but have different link weights.

5 Breadth and Depth of Collaboration in OSM

Using the interlocking measure defined in the previous section, we analyzed OSM data sets to evaluate its relevance. The objective of the evaluation is twofold showing that (1) the depth of collaboration cannot be predicted by a measure which

Table 1 Sample regions

Sample region	Berlin	London	Praha	Franconia
City center	Mitte	Westminster	Praha Old Town	
Urban area	Charlottenburg	Hounslow		Bamberg, Bayreuth
Rural area				Hof

only takes the breadth of collaboration into account, (2) the depth of collaboration cannot be predicted by any measure only taking the history length and the number of referenced objects into account.

To inspect a wide variety of well-mapped areas (as this is the prerequisite for deep interaction), we chose several parts of three major European cities and three smaller German towns. For testing on differences, we grouped main touristic centers and suburban parts of the sample regions. Table 1 provides an overview. The Overpass API[4] provides a read-only interface for bulk access to OSM data and allows fetching data by different criteria including geospatial relations. We use Overpass to get all OSM objects within a given area, e.g. the city borders mentioned above.

For any OSM object (node, way, relation) an edit history, i.e. a list of revisions with user and timestamp, is stored. Unfortunately, Overpass does not provide the revision history for these OSM objects.[5] The OSM API provides history requests but should not be used for large data requests as this would slow down the OSM server and disturb others. Therefore we imported the OSM world history dump (planet file) from November 2013 into a local postGIS database (with indexes: ~ 1.7 TB).

Note that the OSM object history itself is not complete. Edits before 2007 are not available (since object history part of OSM before API 0.5 which was introduced on 2007-10-07[6]), 1 % of OSM data was deleted due to the OSM license change (which took effect on 2012-09-12[7]) as data from users not complying with the change cannot be included in the working OSM dataset. We implemented an analysis tool, the OSM-Explorator, which computes several global statistics (e.g. users with many edits vs. users with few edits). The interlocking approach to the depth of collaboration is compared in two ways with approaches taken from literature: (1) our interaction model which considers implicit shape changes of ways is compared to results on the interaction on plain OSM objects, (2) interlocking measures considering the depth of interaction are compared to approaches which only consider the breadth of interaction.

[4] http://overpass-api.de/ (2014-05-26).

[5] It is announced that it will provide some history information in future (announced at FOSSGIS 2013, Robert Olbricht).

[6] http://wiki.openstreetmap.org/wiki/API_0.5 (2014-05-23).

[7] http://wiki.openstreetmap.org/wiki/Open_Database_License (2014-05-23).

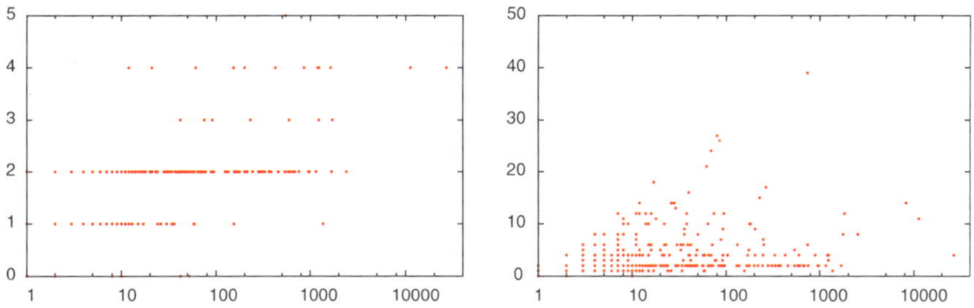

Fig. 6 Comparison of total user edits (x-axis) and maximum interlocking value (y-axis) on nodes and ways (*left*) versus nodes, ways and relations (*right*). Bamberg

In a first step, we check whether aggregated interlocking values (network degree) of OSM objects can be explained by the size of the users' edit history. Figure 6 visualizes a scatter plot for data set of the city of Bamberg. The data shows that the interlocking values of relations outnumber the values on nodes and ways by a factor 10. This can be partially explained by the fact that relations often add some kind of semantic overlay to basic geofeatures as will be discussed below.

As expected, the data further confirms a weak correlation between the edit count and the interlocking value. The scatter plots permit to identify four user types. As expected there are users with high edit count and high degree as well as users with low edit count and low degree. More interesting are the two other types: users with high edit count but low degree can be considered *lonely mappers*; users with relatively low edit count but high degree are intensively collaborating. At this point, we are not able to distinguish between different kinds of collaboration yet.

In a second step, we compare the results provided by collaboration breadth with collaboration depth. This allows distinguishing further user types: A user with high collaboration depth revisits the same OSM objects several times reacting to edits of other users on the OSM objects. Users with high collaboration breadth have interactions with many other users on a large number of OSM objects (but not necessarily recurring interactions). Deep interaction may occur on different OSM objects, if more than two users collaborate with each other.

Senior mappers show a high breadth and depth value and rarely active users show low breadth and depth value. More interestingly, the joint analysis of collaboration breadth and depth reveals the most asymmetric users, i.e. users who interact often, but on different OSM objects (breadth) and users who stay on a small number of OSM objects but edit them very often interactively with other users. Figures 7, 8 and 9 illustrate these considerations as collaboration graphs and two different scatter plots comparing breadth against depth. Note that the plots show the number of interactions on both axes, not simply edits.

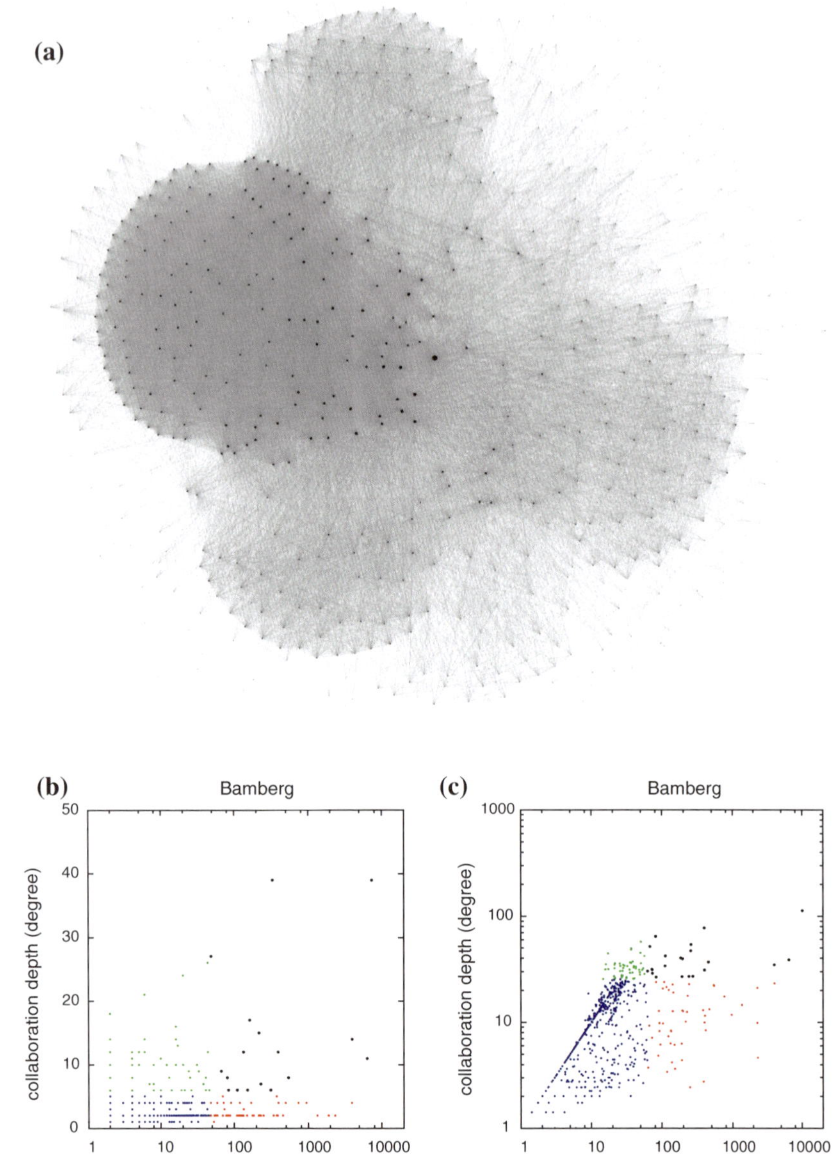

Fig. 7 Interlocking networks for Bamberg, **a** weighted coautorshipgraph ($\mathrm{il}^{\lceil 2}$), zoomed and clipped, **b** maximum interlocking breadth ($\mathrm{il}^{\#}$) to maximum interlocking depth ($\mathrm{il}^{\top}$) for each user, **c** weighted degrees (square root of sum of squares) of maximum interlocking breadth ($\mathrm{il}^{\#}$) to maximum interlocking depth ($\mathrm{il}^{\top}$) for each user

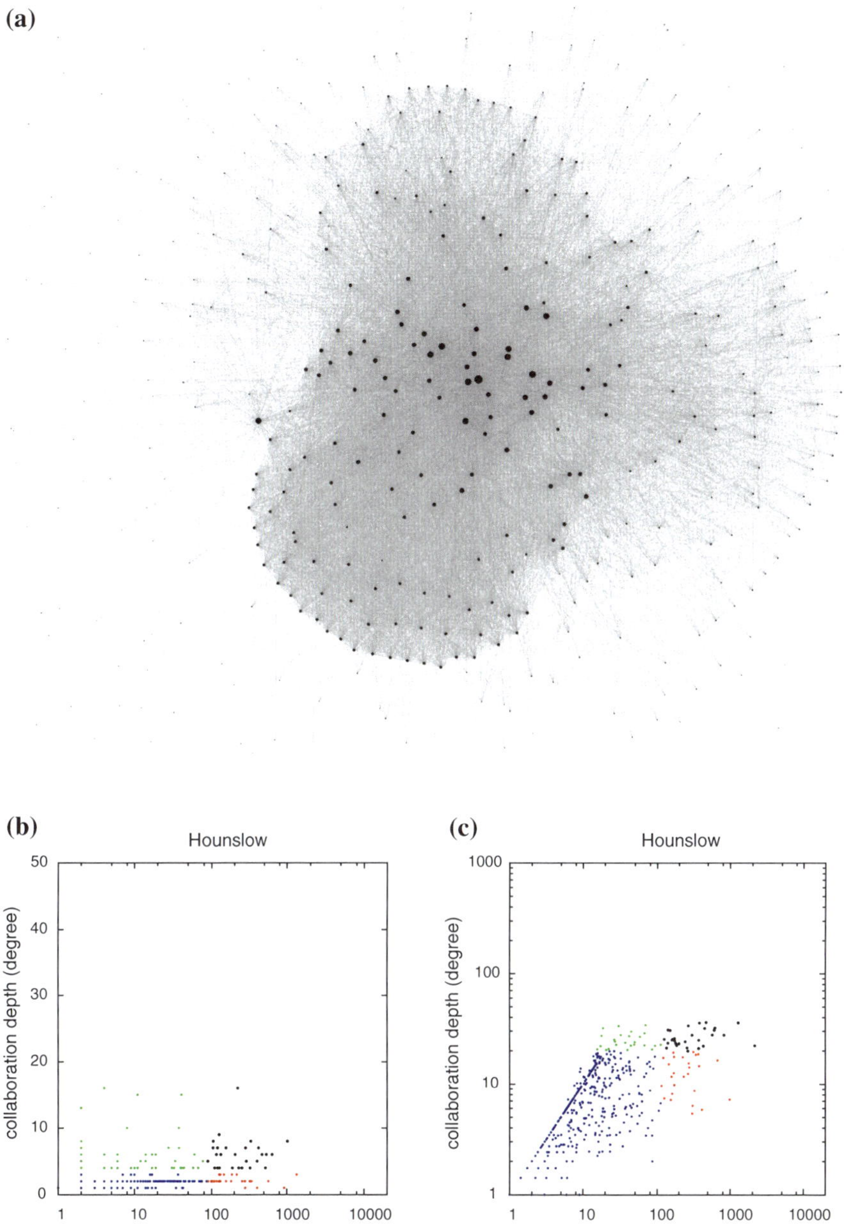

Fig. 8 Interlocking networks for Hounslow, **a** weighted coautorshipgraph ($\mathrm{il}^{\lceil 2}$), zoomed and clipped, **b** maximum interlocking breadth ($\mathrm{il}^{\#}$) to maximum interlocking depth ($\mathrm{il}^{\top}$) for each user, **c** weighted degrees (square root of sum of squares) of maximum interlocking breadth ($\mathrm{il}^{\#}$) to maximum interlocking depth ($\mathrm{il}^{\top}$) for each user

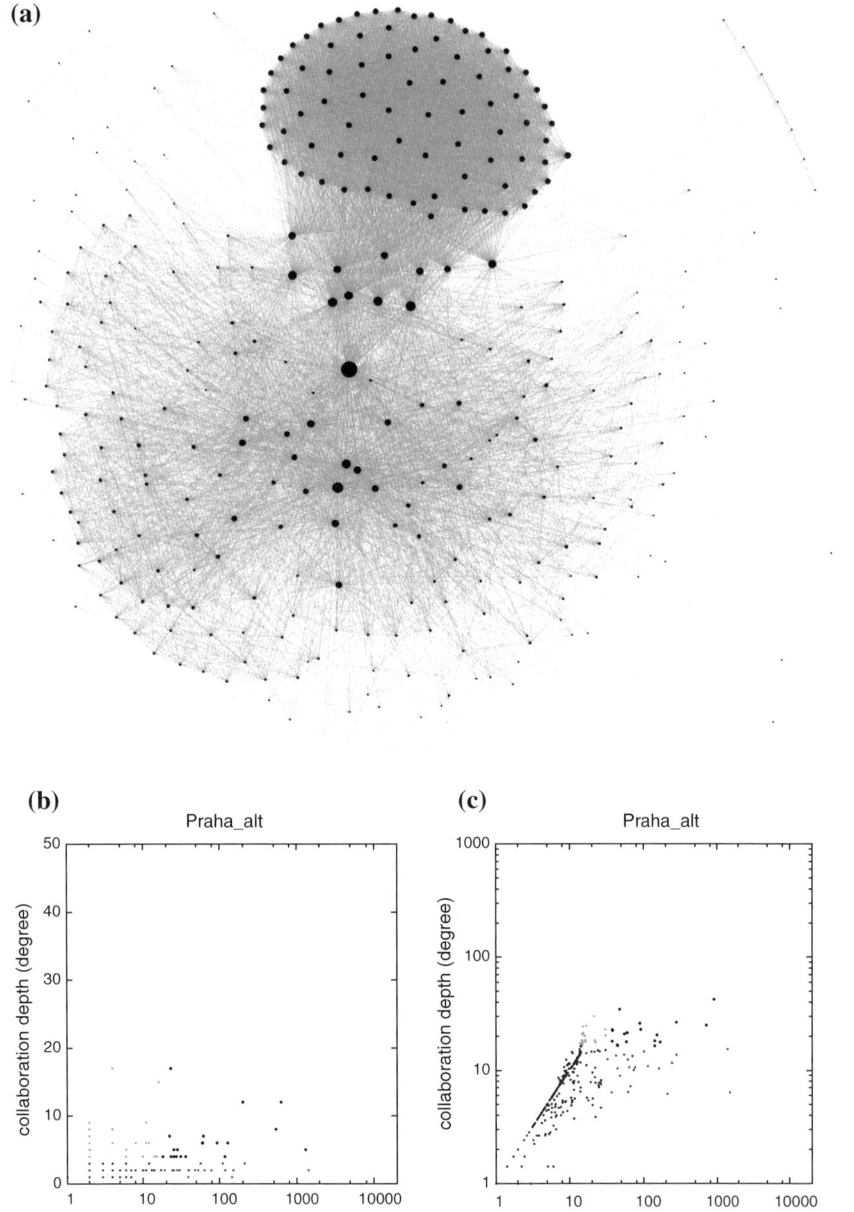

Fig. 9 Interlocking networks for Praha Old Town, **a** weighted coautorshipgraph ($\mathrm{il}^{\lceil 2}$), zoomed and clipped, **b** maximum interlocking breadth ($\mathrm{il}^{\#}$) to maximum interlocking depth ($\mathrm{il}^{\top}$) for each user, **c** weighted degrees (square root of sum of squares) of maximum interlocking breadth ($\mathrm{il}^{\#}$) to maximum interlocking depth ($\mathrm{il}^{\top}$) for each user

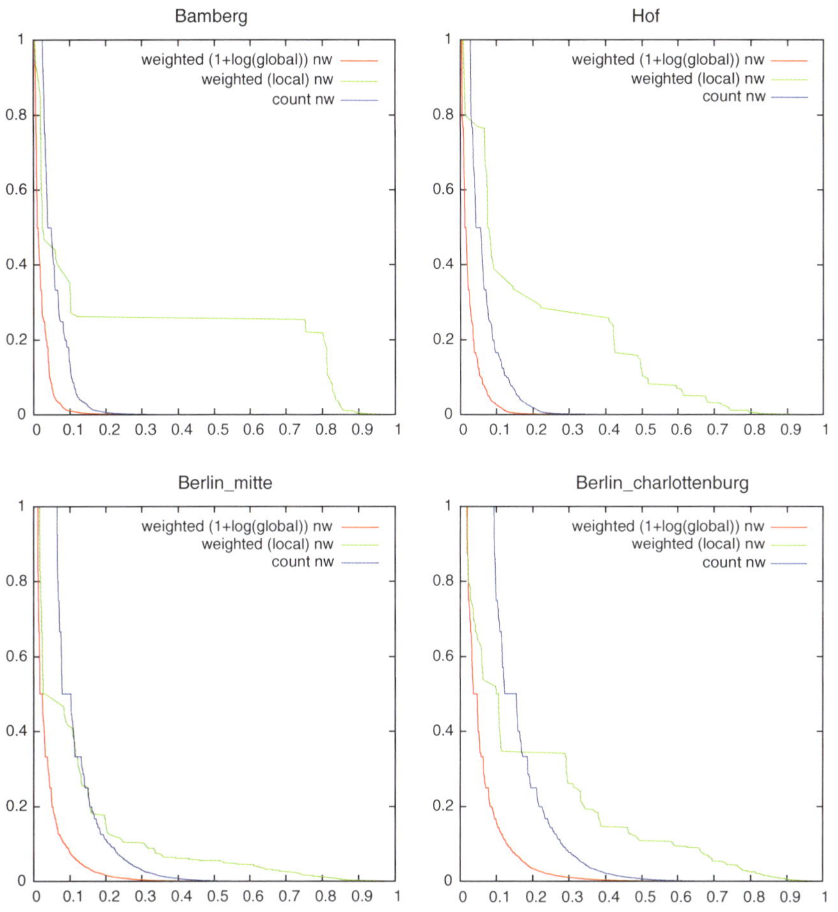

Fig. 10 Locality of mappers, *x-axis* users sorted by the share of their contribution to nodes and ways, *y-axis* fraction of user changesets in region to total user changesets, *blue* all users present in this region, *green* users weighted by the share of their contribution to the region, *red* users weighted by the share of their contribution to OSM

The visualizations of the collaboration graphs provide additional information. Bamberg and Praha each have one prominent mapper (visible in the graph), whereas in Hounslow several high activity mappers are connected to other groups of mappers. A noticeable feature of Praha is a subgroup of strongly interconnected users who connect to the remaining network by only a few gatekeepers.

The plots to the left have been colored to highlight the classification of different kinds of users. In both cases, the threshold has been set to the top 10 %-percentile separating *senior mappers* (black) from the rest. Blue and red mappers are both *not-recurring mappers* as they do not collaborate on an intense level, but red mappers touch a lot of OSM objects. Green mappers focus on few objects, but on an intense level and can be described as *local mappers or experts*.

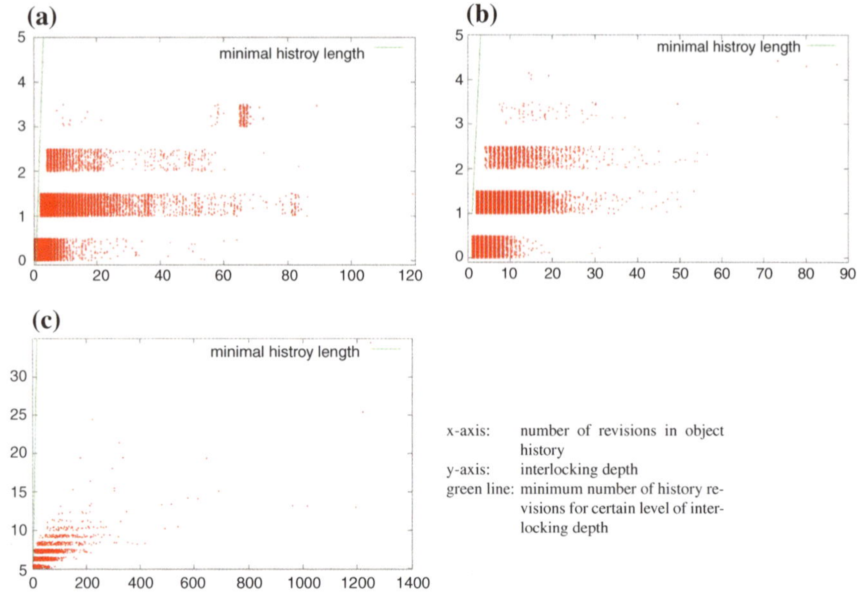

Fig. 11 Comparing history size to interlocking depth. **a** nodes. **b** ways. **c** relations

The interlocking approach allows discriminating the red and black versus the red and green parts of the diagram. Considering only collaboration breadth, local mappers would have been projected on senior mappers, whereas green mappers would have been misleadingly regarded as making no major contribution. As a side-effect, *mass imports* can be separated from non-recurring mappers. Both show a low interlocking value, but mass imports can be identified by zero outgoing links as they never touch existing OSM objects.[8] *External mappers* i.e. users mapping in a previously mapped region for a short time, will also show a small interlocking value with locals, but with a normal out degree.

Figure 10 provides additional insights into the locality of mappers based on the number of change sets. The highest locality of contributing mappers can be measured in Berlin-Charlottenburg with over 20 % of the users dedicating 20 % of their work in that area. In Bamberg and Hof, only about 10 % of active users dedicate 20 % of their work inside that area. Regarding OSM objects, we already saw small interlocking values on nodes (with name tag; max 4) and history on nodes with name tag (max 85) in Fig. 6. On relations we see an interlocking maximum of 35 and a maximum history greater than 1,000. Remember that relations often are some kind of semantic overlay in OSM. The last evaluation step checks whether collaboration depth can be predicted from the number of revisions. Only OSM

[8] There are exceptions: reuse of deleted OSM object ids was not prohibited in the beginning.

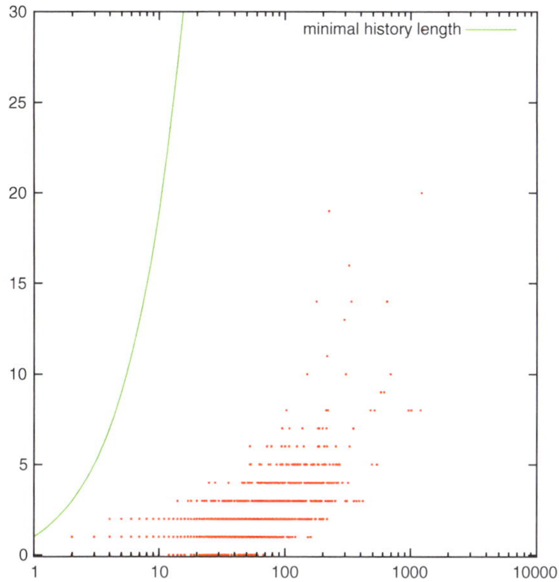

Fig. 12 Interlocking depth versus history size

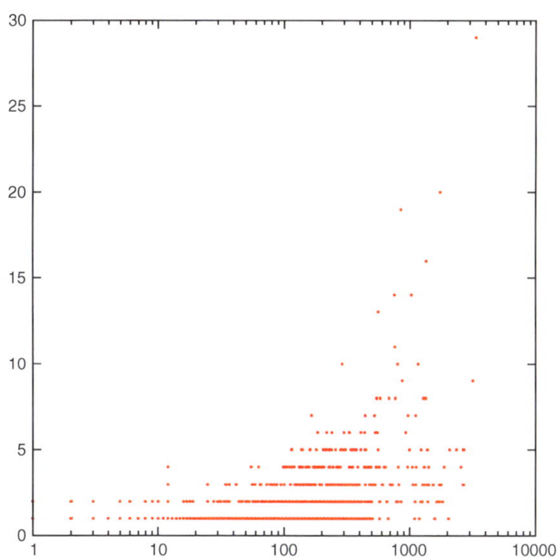

Fig. 13 Interlocking depth versus number of object references on relations

objects that have at least one part in the query regions are considered. Across all query regions and OSM object types, interlocking depth and number of revisions in object history only correlate weakly (Fig. 11). Note that interlocking counts how often two users take turns editing an OSM object, but that these edits do not need to follow directly on each other.

Looking only at relations and substituting the history size with the number of OSM objects referred to by this relation adds no further information. Again, across all query regions and OSM object types, interlocking depth and number of revisions in object history only correlate weakly (Figs. 12 and 13).

6 Conclusion and Outlook

This chapter described a methodical approach for analyzing the depth of user collaboration on OSM data. Transferring the interlocking collaboration measure which had originally been defined for wikis to OSM, gives a detailed view on the collaboration process. The approach was evaluated on OSM data sets collected for that purpose and analyzed by the OSM-Explorator Tool. The evaluation showed that the depth of collaboration cannot be predicted by any measure only taking breadth of collaboration into account. Furthermore, the depth of collaboration cannot be predicted from the history length and number of referenced objects.

Based on the analysis of collaboration depth, we were not only able to identify different kinds of mappers described in the literature but also to refine the types. Collaboration is not necessarily restricted to edits on the same OSM object. Adding a building next to a street created by another user is some form of collaboration because it takes notice of previous work, but leaves it untouched (cf. Keßler and Groot (2013)[9]). Opening the OSM editor for an area shows the features created by other users and the additions of the current user will be influenced by what there is.

For that reason, working on OSM objects which are close to each other could be considered as working together. Our future research will look into efficient methods for computing interlocking collaboration graphs based on close spatial edits. Conceptually, a challenge consists in generalizing the concept of collaboration from the interaction dyad of pairs of users to group-responses.

References

Budhathoki NR, Haythornthwaite C (2013) Motivation for open collaboration crowd and community models and the case of openstreetmap. Am Behav Sci 57(5):548–575

Crandall DJ, Backstrom L, Cosley D, Suri S, Huttenlocher D, Kleinberg J (2010) Inferring social ties from geographic coincidences. Proc Nat Acad Sci 107(52):22436–22441

[9] Who regard it as confirmation and use it to determine trust in and quality of given features.

Cranshaw J, Toch E, Hong J, Kittur A, Sadeh N (2010) Bridging the gap between physical location and online social networks. In: Proceedings of the 12th ACM international conference on ubiquitous computing, ACM, pp 119–128

Freeman LC, Borgatti SP, White DR (1991) Centrality in valued graphs: a measure of betweenness based on network flow. Soc Netw 13(2):141–154

Gibson DR (2005) Taking turns and talking ties: networks and conversational interactionl. Am J Sociol 110(6):1561–1597

Girardin F, Vaccari A, Gerber A, Biderman A, Ratti C (2009) Quantifying urban attractiveness from the distribution and density of digital footprints. Int J Spat Data Infrastruct Res 4:175–200

Haklay M (2013) Neogeography and the delusion of democratisation. Environ Plann A 45(1): 55–69

Haythornthwaite C (2012) Democratic process in online crowds and communities. JeDEM—eJournal of eDemocracy and Open Government 4(2). http://www.jedem.org/article/view/137

Hecht R, Kunze C, Hahmann S (2013) Measuring completeness of building footprints in openstreetmap over space and time. ISPRS Int J Geo-Inf 2(4):1066–1091

Keßler C, de Groot RTA (2013) Trust as a proxy measure for the quality of volunteered geographic information in the case of openstreetmap. In: Geographic Information Science at the Heart of Europe, Springer, Berlin, pp 21–37

Kittur A, Suh B, Pendleton BA, Chi EH (2007) He says, she says: conflict and coordination in wikipedia. In: Proceedings of the SIGCHI conference on Human factors in computing systems, ACM, pp 453–462

Kremer D, Stein K (2014) Ein Analyseansatz für Nutzerverhalten auf Basis von OSM-Daten. Kartographische Nachrichten 64(3):144–152 (German: An approach for analyzing user behavior on OSM data)

Mooney P, Corcoran P (2012a) Characteristics of heavily edited objects in openstreetmap. Future Internet 4(1):285–305

Mooney P, Corcoran P (2012b) How social is openstreetmap. In: Proceedings of the 15th association of geographic information laboratories for europe international conference on geographic information science, Avignon, France, pp 24–27

Mooney P, Corcoran P (2012c) The role of communities in volunteered geographic information projects. In: Krisp J (ed) Proceedings of the 9th symposium on location based services, vol 1, Springer, Berlin Lecture Notes in Geoinformation and Cartography, pp 35–371

Mooney P, Corcoran P (2012d) Who are the contributors to openstreetmap and what do they do. In: Proceedings of GISRUK

Mooney P, Corcoran P (2013a) Analysis of interaction and co-editing patterns amongst openstreetmap contributors. Transactions in GIS

Mooney P, Corcoran P (2013b) Has openstreetmap a role in digital earth applications? Int J Digit Earth (ahead-of-print)

Mooney P, Corcoran P (2013c) Understanding the roles of communities in volunteered geographic information projects. In: Progress in Location-Based Services, Springer, Berlin pp 357–371

Neis P, Zipf A (2012) Analyzing the contributor activity of a volunteered geographic information project—the case of openstreetmap. ISPRS Int J Geo-Inform 1(2):146–165, doi:10.3390/ijgi1020146, http://www.mdpi.com/2220-9964/1/2/146

Neis P, Zielstra D, Zipf A (2013) Comparison of volunteered geographic information data contributions and community development for selected world regions. Future Internet 5 (2):282–300

Newman ME (2001) The structure of scientific collaboration networks. Proc Natl Acad Sci 98 (2):404–409

Opsahl T, Panzarasa P (2009) Clustering in weighted networks. Soc Netw 31(2):155–163

Rehrl K, Gröechenig S, Hochmair H, Leitinger S, Steinmann R, Wagner A (2013) A conceptual model for analyzing contribution patterns in the context of vgi. In: Progress in Location-Based Services, Springer, Berlin pp 373–388

Rhodes C, Keefe E (2007) Social network topology: a bayesian approach. J Oper Res Soc 58(12):1605–1611

Schmidt M, Klettner S (2013) Gender and experience-related motivators for contributing to openstreetmap. In: Mooney P, Rehrl K (eds) International workshop on action and interaction in volunteered geographic information (ACTIVITY), Leuven, Belgium, pp 13–18

Stefanidis A, Crooks A, Radzikowski J (2013) Harvesting ambient geospatial information from social media feeds. GeoJournal 78(2):319–338

Stein K, Blaschke S (2010) Interlocking communication: measuring collaborative intensity in social networks. In: Memon N, Alhajj R (eds) Social network analysis and mining: foundations and applications. Springer, Berlin

Xiao X, Zheng Y, Luo Q, Xie X (2014) Inferring social ties between users with human location history. J Ambient Intell Humanized Comput 5(1):3–19

Zheng Y (2011) Location-based social networks: users. In: Computing with spatial trajectories, Springer, Berlin pp 243–276

Zheng Y, Xie X (2011) Location-based social networks: locations. In: Computing with spatial trajectories, Springer, Berlin pp 277–308

Part III
Network Modeling and Routing

Route Choice Analysis of Urban Cycling Behaviors Using OpenStreetMap: Evidence from a British Urban Environment

Godwin Yeboah and Seraphim Alvanides

Abstract The neglect of non-motorized transportation options in transport planning and demand modelling is gradually being addressed in the United Kingdom. In route choice research there has been, in recent years, a trend away from modelling hypothetical situations towards field testing. This is partly due to the effective use of emerging GPS technologies for gathering travel behavior data in "wild" urban spaces, making it possible to observe realistic situations. Such data on detailed travel behaviors offer possibilities for further research, especially in the non-motorized transportation arena. Globally, there has been progress in the development of cyclists' route choice models using revealed preference GPS data from various geographical and local contexts. However, we have little evidence on detailed cyclists' route choices in the UK in a national and local context. This is particularly the case with low cycling participation cities in North England, where there have been various attempts to increase cycling uptake in recent years. This chapter fills this knowledge gap by undertaking a route choice analysis using the cycling-friendly version of OpenStreetMap (OSM) as the transportation network for analysis, alongside GPS tracks (7 days) and travel diary data for 79 Utility Cyclists around Newcastle upon Tyne in North East England. We examined specific variables as proposed in the relevant cycling literature and used these to develop a model testing the null hypothesis that network restrictions (i.e. one way, turn restrictions and access) do not have any significant influence on the movement of commuter cyclists. The findings suggest that OSM can provide a robust transportation network for cycling research, in particular when combined with GPS track data. The observed routes were significantly longer than their shortest path alternatives, the only exception being the straight-line distance between the observed bike routes and the unrestricted network routes, where the difference was not statistically significant.

G. Yeboah (✉)
The Centre for Transport Research, University of Aberdeen, Aberdeen, UK
e-mail: godwin.yeboah@abdn.ac.uk

S. Alvanides
Faculty of Engineering and Environment, Northumbria University, Newcastle upon Tyne, UK
e-mail: s.alvanides@northumbria.ac.uk

J. Jokar Arsanjani et al. (eds.), *OpenStreetMap in GIScience*, Lecture Notes in Geoinformation and Cartography,
DOI 10.1007/978-3-319-14280-7_10

We conclude that network restrictions for both observed and shortest paths are significant, suggesting that route directness is an important factor to be considered for restricted and unrestricted networks.

Keywords Urban cycling behaviors · OpenStreetMap · Route choice · Bicycle infrastructure · Spatial network analysis · Travel behavior

1 Introduction

The neglect of non-motorized transportation options in transport planning and demand modelling is gradually changing in the UK in a positive way. For example, the Department for Transport (DfT) has recently allocated funding for the improvement of cycling in cities through a cycling ambition program and for improving the transport infrastructure in the UK (DfT 2013a, b). In route choice research there has been, in recent years, a trend away from modelling hypothetical situations towards field testing (Papinski and Scott 2011). This is partly due to the effective use of emerging GPS technologies for gathering travel behavior data in "wild" urban spaces, making it possible to observe realistic situations (Gong and Mackett 2008, p. 3). Such detailed travel behavior data offers possibilities for further research, especially in the non-motorized transportation arena.

Recently, there has been increased interest in the development of cyclists' route choice models using revealed preference GPS data from various geographical and local contexts employing a variety of methodological approaches (Snizek et al. 2013; Ehrgott et al. 2012; Broach et al. 2012; Hood et al. 2011; Larsen et al. 2011; Menghini et al. 2010; Sener et al. 2009; Harvey et al. 2008; Aultman-Hall 1996). However, limited knowledge of detailed cyclists' actual route choice preferences exist within the British geographical and local contexts; especially for low cycling cities in North England where there have been various attempts to promote cycling uptake. This chapter examines cycling behaviors based on recently acquired route choice information by analyzing daily trips from home to work at the network level. The chapter fills a knowledge gap by analyzing actual home-to-work route choices of commuter cyclists around the conurbation of Newcastle upon Tyne (North East England) derived from the primary data collected for this study.

The application of a route choice modelling approach in understanding usage of transport infrastructure is considered quite challenging but very rewarding (Bierlaire and Frejinger 2008; Harvey et al. 2008; Papinski and Scott 2011). The challenges revolve around the lack of detailed information about the users of the infrastructure, in addition to limited availability of accurate data on the transport infrastructure itself. Perhaps a more pertinent emerging concern is how to determine and use the necessary variables to better understand the relationship between the collected data on the one hand and the infrastructure on the other hand. Even when the relevant variables are identified, accurate calculation of their values can be very time consuming, thus adding to the demands of the research. Papinski and Scott (2011) developed a toolkit to generate relevant variables, with the aim of improving variable-generation time. They used 237 home-to-work GPS-based observed routes of car drivers in Halifax (Nova Scotia, Canada) and developed the toolkit in ArcGIS with the *NetworkAnalyst* extension and visual basic for applications (VBA) programming language. Unfortunately, the toolkit is not available to the research community due to licensing issues. However, Papinski and Scott's (2011) study provides useful insights both on the process and on the generated variables themselves which could be applied to non-motorized route choice analysis, as reported in our study.

The aim of this chapter is twofold. First, to identify specific variables tested in cycling-related route choice studies and use them as inputs for route choice analysis with OpenStreetMap (OSM) as the transportation network. Second, we statistically test the hypothesis that network restrictions (i.e. one way, turn restrictions, and access) do not have any significant influence on the movement of commuter cyclists; for this we use both parametric and non-parametric statistical techniques. The remaining chapter is structured into four sections. The next section introduces the contextual background for the study followed by a detailed description of the methods used. This is followed by a section presenting the statistical analysis and results, before discussing our conclusions and the implications of rejecting our null hypothesis in the final section of the chapter.

2 Contextual Background

In the last decade, a number of cycling-related studies based on route choice data and analysis have been undertaken, mostly outside the British context. Aultman-Hall's (1996) study was one of the earliest attempts to look into urban cycling behaviors using a geographic information system (GIS) with time and distance variables as a fundamental part of the investigation. Although the behaviors were described as "*actual*" this was not necessarily the case as participants had to draw their route by memory on a prepared map of the study areas around Guelph, Toronto, and Ottawa (Canada). This prepared map was an integral part of the survey instruments for the data collection and the stated routes drawn by participants were subsequently geocoded, but no GPS devices were used to log their actual movement. The collected routes from participants were then compared with their shortest path alternatives alongside other variables such as distance, time, signage, turns, slope, age, and gender. Using multinomial logistic regression modelling, the study found that age, gender, and winter cycling were significant factors in route preferences of the cyclists sampled and within the study area, while off-road paths, although of high quality, were rarely used. The study suggests that results from the analysis of cyclists' route choice preferences have the potential to assist transportation planners and engineers in addressing the needs of cyclists.

Harvey et al.'s (2008) study was one of the first to use data from a stated preference survey together with a GPS-based survey for 49 cyclists in Minneapolis

(US). They compared preferred routes of commuter cyclists to their shortest path alternatives and found that the mean distance of the preferred routes was significantly longer than the computed shortest possible route on a network without implementing any kind of choice model. For the computation of shortest path alternatives, all roads with restrictions for cycling were removed prior to the computation. They also used network attributes, such as cycling on street, on-street bicycle lane, and off street path, and compared them between preferred and shortest routes for each commuter cyclist. Another early study in Texas (US) based on a web-enabled stated preference survey of cyclists investigated variables influencing route choice preferences of cyclists (Sener et al. 2009). They examined demographic characteristics of cyclists, on street parking, and types of bike facilities along with roadway physical, operational (e.g. travel time), and functional (e.g. traffic volume) characteristics. The findings highlighted the importance of demographic characteristics and route-related attributes in route choice analysis and decision-making. Sener et al. (2009) emphasized that commuting time and motorized traffic volume are the most important variables for consideration, while signage, speed, on-street parking, and availability of bike facilities *en route* were also significant factors.

Unlike the previous two studies, Menghini et al. (2010) used secondary GPS data of cyclists' route choice preferences as an input to route choice analysis. They analyzed a representative sample of 2,435 residents in Zurich, Switzerland, but without recording residents' demographic information. As such, only distance, observed routes, and their related but unique alternatives were analyzed as part of the modelling. Their findings suggest that policies towards the improvement of cycling uptake should aim at the provision of direct and clearly marked bike paths for cyclists within the study area. Data limitations, such as availability of a navigable street-level network, forced the researchers to investigate a rather limited, but useful, number of variables. In addition, Menghini et al.'s (2010) study contributed to existing studies by developing a route choice model based on a large sample of GPS data to address limitations inherent in the stated preference approach. They used the multi-agent transport simulation toolkit (www.MATSim.org) to generate non-chosen (alternative) routes for the origin and destination pairs and considered the length of a route (link) as a cost attribute along with the assumption that "the speed of the cyclists depends in the main on their own choice."

Similar to Sener et al. (2009) and Larsen et al.'s (2011) study used a web-enabled stated preference survey together with other secondary data such as origin–destination information to investigate location-allocation and prioritization of bike facilities based on defined grid-cells across the study area in Montreal, Canada. The choice of a grid-cell approach was based on the biasness of generated routes towards arterial roads. The use of diverse data sources presented many challenges especially when, for example, a network for analysis is not cycling friendly. Both observed and potential trips were generated using a shortest path algorithm and later rasterized to fall within 300 m grid-like corridors. Variables considered in the analysis were time, origin, destination, turns, and weather. Methodologically, Larsen et al. (2011) concluded that the grid-cells method is not appropriate for detailed analysis of cyclists' actual route choice preferences. They also emphasized the importance of cycling infrastructure and the fact that methods assisting objective assessments are essential to provide the evidence needed for effective use of finite resources allocated to improvements of cycling infrastructure.

Hood et al. (2011) designed a revealed preference survey and utilized GPS-enabled smartphones to understand the route choice decisions of cyclists in San Francisco (US). The measured trips were instantly filtered (at the point of logging), for the detection of purpose and mode as well as being map-matched to the transport network. This means that routes not already captured by the network but traversed by the traveler were missed. Shortest path alternatives were also considered as part of the analysis. The study concluded that traffic volume, count of lanes, crime, speed of traffic, and nightfall had no effect on route choice decisions, while length and turns were found to have a negative effect. Also, frequent cyclists did not value bike lanes more than infrequent cyclists.

Broach et al. (2012) study, based on revealed route choice preferences, suggests a significant difference between all other utilitarian purposes and commuting purposes among cyclists in Portland, Oregon (US). Also, distance, frequency of turns, slope, off-street bike paths, traffic signage at intersections, bridges, and volume of traffic played a role in cyclists' route choice decision making. Ehrgott et al. (2012) suggested variables such as time, traffic speed, traffic volume, bike lanes, and gradient among others as inputs to the development of a route choice model. Snizek et al. (2013) used a web-enabled stated preference survey, not GPS-based, for collection of route choice information from the participants. Although similar to Aultman-Hall's (1996) study, they used a web-enabled map rather than a paper map. The study concluded that cyclists enjoy riding near bike facilities, as well as near water bodies or green areas. Almost all of these cycling-related studies have focused only on route choice modelling rather than on the visual aspect of the captured cycling behaviors.

Table 1 lists the most recent cycling-related studies that have analyzed adult cyclists' route choice preferences. Essential explanatory variables used in these studies are cross-referenced in Table 1. The next section discusses these variables further and justifies their selection for analysis in this research.

3 Methods

3.1 Study Area and Sample

Our study area focuses on the city of Newcastle upon Tyne (North East England), which has long been acknowledged as the central place for the Tyneside conurbation (Freeman and Snodgrass 1966, pp. 180–205). The Tyneside conurbation comprises four local authorities: Newcastle upon Tyne, Gateshead, North Tyneside, and South Tyneside. Newcastle upon Tyne, North Tyneside, and South Tyneside have a relatively high prevalence of cycling activity, with the area around the

Table 1 Variables used for cycling-related route choice studies

Study area (authors, year)	Copenhagen (Snizek et al. 2013b)	Auckland Ehrgott et al. 2012)	Portland (Broach et al. 2012)	San Francisco (Hood et al. 2011)	Montreal, Quebec (Larsen et al. 2011)	Zurich (Menghini et al. 2010)	Texas (Sener et al. 2009)	Minneapolis (Harvey et al. 2008)	Ottawa, Guelph and Toronto (Aultman-Hall 1996)
Travel time		✓					✓	✓	✓
Travel distance/origin, destination/bike lane/quantitative information	✓	✓	✓	✓	✓	✓	✓	✓	✓
Nodes/turns			✓	✓	✓				✓
Weather		✓		✓			✓		✓
Safety/crime	✓			✓	✓			✓	✓
Traffic lights/signal	✓	✓	✓			✓	✓		✓
Gradient/slope		✓	✓	✓		✓	✓		✓
Socio-demography		✓			✓		✓	✓	✓
Trip purpose		✓	✓				✓		✓
Route familiarity	✓						✓		✓
Route complexity			✓	✓	✓		✓		✓
Level of service	✓					✓			
Traffic volume		✓	✓	✓			✓		✓
Route detours/directness (route efficacy)	✓							✓	✓

(continued)

Table 1 (continued)

Study area (authors, year)	Copenhagen (Snizek et al. 2013b)	Auckland Ehrgott et al. 2012)	Portland (Broach et al. 2012)	San Francisco (Hood et al. 2011)	Montreal, Quebec (Larsen et al. 2011)	Zurich (Menghini et al. 2010)	Texas (Sener et al. 2009)	Minneapolis (Harvey et al. 2008)	Ottawa, Guelph and Toronto (Aultman-Hall 1996)
intersections	✓	✓	✓						
Comfort	✓	✓						✓	✓
Speed/velocity		✓		✓		✓	✓		✓
Marked bike path						✓		✓	
Bridge			✓						✓
Crossings			✓				✓		✓
Parking	✓	✓					✓		
Natural areas/en-route stops or delays/ congestion	✓								

Quantitative information is usually secondary data. Natural areas may include forest, parks, cemeteries, bodies of water, wetlands, and green spaces central station of Newcastle upon Tyne being more visible (Yeboah 2014). The finding was based on triangulation of 2001 and 2011 UK Census alongside a local Newcastle Cycling Campaign 2010 Petition Survey, the Tyne and Wear Household Travel Survey from 2003 to 2011, and Average Annual Weekday Cycling Traffic in Tyne and Wear (2004–2012) (Yeboah 2014).

The use of GPS in tracking movement behaviors in the built environment is relatively new (Stopher et al. 2008; Gong and Mackett 2008; Van der Spek et al. 2009; Yeboah 2014; Yeboah et al. 2015). The sampling criteria for the study were: any adult *utility cyclist* who is more than 19 years and willing to freely volunteer as a participant; should be a *utility cyclist* and commute by bicycle at least once a day in a week; must have home, work, or school location within the center of Newcastle upon Tyne geographic area; and, must be willing to carry a personal GPS tracker continuously for one week (7 days: Monday–Sunday) along with filling a travel diary. Utility cycling is defined as any cycling not done primarily for fitness, recreation (such as cycle touring), or sport (such as cycle racing), but as a means of transport and covers activities such as traveling to work, to shops, to run errands, to see friends and family at various locations, and to locations of other social activities. The term *cycling behavior* is defined here as the movement characteristics of cyclists together with a number of associated factors, such as perceptions of the built environment, along with relevant characteristics of the cyclists themselves (Yeboah 2014). The research reported here focuses on the subset of commuting (home-to-work) trips, while the way these were extracted from all utility cycling activities will be discussed later.

3.2 OpenStreetMap

OSM is a non-commercial collaborative project to create a freely available repository of a map of the earth which also includes a cycleway network. The cycleway network is editable and publicly accessible either through the main project website (http://planet.openstreetmap.org) or via third party websites such as http://shop.opencyclemap.org or http://download.bbbike.org/osm. More general OSM data can be accessed also from http://www.geofabrik.de/data/download.html or http://market.weogeo.com. Here, the general OSM data comprises trails, roads, railway stations, cafes, layout of urban areas, continental boundaries, as well as other built up features on the surface of the earth. The OSM project is mainly based on local knowledge, community driven, open data, and supported by various academic and non-academic partners (OSM 2014). Although it is free to use for any purpose, it is expected that users credit OSM and its contributors. Earlier studies have used OSM in diverse ways comprising, among others, positional accuracy (Haklay 2010), data completeness of bicycle trail and lane features in OSM for the United States (Hochmair et al. 2014), and evolution of the OSM street network for car navigation in Germany from 2007 to 2011 (Neis et al. 2012). The next section points out further characteristics of OSM data and those considered for our analysis.

3.3 Variable and Route Generation for Route Choice Analysis

Route choice researchers are usually faced with two main challenges in route choice analysis (Papinski and Scott 2011, p. 436). The first challenge is the generation of feasible alternative routes connecting departure and arrival locations to enable comparison with observed routes. The second challenge is the identification and evaluation of network and non-network attributes that influence the route choice of the traveler. This section discusses how contextual variables, alternative and observed routes were generated as an input to the route choice analysis in this study. The observed routes used for the analysis were only the home-to-work commuting trips from the study sample.

Contextual variables were derived from the cycling-related route choice studies identified in the relevant literature as discussed in the previous section. Apart from Harvey et al.'s (2008) study, it was not clear how the shortest path was computed/generated from the network used by the other studies, with respect to whether the restrictions on the network were released or maintained before the computation (Table 2). Generating a shortest path based on a network with restricted paths may be different from one based on a network with unrestricted paths. Also, the measure of directness of a route, route efficacy, is generally important enough to be considered. A cue taken from Harvey et al.'s (2008) findings is that the more experienced bicyclists maneuver from point A to point B, especially in heavy traffic conditions, the less willing they are to travel extra distances. Nevertheless, such findings should be properly considered in relation to the purpose of riding as differences-in-preferences of different types of cyclists (e.g. recreational, commuter, utilitarian) are reported by Harvey et al. (2008). Despite the large number of variables that could be considered as inputs to route choice analysis, the reported studies have demonstrated that the decision to incorporate such variables is somewhat based on data availability and quality, methodology, line of inquiry, and the context of usage of variables. Taking into account the above studies and what could be computed given the available data and tools, 13 variables were selected for analysis in this chapter. Larsen et al. (2011) suggest that characteristics of multiple datasets and methodological issues play an important role in the selection of factors to study cycling infrastructure. The available multiple datasets, chosen methods, and knowledge of the analyst informed the selection of variables for analysis in this research, as shown in Table 2. The OSM transport network was used, rather than the OS MasterMap® Integrated Transport Network Layer™, because of the difficulty in updating the transport network offered by the latter to reflect the captured route choice preferences. OSM was not only easier to update but also reflected more of the captured route choice preferences from the sampled cyclists.

The OSM cycle network infrastructure used for the generation of constrained routes was extracted from the parent/global planet.osm file using the free version of the www.BBBike.org web service (Fig. 1). This service provided a time-efficient way for the extraction of the needed data from the huge layer of OSM covering the entirety of earth. ArcGIS Editor for OSM 2.1 was used for the conversion from

Table 2 Variables used for analysis

Variable	Definition
Distance	Travelled distance of a trip or trip segment
Time	Network travelled time of a trip
Straight line distance	Euclidean distance between the points of departure and arrival
Route efficacy	Actual travelled distance divided by computed straight-line distance
Road types from OpenStreetMap OSM (2012)	
Tertiary	Busy unclassified through roads. Roads wide enough to allow two cars to pass safely and having adequate road markings. *OSM tag: highway = tertiary*
Primary	A roads having OSM tags: *highway = primary; highway = primary_link*
Secondary	B roads having OSM tags: *highway = secondary; highway = secondary_link*
Residential	Residential roads.
	(Used only on roads that have no other function other than for residential purposes).
	OSM tag: highway = residential
Service	Service roads.
	(driveways, carpark entrance roads, private roads, bus-only roads, etc.). OSM tag: *highway = service*
Cycleway	A cycle track which may or may not follow a road. OSM tags used: *highway = cycleway; highway = track;*
	highway = bridleway; highway = path
Footway	A path mainly for walking. OSM tags:
	highway = footway; highway = pedestrian
Unclassified	Country lanes. OSM tag: *highway =unclassified*
Unknown	Unknown categories.
	OSM tags: *highway = null; highway = road*

OSM to ArcGIS Network Analyst data format. To ensure that the converted ArcGIS data was optimally configured for cycling, the OSM network data configuration file "*CycleGeneric.xml*" was used as part of the input in the conversion process. An option that does not require an ArcGIS license is the use of www.cyclestreets.net but this was not explored here.

The file extracted from www.BBBike.org was provided as a downloadable file and is used if one requires a cycle routing network for analytical purposes. Further, the file contains information on the following OSM tag definitions:

highway	*cycle navigable roads*
access	*road access restrictions*
barrier	*cycle barrier restrictions/time penalties*
oneway	*turn restrictions onto oneway roads*
surface	*paving of tracks, cycle ways and footways*
smoothness	*drive time penalty for rough roads*
bicycle	*restrictions on bicycle traffic*

Name: NEEngland

Coordinates: -1.895,54.842 x -1.333,55.145

Script URL: http://extract.bbbike.org/?sw_lng=-1.895&sw_lat=54.842&ne_lng=1.333&ne_lat=55.145&format=osm.bz2&coords=-1.538%2C54.842%7C-1.438%2C54.844%7C-1.333%2C54.867%7C-1.338%2C54.923%7C-1.352%2C54.975%7C-1.394%2C55.022%7C-1.454%2C55.135%7C-1.631%2C55.145%7C-1.788%2C55.106%7C-1.895%2C55.045%7C-1.808%2C54.942%7C-1.78%2C54.905%7C-1.753%2C54.863%7C-1.668%2C54.861%7C-1.606%2C54.842&city=NEEngland

Square kilometer: 1,212

Granularity: 10,000 (1.1 meters)

Osmosis options: omitmetadata=true granularity=10000

Format: osm.bz2

File size: 4.7 MB

SHA256 checksum:
e65b6666c894331ade14af53cdf3740ce4655c000d68909ba78369940f6c01da
Last planet.osm database update: Fri Mar 29 19:31:02 2013 UTC

License: OpenStreetMap License

Fig. 1 Extracted area parameters by www.BBBike.org

The extracted cycle network from www.BBBike.org was subsequently updated using the actual route choices from the primary data along with the OSM background map available in ArcGIS and a cycle street network shape file provided by Newcastle City Council. The update was completed manually by superimposing the various data layers. Each route (i.e. observed GPS track) indicating the movement of the cyclists was inspected with these data layers at the background to make sure that they are completely covered by the cycle network so that the generated routes represent accurately the actual GPS tracks. This update was done to ensure that the network constrained routes could easily be generated using the method proposed by Papinski and Scott (2011). Although the proposed method was originally applied to motorized transport networks, it can be extended for cycle networks if the underlying routes reflect the observed paths. The aim here was to ensure that the provided network reflects the cycle network infrastructure. The outcome of this process was an updated OSM network that accurately reflects cycle path infrastructure provision in the study area, although the update process was very time consuming and tedious.

The four-step method for generating routes, as described by Papinski and Scott (2011), was used to generate routes on the OSM transport network. Their method provides an alternative to the use of map-matching techniques to align the extracted home-to-work trips from the GPS data to the OSM cycle network. The method generates network routes which reflect the GPS measured route choice preferences from the sampled cyclists and allows network level analysis to be performed. For visualization purposes, the observed routes (in black) are superimposed on the OSM network (in grey) as shown in Fig. 2.

In addition, using only the origin and destination information, the shortest path routes were computed for both applied restrictions and released restrictions on the network. This computation was completed using ESRI ArcGIS 10.0 Network Analyst extension 10.0. In total, three sets of network data were used as part of the inputs to the statistical model:

- The observed constrained routes (OCR);
- The network restricted routes (NRR); and,
- The network unrestricted routes (NUR).

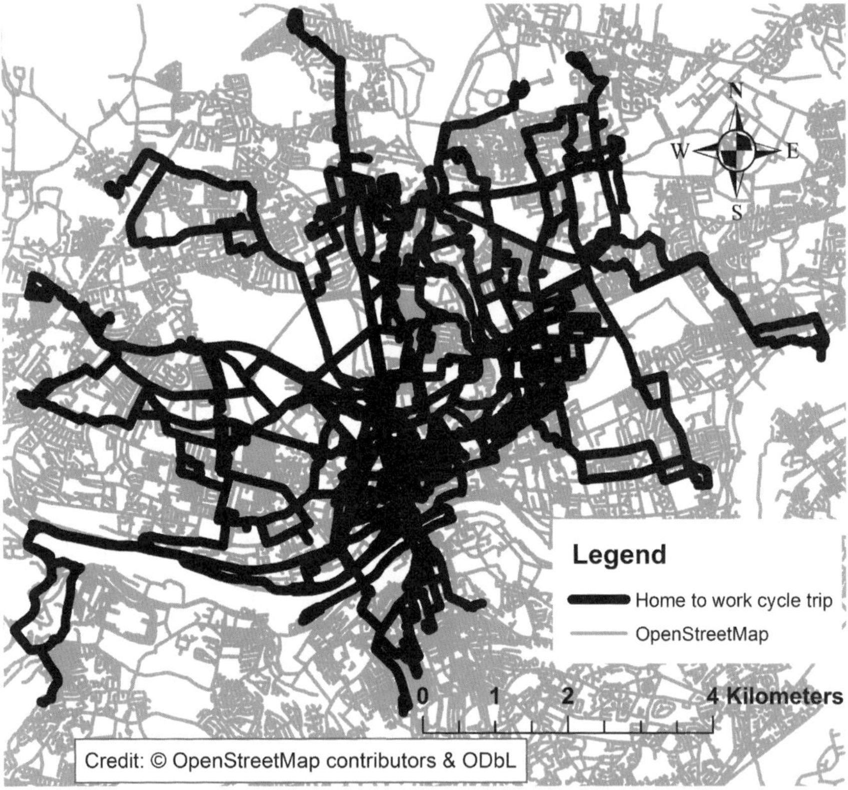

Fig. 2 Home-to-work routes (in *black*) with updated OSM network dataset as basemap

4 Analysis and Results

Little is known about how restrictions on the cycle network influence cycling behaviors, especially in the UK context. The statistical analysis we performed here tests the null hypothesis that urban transport network restrictions (i.e. one way, turn restrictions, and access) do not have any significant influence on the movements of commuter cyclists. Paired samples t-test was used to compare the attributes of the OCR to that of the NRR and NUR. To check robustness for non-normally distributed variables the non-parametric Wilcoxon Signed Ranked t-test was also used for the same datasets and the results compared. Summary statistics of the variables used in this study are presented in Table 3.

Figure 3 shows box plots of the time variable for the three commuting datasets suggesting that there were no instances of extreme outliers. Unlike the first three items from the left side, the last two box plots show the symmetrical variations of time differences between the OCR and the NRR as well as the NUR.

Figure 4 show box plot of commuting distances and their differences. Box plot of commuting Euclidean distances, and their differences, is shown in Fig. 5. More specifically, Fig. 4 shows that the median of the observed routes dataset was higher than their generated shortest path routes but has a slight variation in the differences as the arrival points were moved to the point of restriction towards the destination. For example, if a cyclist cycles from point A to C via B and there was a restriction such that cycling was not possible on the network from B to C, then point B was considered the arrival point. This is to ensure that the effect of network restrictions in movement is properly reflected in the generated routes for NRR. In general, the

Table 3 Comparing observed routes and shortest paths alternatives based on distance impedance and network restrictions (n = 219)

Variable	Observed constrained route (mean ± std.)	Network restricted route (mean ± std.)	Network unrestricted route (mean ± std.)
Time (min)	17 ± 8	15 ± 8	15 ± 7
Distance (m)	4,463 ± 2,100	4,036 ± 1,914	3,926 ± 1,896
Straight line distance (SLD)	3,193 ± 1,554	3,187 ± 1,552	3,192 ± 1,554
Route efficacy	1 ± 0.2	1.3 ± 0.1	1.2 ± 0.1
Percentage of route based on OSM road type			
% of distance on tertiary	70 ± 137	135 ± 162	132 ± 163
% of distance on primary	82 ± 194	114 ± 163	103 ± 167
% of distance on secondary	35 ± 86	100 ± 149	89 ± 141
% of distance on residential	57 ± 95	106 ± 115	108 ± 114
% of distance on service	31 ± 70	36 ± 54	44 ± 61
% of distance on cycleway	70 ± 201	185 ± 319	189 ± 299
% of distance on footway	22 ± 62	63 ± 96	69 ± 129
% of distance on unclassified	28 ± 61	66 ± 72	61 ± 70
% of distance on unknown	141 ± 480	370 ± 649	389 ± 697

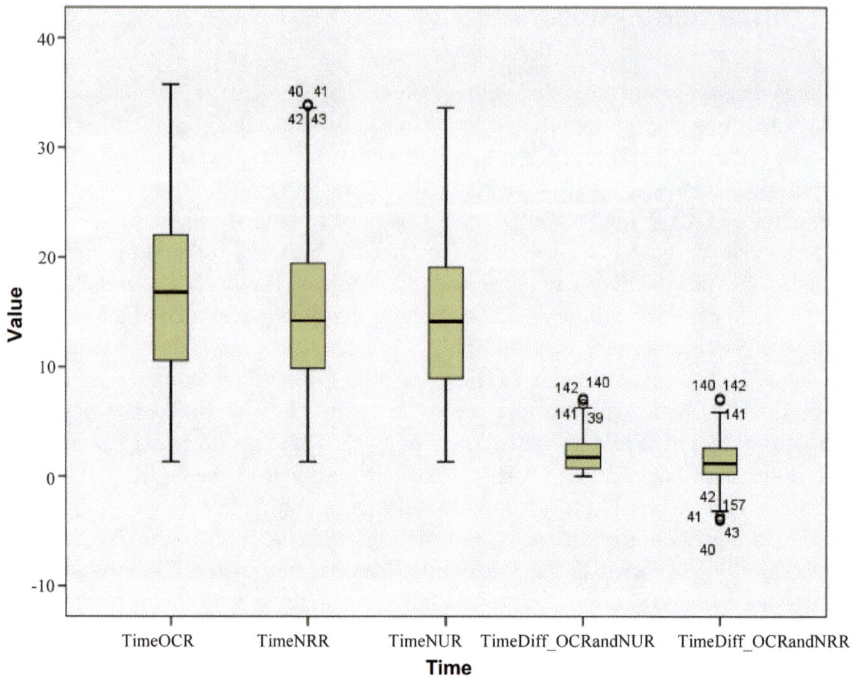

Fig. 3 Box plot of OCR, NRR and NUR commuting time and their differences

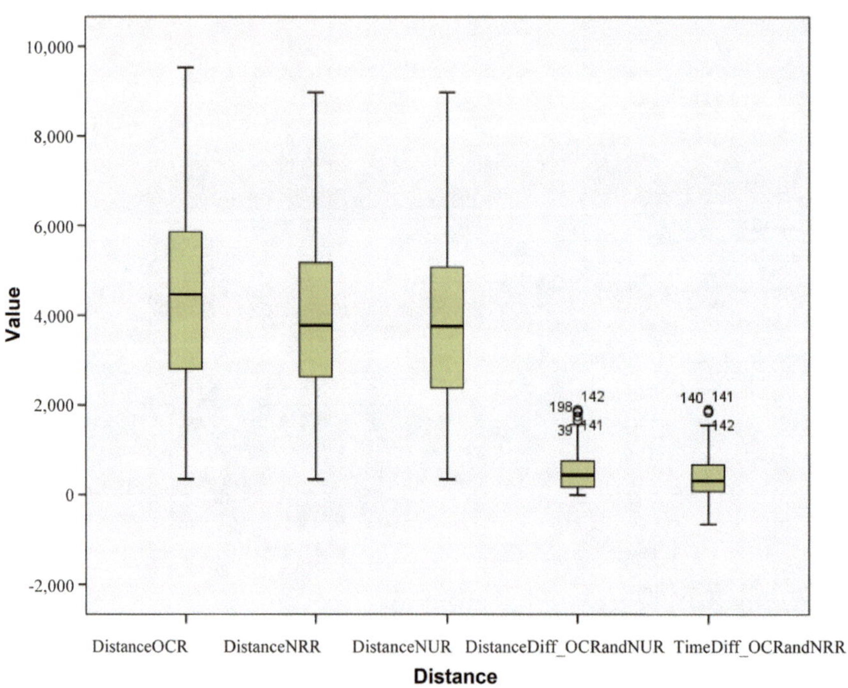

Fig. 4 Box plot of OCR, NRR and NUR commuting distances and their differences

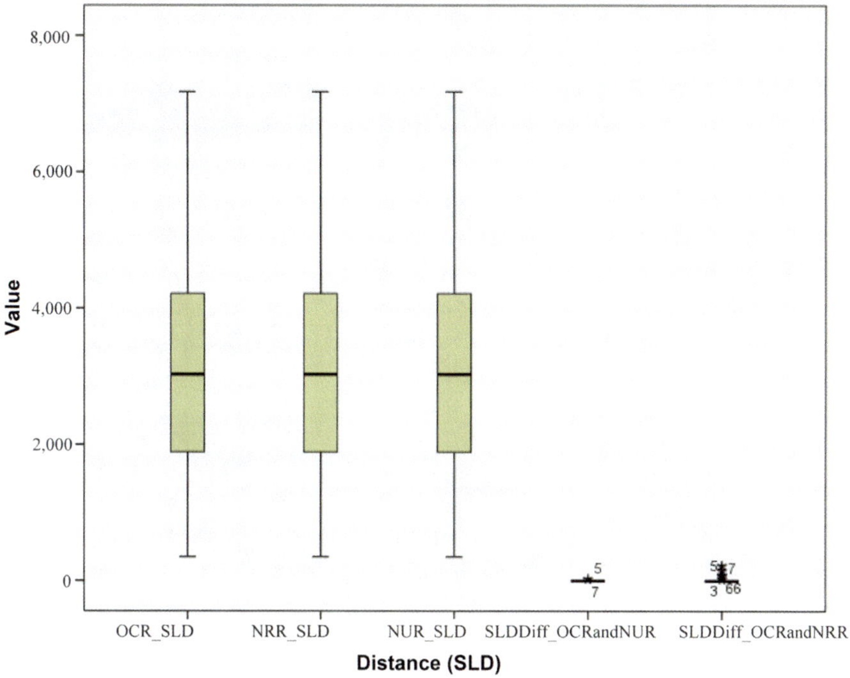

Fig. 5 Straight line Euclidean distance (SLD) variable box plots of commuting datasets

straight-line distances between origin and destination for all three commuting datasets are very close (Fig. 5).

The contextual attributes of observed routes compared to their shortest-path alternatives via paired-samples t-tests are shown in Table 4, using the test statistics and their *p*-values (in brackets). The values in bold indicate significant differences at the 0.05 significance level (n = 219). The contextual attributes of observed routes compared to their shortest-path alternatives, using the Wilcoxon Signed Ranked test, are shown in Table 5 based on the sign of the ranks and their p-values in brackets.

In order to follow the interpretation and discussion of the results from the network level analysis, an example of hypothesis testing is given in Table 6. From this example, the null hypothesis that there is no significant difference between TimeOCR (cycle time from observed routes) and TimeNUR (cycle time from network unrestricted routes) is rejected because the difference between their medians is significant (i.e. $p = 0.001 < 0.05$). Therefore, the Wilcoxon T test indicated a significant difference between the two conditions (i.e. TimeNUR and TimeOCR) with T (N = 219) = t-value indicated, p = p-value indicated.

The statistical findings from the comparison of observed routes against shortest routes based on distance suggest that the observed routes were significantly longer than their shortest path alternatives. The only exception was the straight-line distance between the observed bike routes and the unrestricted network routes, where the

Table 4 Comparison of observed routes compared to their shortest-path alternatives using the paired-sample t-test (n = 219)

Contextual variable	OCR versus NUR	OCR versus NRR
Time (min)	**18.136 (0.000)**	**10.309 (0.000)**
Distance (m)	**17.676 (0.000)**	**13.378 (0.000)**
Straight line distance (SLD)	1.367 (0.173)	**2.849 (0.005)**
Route efficacy	**16.274 (0.000)**	**10.769 (0.005)**
Percentage of route based on OSM road type		
% distance on tertiary	**−4.106 (0.000)**	**−4.338 (0.000)**
% distance on primary	**−3.782 (0.000)**	**−2.097 (0.036)**
% distance on secondary	**−7.943 (0.000)**	**−7.598 (0.000)**
% distance on residential	**−8.990 (0.000)**	**−10.844 (0.000)**
% distance on service	−0.047 (0.963)	−0.504 (0.615)
% distance on cycleway	**−3.882 (0.000)**	**−4.547 (0.000)**
% distance on footway	**−5.329 (0.000)**	**−4.985 (0.000)**
% distance on unclassified	**−6.651 (0.000)**	**−4.579 (0.000)**
% distance on unknown	−1.741 (0.084)	**−2.017 (0.046)**

Note The bolded values shown in Table 4 are significant differences at the 0.05 significance level (n = 219)

Table 5 Comparison of observed routes and their shortest-path alternatives using the Wilcoxon Signed Ranked t-test (n = 219)

Contextual variable	OCR versus NUR	OCR versus NRR
Time (min)	**−12.801b (0.000)**	**−9.381b (0.000)**
Distance (m)	**−12.801b (0.000)**	**−11.187b (0.000)**
Straight line distance (SLD)	−1.342b (0.180)	**−2.934b (0.003)**
Route efficacy	**−12.831b (0.000)**	**−10.525b (0.005)**
Percentage of route based on OSM road type		
% distance on tertiary	**−8.036 b (0.000)**	**−7.531b (0.000)**
% distance on primary	**−10.424b (0.000)**	**−8.437b (0.000)**
% distance on secondary	**−12.176b (0.000)**	**−13.245b (0.000)**
% distance on residential	**−11.291b (0.000)**	**−12.880b (0.000)**
% distance on service	**−2.526b (0.012)**	**−3.057b (0.002)**
% distance on cycleway	**−7.558 b (0.000)**	**−10.253b (0.000)**
% distance on footway	**−10.188b (0.000)**	**−11.294b (0.000)**
% distance on unclassified	**−10.359b (0.000)**	**−9.253b (0.000)**
% distance on unknown	**−2.617b (0.009)**	**−2.483b (0.013)**

Note The bolded values shown in Table 5 are significant differences at the 0.05 significance level (n = 219); b means based on positive ranks

difference was not statistically significant. For all the other contextual variables/conditions, both the parametric and non-parametric outputs from the statistical analysis showed significant differences and the null hypothesis was rejected in favor of the

Table 6 An example of hypothesis test summary using Wilcoxon T-test

Null hypothesis	Test	Sig.	Decision
The median of differences between *TimeOCR* and *TimeNUR* equals 0	Related-samples Wilcoxon signed rank test	0.001	Reject the null hypothesis

alternative. This means that urban transport network restrictions (i.e. one way, turn restrictions and access) appear to have significant influence on the daily movement of commuter cyclists. Therefore, only minor changes in the configuration of the network which in part constitute the built environment may or may not support cycling uptake.

5 Discussion and Conclusion

To the knowledge of the authors, this is the first time this type of analysis has been completed and could serve as a benchmark or evidence base for further studies where the effect of (un)restricted network on cycling behavior could be further examined in different built environments. The closest statement we have found in the literature is "*No studies measured the effects of cut-throughs or right-turn shortcuts*" (Pucher et al. 2010, p. S110). The research reported here partly confirms the findings of Hood et al. (2011) in San Francisco, which suggest that frequent cyclists (in our case experienced cyclists) tend to make full use of cycle lanes. Policy makers in the UK have emphasized that cycling is at the heart of transport and health strategies (DH 2010), but cyclists in many UK cities are yet to experience this despite investment on cycling infrastructure 4 years after such statements (Walker 2013a; Siddique 2013; DfT 2013b). In the UK (excluding Ireland), there seems to be a gradual decline in cycling on main roads, while those using any space other than the main road network to cycle appear to be on the increase from 2002 to 2012 as shown in Fig. 6 (DfT 2013d); could this mean that the road infrastructure is not adequate, while awareness of the usefulness of cycling is on the increase nationally? On the other hand, a downward trend of cycling on main roads in the UK may be due to the generally unsafe and inadequate transport network for cycling. While the number of people cycling on pavements, cycle paths, parks, open country, and private land increased, main-road cycling appears to have decreased during the decade (DfT 2013d). This claim could be supported by the recent announcement by politicians that there will be about £77 million of public money invested into cycling in England with the aim of improving cycling facilities on the transport network at 14 locations (Walker 2013a; Siddique 2013).

Despite the apparent effort and financial commitment by the UK government towards improving cycling, there have been outspoken critics. For example, Walker (2013b) argues that the government is not doing enough towards making the UK one of the "cycling nations" in Europe. The basis of this argument is that the long-awaited DfT response to the "Get Britain Cycling Inquiry" recommendations was far

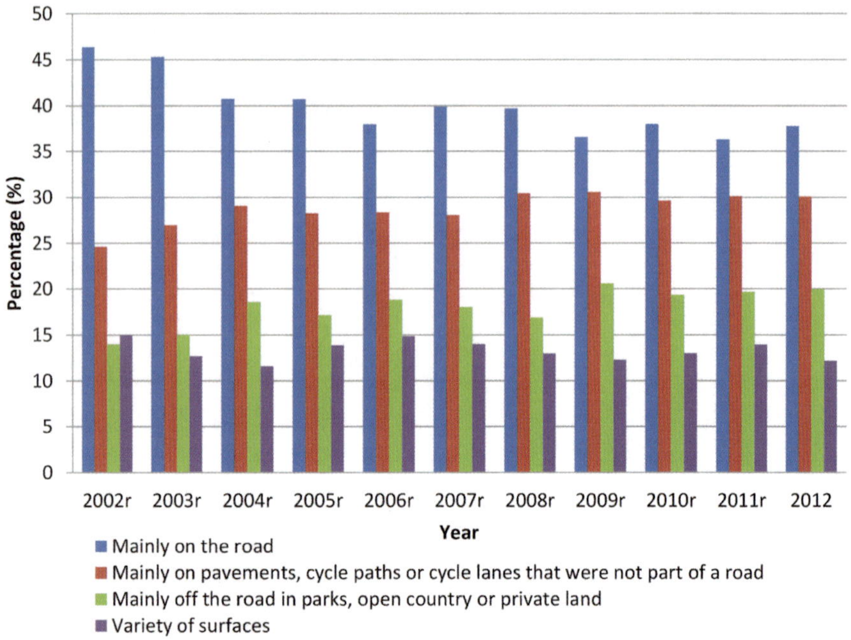

Fig. 6 Where people usually cycled in the last 12 months in the UK (excluding Ireland). *Note* The letter "r" suffixed to years (i.e. from 2002r to 2011r) means revision was made in the estimation by DfT. Graph produced by the authors using data from the National Travel Survey 2002–2012 (DfT 2013d)

below the expectations and seems not to offer any hope by linking responses to already existing nationwide projects (Walker 2013b; DfT 2013c; Goodwin 2013). The MP for Newcastle Central, Chi Onwurah, has also argued for better national leadership and backing for cyclists in the study area during a recently held (Get Britain Cycling) debate in parliament (Pearson 2013). This political campaign suggests that there is more to be done than just a well-formulated local plan as in the Tyne and Wear area. Strategic policies should be put in place to connect on- and off-road cycle lanes by improving the completeness of the cycle network. In addition, key hubs and trip generators should be linked together with the cycle network with the aim of improving the everyday cycling experience (LTP3 2011, p. 160).

In addition, our results suggest that shortest paths do not accurately represent observed bike routes for the home-to-work commute. It is also important to state that the shortest path calculation was based on distance. It was found that using time-impedance for the shortest path with the OSM cycle network was similar to the distance-impedance. This may be due to the lack of posted speed values on the cycle network. It is likely that the shortest path cycle trips generated would have been different given posted speed values integration with OSM. The differences for the paired-samples t-tests are computed as observed route attribute minus the shortest path attribute. This means that if the statistic is positive, the observed route is longer, while if the statistic is negative, the shortest path attribute is longer.

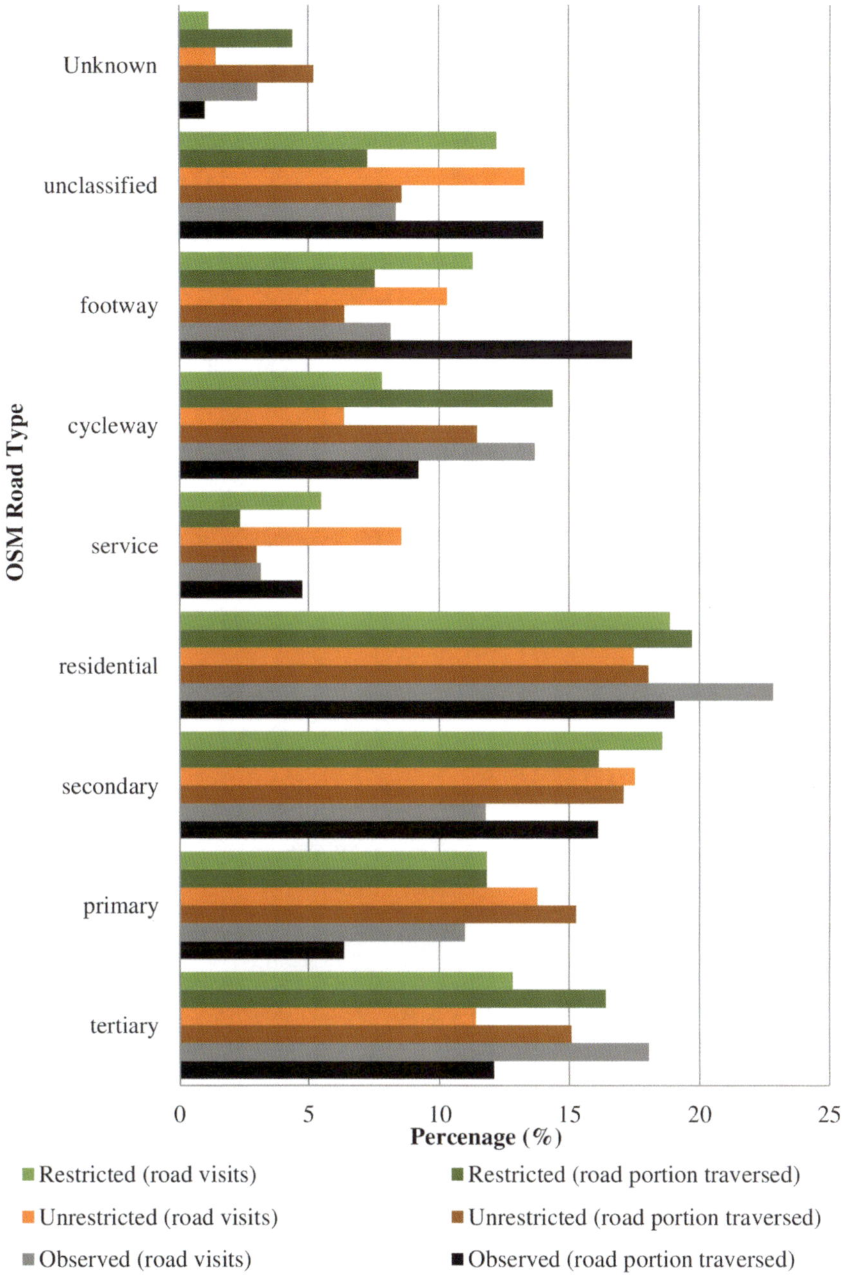

Fig. 7 Percentage of route used based on road type

Our findings suggest that cycling from home-to-work journeys tend to be on residential roads and then footways as shown in Fig. 7. Footways, as the name suggests and according to the coding system of OSM, should not generally be used for cycling. However, they appear to be the best way for cyclists in this study to get to work on time and safely, as they use them in an attempt to minimize exposure to tertiary and primary roads.

Following our statistical analysis, we argue that in order to increase cycling uptake, the network structure needs to support higher destination accessibility (i.e. closer to a value of one for the route efficacy measures) with bike lanes supporting greater connectivity between local streets. The ease-of-travelling to a destination is what is generally considered destination accessibility (Zhao 2014). The route efficacy (i.e. the directness of routes) for both observed and shortest paths was found here to be significant suggesting that route directness is an important factor to be considered for restricted and unrestricted networks.

Acknowledgments The authors would like to thank Northumbria University at Newcastle for sponsoring this research under the Research Development Fund (RDF) and the anonymous participants who helped in collecting the primary data for this research. We also acknowledge the constructive comments made by two anonymous reviewers and the book editors to improve the content.

References

Aultman-Hall LM (1996) Commuter bicycle route choice: analysis of major determinants and safety implications. Doctor of Philosophy (PhD), McMaster University

Bierlaire M, Frejinger E (2008) Route choice modeling with network-free data. Transp Res Part C Emerg Technol 16(2):187–198. doi:http://dx.doi.org/10.1016/j.trc.2007.07.007

Broach J, Dill J, Gliebe J (2012) Where do cyclists ride? A route choice model developed with revealed preference GPS data. Transp Res Part A Policy Pract 46(10):1730–1740. doi:10.1016/j.tra.2012.07.005

DfT (2013a) Guidance: cycle city ambition grants, from https://www.gov.uk/government/publications/cycle-city-ambition-grants. Retrieved 02 June 2013

DfT (2013b) Minister announces record £62 million investment in cycling. GOV.UK, from https://www.gov.uk/government/news/minister-announces-record-62-million-investment-in-cycling. Retrieved 01 June 2013

DfT (2013c) Response to the 'get Britain cycling' report published by the all party parliamentary cycling, from https://www.gov.uk/government/uploads/system/uploads/attachment_data/file/232611/appcg-response.pdf. Retrieved 28 Aug 2013

DfT (2013d) Table NTS0315: where usually cycled in the last 12 months: Great Britain, 2002 to 2012. Department for Transport, from https://www.gov.uk/government/uploads/system/uploads/attachment_data/file/9922/nts0315.xls. Retrieved 31 July 2013

DH (2010) Active travel: the miracle cure? A guide for the NHS on raising physical activity levels through your local transport plan, from http://www.corporatecitizen.nhs.uk/data/files/resources/269/Active-Travel-The-Miracle-Cure.pdf. Retrieved 29 Nov 2012

Ehrgott M, Wang JYT, Raith A, van Houtte C (2012) A bi-objective cyclist route choice model. Transp Res Part A Policy Pract 46(4):652–663. doi:10.1016/j.tra.2011.11.015

Freeman TW, Snodgrass CP (1966) Tyneside. In: The conurbations of Great Britain, 2nd edn. Manchester University Press, Manchester

Gong Y, Mackett R (2008) Visualizing children's walking behavior using portable global positioning units (GPS) and energy expense monitors. In: Paper presented at the virtual geographic environments, The Chinese University of Hong Kong, China, 7–8 Jan 2008, from http://www.iseis.cuhk.edu.hk/downloads/vge/61.pdf. Retrieved 03 Aug 2013

Goodwin P (2013) Get Britain cycling—full report from the inquiry, from http://allpartycycling.files.wordpress.com/2013/04/get-britain-cycling_goodwin-report.pdf. Retrieved 25 Apr 2013

Haklay M (2010) How good is volunteered geographical information? A comparative study of OpenStreetMap and ordnance survey datasets. Environ Plan B Plan Des 37(4):682–703

Harvey F, Krizek KJ, Collins R (2008) Using GPS data to assess bicycle commuter route choice. Transportation Research Board, 20p

Hochmair HH, Zielstra D, Neis P (2014) Assessing the completeness of bicycle trail and lane features in OpenStreetMap for the United States. Trans GIS doi:10.1111/tgis.12081

Hood J, Sall E, Charlton B (2011) A GPS-based bicycle route choice model for San Francisco, California. Transpo Lett Int J Transp Res 3(1):63–75. doi:10.3328/tl.2011.03.01.63-75

Larsen J, Patterson Z, El-Geneidy A (2011) Build it. But where? The use of geographic information systems in identifying locations for new cycling infrastructure. Int J Sustain Transp 7(4):299–317

LTP3 (2011) LTP3: the third local transport plan for Tyne and Wear strategy 2011–2021. Tyne and Wear Integrated Transport Authority, Tyne and Wear, from http://www.tyneandwearltp.gov.uk/wp-content/uploads/2011/04/TW-LTP3-Strategy-Mar-2011-for-upload.pdf. Retrieved 02 Aug 2013

Menghini G, Carrasco N, Schüssler N, Axhausen KW (2010) Route choice of cyclists in Zurich. Transp Res Part A Policy Pract 44(9):754–765. doi:http://dx.doi.org/10.1016/j.tra.2010.07.008

Neis P, Zielstra D, Zipf A (2012) The street network evolution of crowdsourced maps: OpenStreetMap in Germany 2007–2011. Future Internet 4(1):1–21. doi:10.3390/fi4010001

OSM (2014) OpenStreetMap contributors, from http://www.openstreetmap.org/about. Retrieved 28 Aug 2014

Papinski D, Scott DM (2011) A GIS-based toolkit for route choice analysis. J Transp Geogr 19 (3):434–442

Pearson A (2013) Newcastle MP calls for national backing for city's cyclists. Chronicle Live, from http://www.chroniclelive.co.uk/news/local-news/call-greater-cycling-support-newcastle-5837012. Retrieved 06 Sept 2013

Pucher J, Dill J, Handy S (2010) Infrastructure, programs, and policies to increase bicycling: an international review. Prev Med 50(Supplement 1):S106–S125

Sener IN, Eluru N, Bhat CR (2009) An analysis of bicycle route choice preferences in Texas, US. Transportation 36(5):511–539. doi:10.1007/s11116-009-9201-4

Siddique H (2013) David Cameron to announce largest ever investment in cycling. The Guardian, from http://www.theguardian.com/lifeandstyle/2013/aug/12/david-cameron-largest-investment-cycling. Retrieved 14 Aug 2013

Snizek B, Sick Nielsen TA, Skov-Petersen H (2013) Mapping bicyclists' experiences in Copenhagen. J Transp Geogr 30(0):227–233. doi:http://dx.doi.org/10.1016/j.jtrangeo.2013.02.001

Stopher P, FitzGerald C, Zhang J (2008) Search for a global positioning system device to measure person travel. Transp Res Part C Emerg Technol 16(3):350–369

Van der Spek S, Van Schaick J, De Bois P, De Haan R (2009) Sensing human activity: GPS tracking. Sensors 9(4):3033–3055

Walker P (2013a) David Cameron's 'cycling revolution' is barely off the starting blocks. The Guardian, from http://www.theguardian.com/environment/bike-blog/2013/aug/12/david-cameron-cycling-revolution-investment. Retrieved 14 Aug 2013

Walker P (2013b) Despair, cyclists: Britain will not be a 'cycling nation' in your lifetime. The Guardian, from http://www.theguardian.com/environment/bike-blog/2013/aug/30/government-cycling-policy-failed. Retrieved 02 Sept 2013

Yeboah G (2014) Understanding urban cycling behaviors in space and time (Unpublished Doctoral Thesis) Northumbria University at Newcastle, Newcastle upon Tyne

Yeboah G, Alvanides S, Thompson EM (2015) Everyday cycling in urban environments: Understanding behaviors and constraints in space-time. In: Helbich M, Jokar Arsanjani J, Leitner M (eds) Computational approaches for urban environments. Geotechnologies and the environment. Springer, New York

Zhao P (2014) The impact of the built environment on bicycle commuting: evidence from Beijing. Urban Studies 51(5):1019–1037. doi:10.1177/0042098013494423

The Next Generation of Navigational Services Using OpenStreetMap Data: The Integration of Augmented Reality and Graph Databases

Pouria Amirian, Anahid Basiri, Guillaume Gales, Adam Winstanley and John McDonald

Abstract The OpenStreetMap (OSM) project is the most successful collaborative geospatial content generation project. The distinguishing attribute of OSM is free access to huge amounts of geospatial data, which has resulted in hundreds of commercial and non-commercial web and mobile applications and services. The OSM data is freely available and that is why the data can be used within many data infrastructure applications and value-added services. In addition, the free access to data has led to the growth of OSM as a replacement of propriety systems in academic and business environments. This chapter describes the implementation of a navigational application using OSM data as part of the eCampus project in Maynooth University (formerly known as National University of Ireland Maynooth or NUIM). The application provides users several navigation services with navigational instructions through standard textual and cartographic interfaces and also through augmented images showing way-finding objects. There are many navigation services available over the internet; however, the navigation services in this chapter are implemented using a graph database which can be used in connected as well as disconnected modes (online and offline). In addition to the graph database, there is a spatial database for storage and management of images in the system. In other words, the implemented eCampus uses polyglot geospatial data persistence in order to get the best features of several storage systems in a single system in contrast to many traditional storage systems in which all data is stored in a single storage system. The evaluation of the eCampus application by the target users of university students and staff indicated that the visual navigation service using augmented reality provides an intuitive interface that could be integrated into augmented reality systems.

P. Amirian (✉)
The Global Health Network, The University of Oxford, Oxford, UK
e-mail: Pouria.Amirian@ndm.ox.ac.uk

A. Basiri
Nottingham Geospatial Institute, The University of Nottingham, Nottingham, UK

G. Gales · A. Winstanley · J. McDonald
Department of Computer Science, Maynooth University, Kildare, Ireland

J. Jokar Arsanjani et al. (eds.), *OpenStreetMap in GIScience*,
Lecture Notes in Geoinformation and Cartography,
DOI 10.1007/978-3-319-14280-7_11

Keywords OpenStreetMap • Graph database • Navigation services • Augmented reality • Polyglot geospatial data persistence

1 Introduction

The OpenStreetMap (OSM) project is the most successful collaborative geospatial content generation project. The distinguishing attribute of OSM is free access to huge amounts of geospatial data, which has resulted in hundreds of mobile apps and web applications and services. There are several similar commercial services from Google, Microsoft, Yahoo, Esri, TomTom, and many other IT, navigation or mapping companies. However, none of the mentioned commercial services provide access to the data at its lowest level (feature-level). However, this does not mean that it is impossible to implement customized applications with commercial online mapping services. The commercial services provide access to the data using Application Programming Interfaces (APIs) which are mostly just limited to the image of the data (map or more accurately map tiles) and other services that use the data behind the scenes. In other words, the commercial services do not provide direct access to raw geographical data. This limited (or no) access to raw geographical data is a major barrier in front of innovative and creative location-based application development.

In contrast, the OSM project provides access to raw data for anywhere around the world and this is the unique attribute of OSM. Free data from OSM have been used in many applications and services from online gaming to rescue missions and it provides unprecedented opportunities to use data in unforeseen ways. In addition, with the mentioned unique attribute of OSM it is possible to combine data from different sources and implement many innovative and creative applications.

The most widely used service in location-based applications is navigation (Bentley et al. 2014). Typically, navigation applications provide several services for finding shortest path, low-cost path, shortest-travel time path and so on in streets and roads. However, there are many places that need navigation services as well. In general, in many large and complex buildings, markets, university campuses, arenas, stadiums, etc., navigation services are necessary for both indoor and outdoor environments. There are many academic and commercial research projects in this area (Afyouni et al. 2014; Fallah et al. 2013; Strenberg et al. 2013; Purohit et al. 2013). The open and free access to data and the ability to edit data are key advantages of using OSM over all the other services in this area.

Campuses of Maynooth University, like many universities in the world, consist of several buildings, schools, departments and faculties. In addition, Maynooth University is host to many international events and conferences every year. Finding the way between different points of interest (especially different venues) is difficult for new students and visitors. Therefore, the eCampus project started with the following high-level requirements:

- Several kinds of navigation services for Cars, Pedestrians and Wheelchairs with various text, audio and map instructions.
- Augmented Reality to enhance the navigation services.
- Easy to access data for connected and disconnected devices.
- Portable to several platforms such as smart-phones, Web and Cloud.
- Integrated services for indoor and outdoor navigation.
- Extendable for future services.

At the core of the above high-level requirements (and even prerequisite for all other requirements) is open access to geographical data. More specifically, the OSM data is at the heart of the eCampus project. The open access to OSM data makes it possible to make a network (graph) data structure out of street network and enrich the data structure using social networks, business networks, personal preferences, and so on for future use cases.

There are several technical challenges in implementing the eCampus application. Data storage and structure have to be as similar as possible in order to avoid in-memory mapping procedures. In addition, eCampus needs to work in connected and disconnected modes, it needs to be implemented as set of online services with a web interface for connected users and as a standalone mobile app for disconnected users. In addition, it needs an image database to store the physical address (URL) of the images based on geographic data (points). Finally, the augmented reality feature of eCampus needs special algorithms, which are computationally expensive.

The resulting eCampus application provides four navigational services for students, staff, and visitors of the NUIM. Three services (for cars, pedestrians, and wheelchairs) are implemented based on standard shortest-path algorithms. In addition, a fourth navigation service provides augmented images along the path to guide users visually (Fig. 1).

The next section first explains storage and structure of highly connected data (like street network or social network data) in databases. The subsequent section provides implementation details of several navigation services and augmented reality service and finally the chapter explains future directions.

2 Storage and Management of Highly Connected Data in SQL and NoSQL

Highly connected network data are at the heart of most LBS (Location-Based Services) applications. Elements in location-based social networks, road networks, and in general any producer and consumer network constitute highly connected data. Storage and processing of such highly connected and interrelated data can be problematic for most conventional and modern storage systems such as Relational Database Management Systems (RDBMS) and most types of NoSQL databases. The mentioned issue gets worse if the LBS applications need to provide responsive interactivity to the end users when the modification of network data is done in real

Fig. 1 Navigation services in the eCampus web application. A *3D arrow* shows the direction of a path in the NUIM campus

time. By nature, LBS applications need to be highly responsive in real time or near real time. In addition, since LBS applications usually have huge number of users, LBS applications must provide appropriate scalability and performance measures. In this case, many LBS applications resort to proprietary network processing technologies. The mentioned technologies usually build a network representation on top of the physical storage model and keep the network structure in the main memory. In other words, in the mentioned proprietary network processing approaches there is mapping layer between the physical storage layer and the in-memory graph data structure. The mapping layer has some negative effects on performance, consistency, and scalability. In addition, some of the network processing technologies process widely used network analysis and store the results beside the network data itself. However, when additional network elements (such as new landmarks, new roads, and new friends) have to be added to the data, all the representation and built network must be rebuilt and recompiled. This recompilation process would be a serious processing task especially when the newly added elements have lots of connection to the existing elements. In order to prevent being faced with such issues, it is a good idea to store connected data in their natural representation and this is where the graph database comes into play (storing data elements as well as their relationships as graphs in the graph database). In summary, graph databases do not need a mapping layer between the physical storage layer and the application logic layer (Basiri et al. 2014a). The mentioned integration of the storage and application layers also results in flexibility for handling consistency of the data when the size of the data is very large and as a result cannot be stored in the main memory of a single server.

Graph theory was pioneered by Euler in the 18th century, and has been actively researched and improved by mathematicians, sociologists, anthropologists, and others ever since. However, it is only in the past few years that graph theory and graph thinking have been applied to information management (Robinson et al. 2013). Graph databases can be categorized as one of several models of modern NoSQL databases. All the other models of modern NoSQL databases lack the capabilities to handle huge volume of highly connected data (Amirian et al. 2013). On the other hand, conventional RDBMS systems can handle relationships, but they are not designed with highly connected data in mind.

3 SQL (Relational) Database and Managing Highly Connected Data

It has been more than 30 years since RDBMSs (SQL DBMSs) became the major solution for storing all kinds of data in many types of applications (Fowler and Sadalage 2012). They manage data using relational algebra and relational calculus as their theoretical foundation. They use tables, relationships, keys, and Structured Query Language (SQL) to perform all sorts of tasks with data. One of the important advantages of SQL systems is the normalization process which ensures that the data is stored in separate tables and only once in the whole database. SQL databases are usually the best solutions when the schema of data is fixed and predefined. In other words, SQL databases are ideal solutions for managing structured data (Amirian et al. 2010). SQL systems can be effectively used in many common geospatial-related workflows. Since they support transaction and locking features, they provide robust consistency and backend for enterprise GIS systems. Usually geospatial data have a fixed schema and, in most cases, they are not used in isolation. That is, a join of two or more datasets and connecting data through spatial operations is needed in most GIS workflows. For this reason, managing fixed schema geospatial data and using them in GIS workflows can usually be done effectively through SQL systems.

The SQL databases manage connected data using relationships and they retrieve connected data using joins. However, joins are one of the most computationally expensive processes for SQL databases (Fowler and Sadalage 2012). In most cases, joins are the bottleneck of SQL databases. In order to avoid many joins (which are needed in handling highly connected data in SQL databases) a denormalization process can be used to store data items several times in order to avoid joins. However, there are several issues associated with the denormalization process, especially with providing consistency in large datasets. In addition to issues related to handling highly connected data, some other problems arise with SQL systems when scalability is needed by adding more servers and technologies to bind them together (Chang et al. 2006). With more load on a SQL system, vertical partitioning, denormalization, and removal of the relational constraints comes into play. In summary, in order to handle highly connected data with scalability and availability, the normalized relational model of data storage has to be compromised and deviated from the relational model.

4 NoSQL Databases and Handling Highly Connected Data

The Not only SQL (NoSQL) DBMSs are a broad class of DBMSs identified by non-adherence to the SQL (relational) model. There are different types of NoSQL databases, each with distinct sets of characteristics, but they all can deal with large amounts of (semi-structured and structured) data, are able to support a large set of read and write operations, and are designed with scalability and distribution of data in mind (Amirian et al. 2014). For this reason, NoSQL and relational models are not in contrast with each other, rather they are complementary. The most widely accepted taxonomies of NoSQL databases are: key-value, document, columnar, and graph (McCreary and Kelly 2013).

The key-value database is the simplest type of NoSQL database. As the name implies, this type of database stores schema-less data using keys. The key is usually a string and the stored values can be of any valid data type such as a primitive programming data type (string, integer etc.) or a Binary Large OBject (BLOB) without any predefined schema. It provides a simple API to access stored data. In most cases, this type of NoSQL database solution provides very little functionality beyond key-value storage. There is no support for relationships in key-value databases. Usually there is no apparent usage of Key-Value databases in LBS applications. However, this type of NoSQL database can be used efficiently for retrieve and search for geographical names, landmarks, and points of interest in addition to low-level usage for caching and lookup.

A document database in its simplest form is a key-value database in which the database understands its values (Lawrence 2014). In other words, values inside the database are based on predefined formats such as XML, JSON, or BSON (Binary JSON). This feature of document databases provides many advantages over key-value databases. Instead of putting too much logic in the application layer, many operations can be done by the database itself. Queries in this type of NoSQL database are quite flexible and some document databases have their own query languages. Similar to key-value databases, there is no need to adhere to a predefined schema to insert data. There is only limited support for relationships and joins, as each document is a standalone data component. Relationships and joins are not supported the way they are supported in relational databases. Often, relationships and joins have to be used in common GIS workflows. However, the document-oriented nature of the system has some major effects on the way that data can be retrieved. For example, if the application needs data items from the same documents it would be very fast. However, whenever the data items are part of different types of documents there is no efficient approach to reduce the number of index lookups.

In summary, indexing is just based on documents and there is no notion of relationships in document databases.

The columnar (or column family or wide-column) databases store data in sets of columns and distribute data based on columns (rather than rows in SQL databases). The column is the smallest unit of data and it is a triplet that contains a key, value, and timestamp (Abadi et al. 2014). Columnar databases store all values beside the name of the columns and store null values simply by ignoring the column. Usually, related columns compose a column family. All the data in a single column family will be stored on the same physical set of files. This feature provides higher performance for search, data retrieval, and replication operations. Queries in this type of NoSQL databases are limited to keys and in most cases they do not provide a way to query by column or value. By limiting queries to just keys, columnar databases ensure that the procedure for finding the machine containing the actual data is quite fast. There is no join capability in most columnar databases, which makes them less useful for managing connected data.

As the name implies graph databases are based on graph theory and employ nodes, properties, and edges as their building blocks. The nodes and edges can have properties. In the graph databases, various nodes might have different properties. The graph databases are well suited for data that can be modelled as networks such as road networks, social networks, biological networks, and semantic webs (Yang et al. 2014). Graph databases have been growing in popularity and the most significant reason for this growth is the importance of graph data structure in the fields of social, information, and biological sciences as well as in the development of computer hardware. The graph data structure fits naturally for modelling of world objects, entities, and relationships between them (Ciglan et al. 2012).

According to Angles and Gutierrez (2008), graph database models are defined as "those in which data structures for the schema and instances are modelled as graphs or generalizations of them, and data manipulation is expressed by graph-oriented operations and type constructors." In other words, a graph database is a database management system that is based on the graph theory introduced in 1736 by Euler. Graph theory uses nodes for storing entities and edges for relationships among them. Graph databases emphasize the relations among entities rather than entities themselves (Dominguez-Sal et al. 2010a). Any model can be thought of as a representation of reality. A graph model can be thought of as a collection of objects such as people or places and the relationship between them such as "friend" or "living in". Those objects and relationships form a network or a graph (Miler et al. 2014).

5 What Is Special About Graph Databases?

The main feature of graph databases is the fact that each node contains a direct pointer to its adjacent node, so no index lookups are necessary for traversing connected data (which is really valuable in navigation services in LBS). As a result, they can manage huge amounts of highly connected data since there is no need for expensive join operations. Some graph databases support transactions in a way that relational databases support them. In other words, the graph database allows the update of a section of the graph in an isolated environment, hiding changes from other processes until the transaction is committed. Geospatial data, especially street network data, can be modelled as graphs. Since graph databases support topology natively, topological relationship (especially connectivity) between geospatial data can be easily managed by this type of NoSQL database. In most GIS workflows, topological relationships play a major role. In addition, since graph databases are ideal for managing data with evolving schema, they can be effectively used in Volunteered Geographic Information (VGI) and crowd sourcing applications. Also, graph databases are the best choice for managing huge linear networks (such as roads) and for routing and navigation applications. In addition, since each edge in a graph database can have different set of properties, they provide flexibility in the traversal of networks based on various properties. For example, it is possible to combine time, distance, number of points of interest, and user preferences in finding the best path and the mentioned path would be unique for each user.

6 Related Work and Benchmarks

At the core of navigation services are graph traversal algorithms and one of the most important graph traversal algorithms in LBS is the shortest path. The most popular shortest path implementation on a relational database is pgRouting, which is implemented on top of the PostgreSQL SQL (Relational) database. Graph databases already have the shortest path implementations in their core. Neo4j is one of the graph database systems used for semi-structured and network-oriented data and has been in production since 2003. It was developed in Java programming language and is free for non-commercial use. Neo4j can be used as an embedded or server database.

The first full graph database performance analysis was presented in a paper by Dominguez-Sal et al. (2010b) which implemented queries of High Performance Computing Scalable Graph Analysis Benchmark v1.0 (HPC-SGAB) (Bardar et al. 2009). HPC-SGAB was designed by several leading researchers from academia and industry. In their paper, the authors tested the performance of four graph database systems on a synthetic generated graph: Neo4j, Jena, HypergraphDB, and DEX. The test was composed of four kernels: edge and node insertion performance, measuring time needed to find a set of edges that meet a condition, measuring time spent on building a subgraph, and traversal performance over the whole graph. The results showed that DEX and Neo4j were the most efficient graph databases at the time of writing.

Recently Miler et al. (2014) demonstrated that the Neo4j graph database outperforms pgRouting in calculation of the shortest path and graph traversal in a transportation network. The test was performed using a huge transportation network (around 630,000 nodes and 750,000 edges) based on Austrian OSM data. The benchmarks were performed with two modes (cold and hot) to determine the time needed for a database to load data into memory and perform the shortest path analysis between about 200 pair of nodes (as source and destination). Results of the benchmarks demonstrated that the Neo4j graph database is 30–35 % faster than pgRouting in calculation of the shortest path in a transportation network.

In another research, Baas (2012) illustrated the usage of Neo4j in the context of spatial databases and compared its performance with PostgreSQL and pgRouting. Again, OSM data was used as the test data for benchmarks in Bass' work 2012. The results of the mentioned research showed that the graph database is most beneficial when queries can be expressed as traversals over local regions of a graph. Queries that are well suited to this approach are, for example, shortest path analyses or connectivity queries. In a related research paper, De Souza et al. (2014) performed a set of experiments on several spatial extensions of NoSQL databases and they concluded that Neo4j currently provides the most complete support for spatial data.

Graph database management systems are not just routing engines. Their primary purpose is local graph traversal based on the property or hyper-graph graph model[1] in use cases such as social networks, fraud detection, recommendation engines, resource authorization, or computer network management.

In the work of Partner et al. (2014), sample social network data was loaded in the graph database (Neo4j) as well as in MySQL RDBMS. Experiments composed of several queries for friend recommendation and friend finding based on level of friendship up to the 5th level; for example, someone's friends are considered as level 1 and friends of friends considered as level 2. The results of the experiments strongly suggest that graph databases are the best choice for highly connected data like social networks (Partner et al. 2014).

The authors of this chapter performed a set of similar experiments in a location-based social application. They implemented a location-based social network application using Neo4j as well as using Microsoft SQL Server 2012 and measured the performance of queries for finding nearest friends or nearest friends of friends for both databases. In the mentioned research, OSM data for Dublin, Ireland was loaded in Neo4j and SQL Server databases. Table 1 illustrates the results of the experiments which indicate the superiority of graph databases over relational models for handling highly connected data.

In all the aforementioned resources, experiments, and benchmarks the results proved that graph databases are the best choice for management of highly connected data and Neo4j is the most efficient graph database at the time of writing. The results of the mentioned resources were the basis for choosing a graph database for this research. In addition, since Neo4j can be used in server mode as well as embedded mode, it can be efficiently used in online as well as offline (disconnected) applications and services.

---

[1] In the property graph model each node and each edge can have different sets of properties and each edge is between exactly two nodes. In other models of graphs like the hyper-graph model a single edge can be shared between many nodes. Neo4j implements the property graph model. In the hyper-graph model it is possible that an edge is shared between more than two nodes. HypergraphDB uses this model.

Table 1 Performance (response time in seconds) of finding various level of friendship relationships in certain geographic areas using SQL Server and Neo4j

Depth	Spatial (SQL server 2012)	Graph (Neo4j)
2	0.018	0.012
3	0.671	0.198
4	198.675	1.759

7 Implementation

Figure 2 illustrates the high-level conceptual architecture of the eCampus applications. An instance of the Neo4j graph database is used at the server-side to store and manage the highly connected OSM data as well as to process the graph structure of the OSM data. There are several navigation web services (on top of the graph database) which are available for clients to use. The mentioned web services can be accessed using REST APIs or simple HTTP GET requests. In order to use the navigation services, the client (desktop web browsers and mobile web browser) needs to be connected to the internet.

For disconnected clients, the Neo4j is used in embedded mode; the Neo4j engine is deployed on smartphones (Android platform) along with the eCampus native

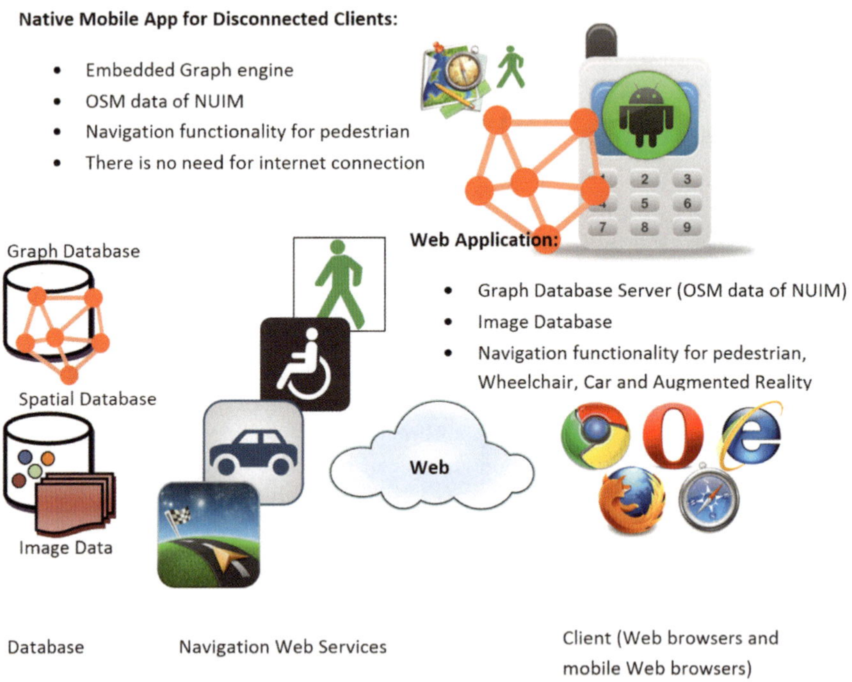

Fig. 2 High-level conceptual architecture of eCampus applications

mobile application. The embedded graph database, which has the same engine as the server-side graph database, includes OSM data for the NUIM. In other words, the native mobile app contains OSM data of the NUIM and the mentioned data is managed using an embedded Neo4j engine.

In addition to the graph database, there is a spatial database for storage of geographical coordinates of the landmarks in the image and the coordinates of the camera (mobile device) when the image was taken. In order to provide the highest retrieval speed for images and since the image data are just read-only for users, the image data are stored as separate files in a public directory on a web server. In other words, the implemented eCampus uses polyglot geospatial data persistence in order to get the best features of several storage systems in a single system; graph of street data in a graph database, point data in a spatial database, and images in the file system.

There are three navigation services that provide the shortest path as well as instructions for Car, Pedestrian and Wheelchair navigation modes in KML format. In addition, a fourth navigation service (Augmented Reality Navigation service, ARN) provides augmented images along the path to guide pedestrian users. Next section describes the implementation details of the ARN service.

8 Augmented Reality Navigation Service Implementation

The Augmented Reality Navigation service takes advantages of geolocation capabilities of mobile devices and/or the HTML5 geolocation API and uses the pedestrian navigation web service in order to provide visual navigation on the NUIM campus. The position of the user is retrieved from the GPS of the mobile device or geolocation API (or manually selected by the user) and a desired destination is selected. The path between these two positions is retrieved using the pedestrian navigation web service. Then a spatial database of images of the campus is queried to retrieve a sequence of images taken from the desired path in the direction the user should take to reach his destination. This retrieval is made possible by storing, in the database, the position of the camera when the image was taken. Then, these images are augmented by overlaying a 3D arrow pointing at the following direction to take, helping the user to visually interpret his position and get his bearings. The ARN uses two services behind the scenes:

- Simple image retrieval: This service takes geographical coordinates as an input and returns the URL of the image with the closest registered coordinates within the database.
- Image sequence retrieval with visual navigation information: This service takes two GPS coordinates as input, calls the pedestrian navigation web service to get the pedestrian path between the two given positions, and returns the sequence of images from the database taken along the path. Finally, the images are augmented with a 3D arrow pointing at the next direction in the path.

Table 2 Schema of image data table

Name	Type	Description
id	Serial	Image unique id number
img_path	Character	Image URL in the public web server
gps_coord	Geography	GPS coordinates of what is shown in the image
img_timestamp	Timestamp	Timestamp of when the image was uploaded
cam_gps	Geography	GPS coordinates of the camera when the image was taken
cam_heading	Real	Heading of the camera when the image was taken (0 for north, 90 for east, ±180 for such, −90 for west)

The mentioned spatial database is implemented using PostgreSQL/PostGIS. Table 2 illustrates the schema of data in the spatial database.

In order to capture images, a native mobile app is implemented which captures images along with the coordinates and heading of a mobile device. Then a script is used to upload data from the mobile device to the spatial database on the web server as well as to transfer the images to a public directory on the web server.

Simple image retrieval is implemented using php scripting language. To use this service, a GET request can be sent to its URL with the latitude and longitude parameters (a geographical position). It returns a JSON string with the URL of the retrieve image in the public directory. Figure 3 illustrates a simple web application for testing this service.

If no image can be found, it returns a 204 HTTP status code, which means "No Content" is available. As mentioned before, the image sequence retrieval service takes two GPS coordinates as input, calls the pedestrian navigation web service to get the pedestrian path between the two given positions (origin and destination), and returns the sequence of images from the file system taken along the path. The images are augmented with a 3D arrow pointing at the next direction in the path and returned to the requester. The implementation detail of the augmented reality functionality is described in the next section.

For each point in the KML in the pedestrian navigation path, the direction, d, between the current point position (index i) and the following one (index i + 1) is calculated using Eq. (1):

$$d = \frac{\pi}{2} - \tan_2^{-1}(\mathrm{lat}_{i+1} - \mathrm{lat}_i,\ \mathrm{lon}_{i+1} - \mathrm{lon}_i). \tag{1}$$

Equation 1: Direction formula

Then, the image with the closest camera position to point i and the closest heading to d is retrieved. To find the mentioned image, a score value s is calculated using

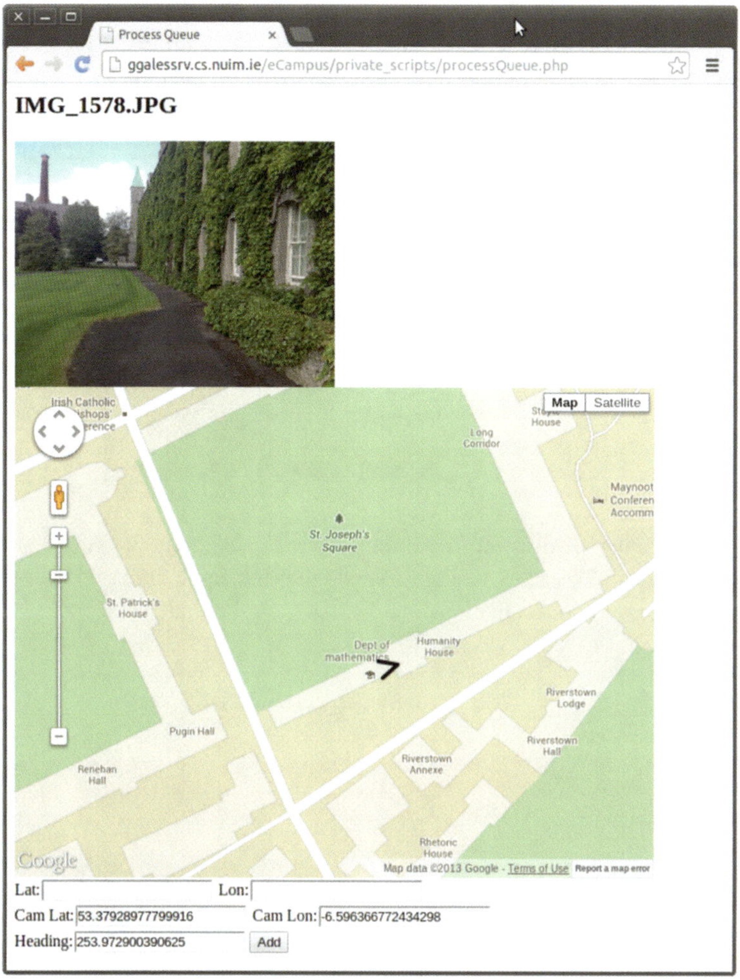

Fig. 3 Simple image retrieval service example in the NUIM

location and heading. A score value s (which is between 0 and 1) is calculated from the Euclidean distance. A sigmoid function is used to set a score close to 0 for near images and 1 for the others ones. A trigonometric function is used to get a similar score for the heading:

$$s = \left(1 - s_{pos}\right)\left(1 - s_{head}\right) \qquad (2)$$

Equation 2: Score formula

In Eq. (2), the s_{pos} is the score given by the Euclidean distance, p, between the position of the point and the position of image (camera position):

$$s_{pos} = \frac{1}{1 + e^{-\frac{2.25(p-15)}{5}}}. \tag{3}$$

Equation 3: Formula for S_{pos}

The constants of the sigmoid function are set experimentally (they represent the threshold distance from which it can be said, "this is too far away and cannot be a match"). Figure 4 shows a plot of this function. The higher the distance p is, the higher the cost.

The heading score s_{head} is calculated by Eq. 4:

$$s_{head} = |\sin(h_{cam} + d)|. \tag{4}$$

Equation 4: Score of heading formula

Figure 5 shows a plot of this function. The higher the angle difference between the camera heading, h_{cam}, and the heading direction, d, the higher the cost.

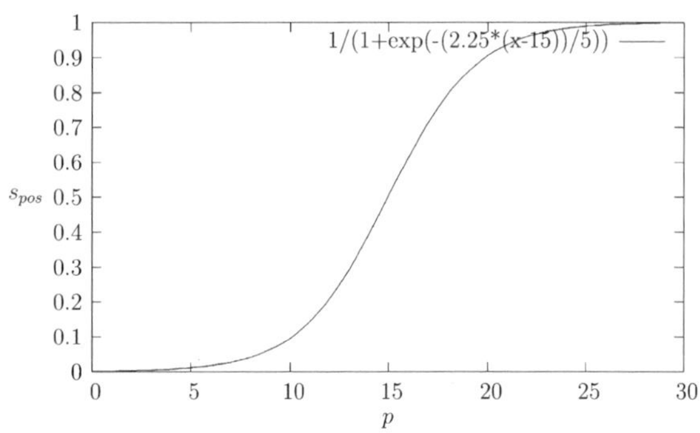

Fig. 4 Position cost function where p is the Euclidean distance between the point positions (in the path) and the image camera position

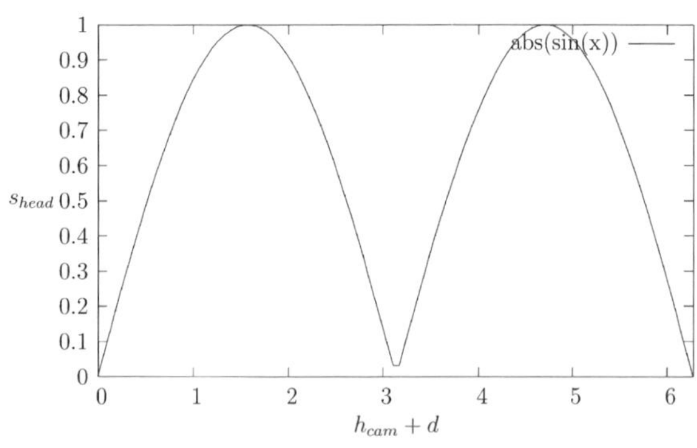

Fig. 5 The heading score is high when the angle difference between the image camera heading, h_{cam}, and the heading direction, d, is high

Fig. 6 Augmented images from the image sequence retrieval service between John Hume building and Chill (restaurant) in the NUIM

Figure 6 illustrates some images from the image sequence retrieval service. The web interface of the application can be found at http://win02.cs.nuim.ie/ws and the mobile web version of the application is located at http://win02.cs.nuim.ie/mobile.

9 Conclusion and Future Work

Graph databases are suitable systems for managing and analyzing highly connected data and, as mentioned in this chapter, there are many experiments and benchmarks that illustrate the efficiency of graph databases in this area. Graph databases provide very efficient tools for modeling, storing, and analyzing spatio-temporal data in LBS applications. Besides, graphs are naturally multi-dimensional so it is quite straightforward to join graphs from several domains (street networks, social networks, professional business networks, and so on) together and ask more sophisticated multi-dimensional questions. There are many navigation services available over the internet; however, the navigation services in this chapter are implemented using a graph database, which can be used in connected as well as disconnected modes. In addition, the evaluation of the eCampus applications by the target users of university students and staff indicated that the visual navigation service using augmented reality provides an intuitive interface that could be integrated into augmented reality systems. With the proliferation of augmented reality glasses and gadgets in the very near future, and with the availability of reliable OSM data coupled with the multi-dimensional nature of graph databases, there will be huge opportunities for innovative and creative applications in the next generation of LBS.

Although the system presented in this chapter has been successfully implemented and tested, the system needs further improvements and research. There are some issues with the GPS/heading accuracy that need to be addressed. In order to improve the augmented reality feature, further research and benchmarks are needed for small areas; for example, if points are too close in the retrieved path, the same image will be retrieved and displayed many times in the final application, which must be avoided. In addition, the integration of indoor maps and OSM data is not yet finished (but this feature is working in certain buildings that have QR codes affixed) (Basiri et al. 2014b). Finally, at the moment the native mobile application is just limited to the Android platform. Improvements in the eCampus system, finding optimized solutions to the above issues, and integration of the eCampus with cutting-edge augmented reality systems like Google Glass are the future direction for this research project.

Acknowledgments Research presented in this paper was funded by a Strategic Research Cluster grant (07/SRC/I1168) by Science Foundation Ireland under the National Development Plan. The authors gratefully acknowledge this support.

References

Abadi D, Boncz P, Harizopoulos S, Madden S (2014) The design and implementation of modern column-oriented database systems (No. 136711)

Afyouni I, Ray C, Claramunt C (2014) Spatial models for context-aware indoor navigation systems: a survey. J Spat Inf Sci 4(2014):85–123

Amirian P, Alesheikh AA, Basiri A (2010) Standard-based, interoperable services for accessing urban services data. Comput Environ Urban Syst 34(4):309–321

Amirian P, Basiri A, Winstanley A (2013) Efficient online sharing of geospatial big data using NoSQL XML databases. Computing for geospatial research and application (COM. Geo). In: Fourth international conference on computing for geospatial research and application (COM. Geo). IEEE, New york, pp 150–152

Amirian P, Basiri A, Winstanley A (2014) Evaluation of data management systems for geospatial big data. In: Proceedings of the 14th international conference on computational science and its applications (ICCSA 2014), ICCSA 2014, Part V. LNCS 8583, pp 678–690

Angles R, Gutierrez C (2008) Survey of graph database models. ACM Comput Surv 40(1):1–39

Bader D, Feo J, Gilbert J, Kepner J, Koester D (2009) Hpc scalable graph analysis benchmark. Citeseer. Citeseer 2009:1–10

Basiri A, Amirian P, Winstanley A (2014a) Use of graph databases in tourist navigation application. In: The 14th international conference on computational science and its applications (ICCSA 2014), ICCSA 2014, Part V. LNCS 8583, pp 663–677

Basiri A, Amirian P, Winstanley A (2014b) The use of quick response (QR) codes in landmark-based pedestrian navigation. Int J Navig Obs (1)

Bass B (2012) NoSQL spatial: Neo4j versus PostGIS. Geographic information management and applications (GIMA). M.S. Thesis, Delft University of Technology, Delft, The Netherlands

Bentley F, Henriette C, Müller J (2014) Beyond the bar: the places where location-based services are used in the city. Pers Ubiquit Comput 1–7

Chang F, Dean J, Ghemawat S, Hsieh W, Gruber R (2006) Bigtable: a distributed storage system for structured data. In: Proceedings of Seventh symposium on operating system design and implementation

Ciglan M, Averbuch A, Hluchy L (2012) Benchmarking traversal operations over graph databases. In: Proceedings of 28th international conference on data engineering workshops. IEEE, New York, pp 186–189

De Souza B, Cláudio C, Daniel F, Maxwell G (2014) NoSQL geographic databases: an overview. Geograph Inf Syst Trends Technol 2014:73–103

Dominguez-Sal D, Martinez-Bazan N, Muntes-Mulero V, Baleta P, Larriba-Pay JL (2010a) A discussion on the design of graph database benchmarks, 13 Sept 2010, pp 25–40

Dominguez-Sal D, Urbón-Bayes P, Giménez-Vañó A, Gómez-Villamor S, Martinez-Bazán N, LarribaPey J (2010b) Survey of graph database performance on the hpc scalable graph analysis benchmark. Web-Age Inf Manage 2010:37–48

Fallah N, Apostolopoulos I, Bekris K, Folmer E (2013) Indoor human navigation systems: a survey. Interact Comput 25(1):21–33

Fowler M, Sadalage P (2012) NoSQL distilled: a brief guide to the emerging world of polyglot persistence. Addison-Wesley publication, Boston

Lawrence R (2014) Integration and virtualization of relational SQL and NoSQL systems including MySQL and MongoDB. In: Proceedings of international conference on computational science and computational intelligence (CSCI), vol 1. IEEE, New York

McCreary D, Kelly A (2013) Making sense of NoSQL. Manning Publications, Greenwich

Miler M, Damir M, Dražen O (2014) The shortest path algorithm performance comparison in graph and relational database on a transportation network. PROMET-Traffic Transport 26 (1):75–82

Partner J, Vukotic A, Watt N (2014) Neo4j in Action. Manning Publication, Greenwich

Purohit A, Sun Z, Pan S, Zhang P (2013) Indoor navigation in retail environments without surveys and maps. In: Proceedings of 10th annual IEEE Communications Society conference on sensor, mesh and ad hoc communications and networks (SECON), 2013. IEEE, New York (2013), pp 300–308

Robinson I, Webber J, Eifrem E (2013) Graph databases. O'Reilly Media, Sebastopol

Sternberg H, Keller F, Willemsen T (2013) Precise indoor mapping as a basis for coarse indoor navigation. J Appl Geodesy 7(4):231–246

Yang S, Yinghui W, Sun H, Xifeng Y (2014) Schemaless and structureless graph querying. In: Proceedings of the VLDB endowment 7(7)

Building a Multimodal Urban Network Model Using OpenStreetMap Data for the Analysis of Sustainable Accessibility

Jorge Gil

Abstract This chapter presents the process of building a multimodal urban network model using Volunteered Geographic Information (VGI) and in particular OpenStreetMap (OSM). The spatial data model design adopts a level of simplification that is adequate to OSM data availability and quality, and suitable to the measurement of the sustainable accessibility of urban neighborhoods and city-regions. The urban network model connects a private transport system (i.e. pedestrian, bicycle, car), a public transport system (i.e. rail, metro, tram and bus) and a land use system (i.e. building land use units). Various algorithmic procedures have been developed to produce the network model, supporting the reproducibility of the process and addressing the challenges of using OSM data for this purpose. While OSM demonstrates great potential for urban analysis, thanks to the detail of its attributes and its open and universal coverage, there is still some way to go to provide the data quality and consistency required for detailed operational urban models.

Keywords Multimodal networks · Street networks · Network model · Spatial network analysis · Sustainable accessibility

1 Introduction

Sustainable accessibility policy goals are concerned with all modes of travel, and give particular attention to walking and cycling for local travel, and to multimodal travel using public transport over longer distances (Van Nes 2002). Studies on the relation between land use and travel patterns have, in the past, had a series of shortcomings (Stead and Marshall 2001; Van Wee 2002), namely the use of

J. Gil (✉)
Faculty of Architecture, Department of Urbanism, Delft University of Technology, Julianalaan 134, 2628 BL Delft, The Netherlands
e-mail: j.a.lopesgil@tudelft.nl

J. Jokar Arsanjani et al. (eds.), *OpenStreetMap in GIScience*,
Lecture Notes in Geoinformation and Cartography,
DOI 10.1007/978-3-319-14280-7_12

aggregate descriptions of urban form, low spatial resolution, a small number of case studies and the difficulty in comparing the results and methods used. These can in part be explained by data sets' reduced geographic coverage or detail, by the cost of acquiring new or existing data, and by modelling and analytical constraints imposed by both software and hardware. Advances over the last decade in open geospatial standards, data collection and distribution, and in open Geographic Information System (GIS) and spatial analysis technologies allowed some of these shortcomings to be addressed, building urban models that are disaggregated, detailed, large scale, relational and reproducible (Jiang 2011, 2013). Recent studies measure sustainable accessibility at the city and regional level by private and public transport modes (Le Clerq and Bertolini 2003; Bertolini et al. 2005; Yigitcanlar et al. 2007; Scheurer and Curtis 2008; Mavos et al. 2012; Hadas 2013) using rich GIS data sets. This type of measurement suits a map view (Goodchild 2000): a static network model requiring a high level of accuracy in the location of and relation between features, consistent coverage and rich attribute data.

GIS-based multimodal network models are data models representing the mobility infrastructure of an urban area as a network, laying out the affordances available for travel using all modes of travel. Goodchild (2000) and Miller and Shaw (2001) have highlighted the challenges and laid out the principles of developing multimodal transportation data models for GIS. Some challenges specific to multimodal network representation relate to the data availability and its multiple formats, overlapping routes of similar or different modes on the same network, the modelling of transfers between modes, and the integration of travel costs on different modes.

There are numerous examples of detailed GIS multimodal network models that include data on public transport lines, services and timetables (Friedrich 1998; Butler and Dueker 2001; Li 2007; Liu 2011; Hadas 2013), and even the internal layout and access points of public transport stations (Chen et al. 2011). These models are usually developed for navigation and route choice modelling, and their level of detail is higher than required to understand the general structure and configuration of multimodal networks (Van Nes 2002). Nevertheless, they offer approaches to address the connection between different modes, for example adding 'abstract connectors' in a single layer (Ismail and Said 2014), or connecting multiple layers and systems for each mode (Mouncif et al. 2006), offering a robust and flexible data model that allows the analysis of each system in isolation (Galvez-Fernandez et al. 2009). Existing studies also highlight the importance of a correctly represented pedestrian network for measuring walking affordances, and propose methods for extending existing street network representations (Kim et al. 2009; Ballester et al. 2011).

This chapter presents the process of building a multimodal urban network model using the OSM street network data set as its main structure, onto which additional public transport and land use data sets are connected, using the case of the Randstad region of the Netherlands. The following sections describe the multimodal urban network model's structure, the selection and pre-processing of the various data sets used, and the procedures developed to integrate the various elements into a topologically consistent model. It concludes with results and reflections on lessons learned from the process, highlighting some of the current challenges in building urban analytical models using OSM data.

2 The Structure and Infrastructure of a Multimodal Urban Network Model

The multimodal urban network model represents the multimodal mobility infrastructure and land use, integrating a private transport system (car, pedestrian and bicycle), a public transport system (rail, tram, metro and bus), a land use system, and interfaces interlinking all three (Fig. 1). The smallest spatial unit of each system is the street segment, the transit stop area and the individual building, respectively. The aim of this model is the analysis of the spatial characteristics of neighborhoods in the region, and of the structure of the region as a whole. This quantitative analysis aim imposes requirements on the data that, in particular regarding its attributes and topology, are different from the requirements for cartographic visualization. At the same time, there is an interest in building a platform that supports the automation of model maintenance, its reproducibility in different geographic contexts, and is accessible to all. These aims inform the subsequent choices of data set and technology stack.

2.1 Data and Software Stack

The multimodal urban network model is built using various VGI and open data sources, taking advantage of the benefits, and attempting to address the issues, of integrating different data sets (Sester et al. 2014). The OSM data set forms the backbone of the model, because of its open access nature, universal coverage and standard, and rich feature set covering all modes of transport. In the case of the Netherlands, when OSM is compared with official street network data (Nationaal Wegen Bestand) according to various categories (Girres and Touya 2010; Jokar Arsanjani et al. 2013a), it presents itself as the most appropriate choice for a multimodal urban network model. OSM offers better semantic accuracy because it includes a representation of soft modes (i.e. walking and cycling) which is essential for the analysis of the walking and cycling environment (Chin et al. 2008); OSM's positional accuracy is not substantially different from official street network data (Girres and Touya 2010; Haklay 2010; Neis et al. 2011; Graser et al. 2014); OSM can have a very good level of completeness: in the Netherlands a contributed road dataset was imported in early 2007[1] and OSM contributors have been

[1] For details refer to the maps on http://wiki.openstreetmap.org/wiki/WikiProject_Netherlands.

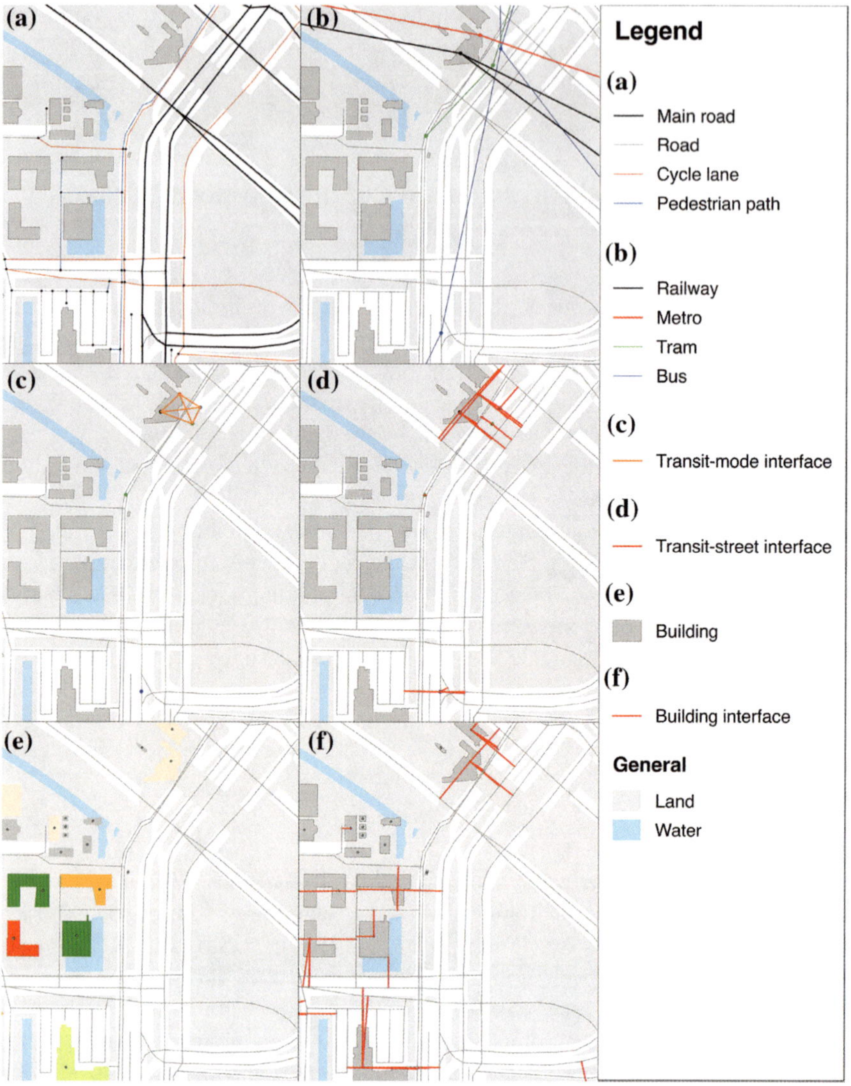

Fig. 1 Structure of the multimodal urban network model's systems—**a** private transport system, **b** public transport system, and **e** land use system—and of the interfaces between all three **c**, **d**, **f**

supplementing and correcting this data since. In addition, the public transport networks can be partly derived from the same OSM data. However, in the case of the Netherlands public transport data has to be complemented with data from OpenOV[2] timetable data, and verified against route maps from local network

[2] http://www.openov.nl/.

operators. Detailed land use data is still very incomplete and inconsistent in OSM (Jokar Arsanjani et al. 2013b; Estima et al. 2013; Jokar Arsanjani and Vaz 2015), both in terms of building footprints and points of interest. In the case of the Netherlands, we used land use data extracted from the public data set Basisregister Addressen en Gebouwen (BAG),[3] which is currently in the process of being imported into the OSM data set.

In line with the chosen data sets, the multimodal urban network model described uses Free and Open Source Software (FOSS) GIS platforms. Osmosis is used to load the OSM data dump into a PostGIS database, where it is maintained together with the other datasets, the model is analyzed using pgRouting, and the results are visualized using QGIS. This software stack offers scripting capabilities in SQL, PL/SQL and Python, thus allowing the model building procedures to be reproduced and adapted to other cases.

2.2 Spatial Data Model

The OSM data set is composed of nodes (points) and ways (polylines), populated by a rich and open set of attribute tags, made up of keys with one or more values. It also supports relations (topologies) combining ways and nodes to describe larger entities, such as routes or named areas (polygons). Polygons can also be defined in closed ways, with a tag identifying it as an area. This data model is compact and flexible, combining all the different features based on their ids. However, this data model also makes data querying and manipulation a non-trivial task. In order to facilitate the analysis and display of the multimodal urban network model, it is based on a more conventional relational spatial data model. In the following sections we dive into the details of building the multimodal urban network model's systems, presenting the procedures for processing and integrating the various data sets, that result in the spatial data model illustrated in Fig. 2.

3 Creating the Private Transport System

The private transport system is the backbone of the multimodal urban network model and, for this reason, it is the first system to be created. It defines the network of streets and paths for cars, bicycles and pedestrians onto which the public transport nodes and the land use activities are connected. The private transport system uses a standard road center line representation and data model, with a table of street segments and one of intersection nodes, populated with attribute fields defined specifically for multimodal urban network analysis (Fig. 2). The following procedure, illustrated in Fig. 3, describes how these features and attributes are

[3] http://bag.vrom.nl/.

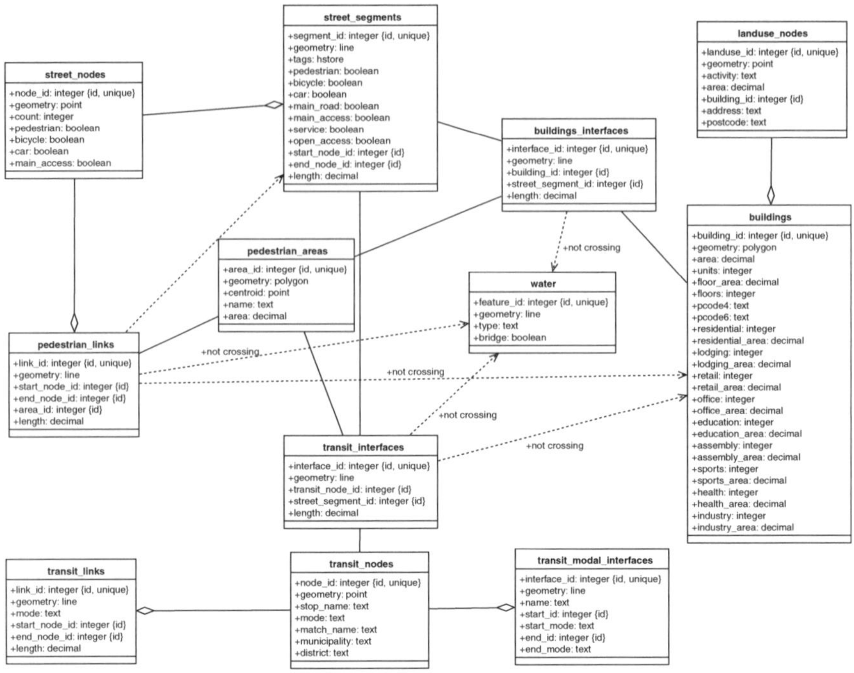

Fig. 2 Diagram of the multimodal urban network spatial data model

created from the original OSM data. The main stages include the preparation of the street network segments, followed by the preparation of the street intersection nodes, concluding with the creation of pedestrian area links.

3.1 Street Network Segment Classification and Correction

There are several steps required to produce the street network segments. Firstly, the OSM ways that contain a 'highway' tag are extracted and inserted into a street segments table, keeping the complete tag contents. Secondly, the street segments are filtered and classified according to mode permissions and restrictions, and road hierarchy using the associated tags. Thirdly, the street segments need to be corrected for geometry and topology problems.

3.1.1 Street Network Segment Classification

The street segments in the multimodal urban network model are assigned attributes that indicate permission or restriction of the private modes (i.e. 'car', 'bicycle' and

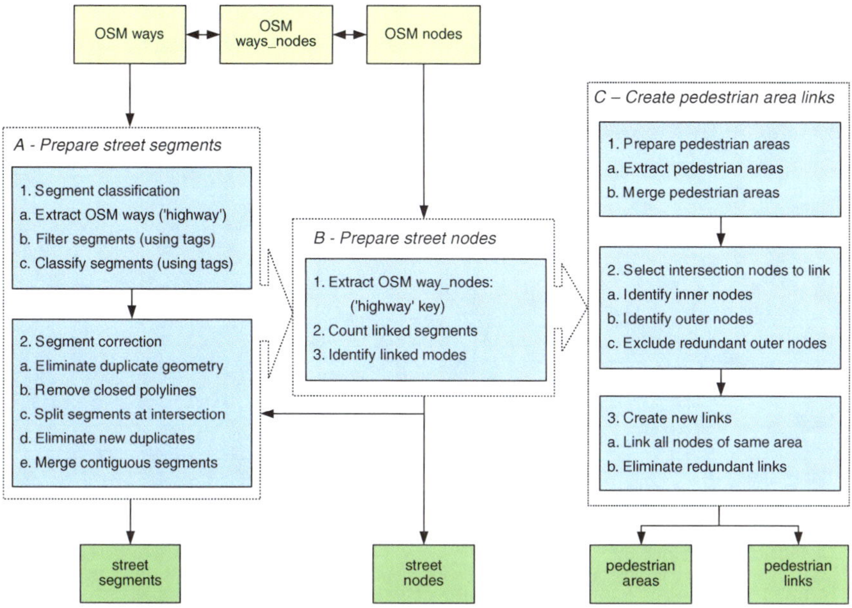

Fig. 3 Procedure to create the private transport system data classes, from OSM data

'pedestrian'), based on observable affordances (Scheider and Kuhn 2010) extracted from a wide range of tags available in the OSM data set (step A1 in Fig. 3). The level of completeness of attributes is critical to this type of network model (Neis and Zielstra 2014), and one has to adopt a simpler model specification (e.g. ignoring turn and speed restrictions) that is consistent throughout the study area, extracting as much information from additional keys as possible.

The first step is to identify the keys and key values available in the data set, their frequency, and to define their relevance to different modes (Table 1), supported by examples mapped over aerial photography. The main key for classifying the street network is 'highway', with values directly related to mode (e.g. 'cycleway', 'footway', 'pedestrian', 'track', 'path', 'steps'), and to road hierarchy (e.g. 'motorway', 'trunk', 'primary', 'secondary', 'tertiary', 'residential'). In the case of the 'highway' key, the most common value is 'unclassified', which can be complemented by information in other mode related keys, namely 'motorcar', 'motor_vehicle', 'bicycle', 'cycleway', 'foot', 'footway' and 'route'. These specify if the segment is designated, accessible (e.g. yes) or restricted (e.g. no) for the given mode. Street segments with the 'construction' value should be ignored, and street segments with the 'service' value are ignored if their access is restricted (e.g. controlled by a gate), they are private, or are exclusively for public transport use. Once we have identified the relevant keys and values, we run a series of queries to populate the private mode attributes with Boolean values, where true is used when the segment is designated, false when the segment is forbidden, and null if the

Table 1 Summary of relevant "highway" key values in the OSM data set, with their frequency count, and indication of the mode that they are accessible to ('T', 'F' or blank)

Key value	Frequency	Car	Bicycle	Pedestrian
Unclassified	652,440	( )	( )	( )
Cycleway	112765	F	T	
Tertiary	99,328			
Footway	73,425	F		T
Residential	68,182			
Service	52,836	(F)	(F)	(F)
Secondary	47,116	T		
Track	34,524	F		
Pedestrian	31,786	F		T
Primary	27,986	T	(F)	(F)
Path	18,914	F		
Living_street	7,566		T	T
Motorway	7,406	T	F	F
Motorway_link	5,221	T	F	F
Trunk	3,794	T	F	F
steps	3,134	F		T
Construction	1,250	F	F	F
Trunk_link	1,030	T	F	F
Primary_link	776	T	(F)	(F)
Bridleway	646	F		

Brackets represent states that depend on the value of additional mode-specific keys

segment is simply accessible to that mode. The latter is the state that applies to the majority of street segments, which can be shared by all three modes.

The open nature of the tagging system in OSM allows contributors to introduce rich detail in feature attributes, but it also adds heuristic complexity. In the data set for a given region one might find keys or values spelt differently (i.e. yes, Yes, YES, y), using regionalisms (i.e. underground, tube, metro), in different languages, or simply misspelled. One has to analyze and consider these specificities when using a data set for the first time, and develop queries to classify the segments based on a broad set of key-value pairs.

3.1.2 Street Network Segment Correction

Regarding the OSM street segment geography and topology, there is a range of possible problems (Girres and Touya 2010): duplicate segments, where geometry is exactly the same; overlapping segments, where geometry partially coincides; missing segments; closed segments, representing areas; orphans, unconnected segments from the rest of the network; segments without segmentation, where intersection nodes exist; contiguous segments, separated where no intersection nodes exist; missing intersection nodes; intersections at bridges or tunnels. One has to identify which problems are present, to what extent, how critical they are for the intended use of the data set, and ultimately decide the degree of error that is acceptable, making corrections accordingly. Considering that the aim is to produce a multimodal urban network model representing a large region, resulting from an automated procedure towards reproducibility and operationalization, we should not introduce steps that depend on the individual identification and correction of mistakes, such as missing data or incorrectly classified data. When creating a smaller network model, one should consider scanning and correcting any additional problems.

The correction steps proposed convert the OSM street segments into a standard road center line representation (A2 in Fig. 3). Firstly, we eliminate duplicate geometry, choosing to remove the segment with greater access restrictions. Secondly, we remove closed polylines that represent pedestrian zones, and do not correspond to road center lines. Thirdly, we split segments that continue over intersections where there is an intersection node, and eliminate any new duplicates resulting from overlapping geometry. Finally, multiple contiguous segments between crossings are merged together into a single polyline.

3.2 Street Intersection Node Classification

The standard road center line representation uses nodes at the endpoints of street segments to indicate level crossings. Where two segments intersect without a node, this indicates the absence of a level crossing, as in the case of bridges and tunnels. The nodes layer is selected from the OSM 'nodes' data set, namely those with an id in the 'node_id' attribute of the 'way_nodes' table, whose 'way_id' is in the extracted street segments table. The intersection nodes can be used to quantify morphological characteristics of the street network, namely the typology of crossings and cul-de-sacs, and to provide this typology, each node has a 'count' attribute with the number of street segments that share it.

3.3 Pedestrian Areas Links Generation

The OSM street network includes pedestrian paths and cycle lanes, however some public open spaces, such as squares, parks or urban block interiors, are represented by closed polylines disconnected from the street network, with the 'highway' = 'pedestrian' value, and a 'area' key with value 'yes'. We create pedestrian area links to represent routes that pedestrians and cyclists can take across these open spaces. These are important shortcuts that affect the measurement of local neighborhood characteristics, because the routes around the perimeter of blocks and open spaces can represent a considerable increase in walking distance.

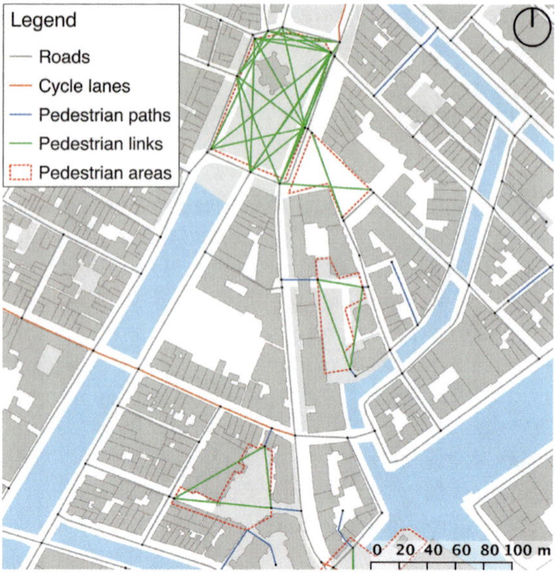

Fig. 4 Map with examples of pedestrian areas, with pedestrian area links complementing the connectivity of the street network

The procedure (C in Fig. 3) generates pedestrian links in an extension of the street network, using the existing street segments, intersection nodes and open areas, without changing the street network geometry or affecting its topology. The first stage is the selection of the pedestrian area polygons, merging together pedestrian areas that intersect, are adjacent, have the same name key, or have the same OSM id. The second stage is the selection of intersection nodes to be connected, identifying 'inner nodes' that belong to street segments intersecting the pedestrian areas and are located on or inside the areas' perimeter, and 'outer nodes' that are located outside but immediately adjacent to the pedestrian areas, within a 25 m buffer. Finally, we exclude 'outer nodes' that are directly connected to 'inner nodes'. The third stage is the creation of new links between the inner and outer nodes belonging to the same pedestrian area. We simplify this procedure by ignoring the pedestrian area's shape and allowing links to cross the perimeter of concave shapes, because the link does effectively exist, albeit not so direct in terms of geometry. The final step eliminates excessive links that cross external buildings, cross external street segments, are entirely outside the pedestrian area, or are duplicates of existing street segments. The result is illustrated in Fig. 4.

4 Creating the Public Transport System

The next stage in building a multimodal urban network model is to build the public transport network and to connect it to the private transport network (Fig. 5). The public transport network is also multimodal, including rail, metro (or light rail), tram,

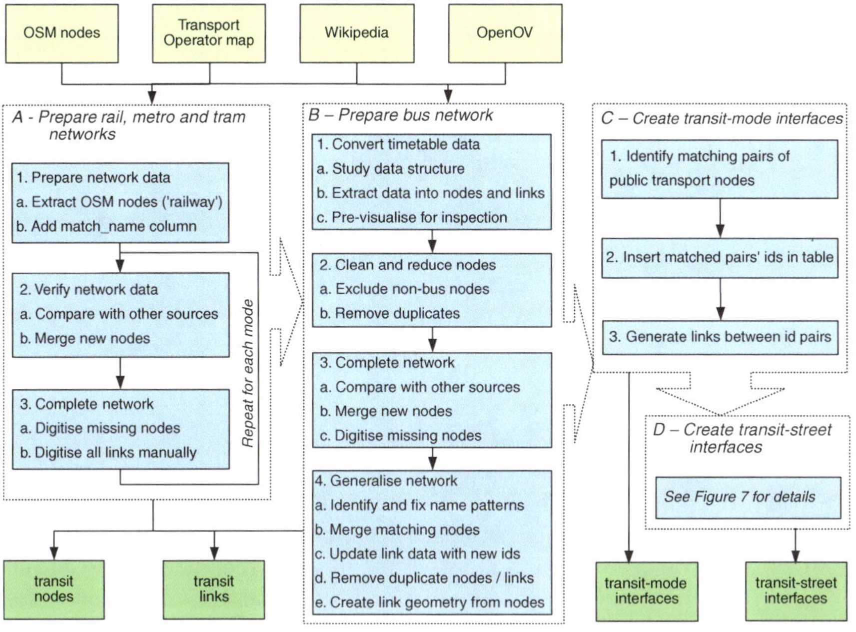

Fig. 5 Procedure to create the public transport system data classes, from OSM data and other sources

and bus networks. We have opted for a simplified representation of the network focusing on the connectivity of the infrastructure, rather than on the service provision, i.e. the lines, services, and their frequency. This simplified representation has lower data requirements, and is easier to produce, verify and explain, retaining an adequate level of detail for the measurement of multimodal accessibility and urban structure (Van Nes 2002). It is made up of nodes representing stations or stops; lines representing the links between stops of the same mode; 'transit-mode interfaces' representing the transfer between stops of different modes; and 'transit-street interfaces' connecting the stops to the streets of the private transport network (Fig. 2).

4.1 Public Transport Networks Preparation

The public transport networks consist of a single node of each mode at a given location, independent of the number of physical stops, platforms, or station entrances that exist; and one line segment between nodes of the same mode, where a service exists connecting them. The OSM dataset has the required tags to construct such a network, however the level of coverage and detail of the data varies across regions and across transport modes. This can lead to different approaches to produce the multimodal network model, which is less systematic than desired, but is useful to understand different methods and issues for different data scenarios.

Table 2 Summary of railway key values of the nodes and ways data

Node key values	Count	Mode	Way key values	Count	Mode
Tram_stop	1,019	Tram	Rail	12,306	Railway
Station	518	Railway	Tram	2,078	Tram
Buffer_stop	184	Ignored	Platform	1124	Ignored
Subway_entrance	83	Ignored	Subway	388	Metro
Halt	60	Metro	Light_rail	309	Metro and railway
			Preserved	180	Ignored
			Disused	112	Ignored
			Narrow_gauge	60	Ignored
			Abandoned	57	Ignored

4.1.1 The Rail, Metro and Tram Networks

The public transport network nodes for the modes that use rail tracks were extracted from the OSM nodes data, where the node tag has a 'railway' key. The values of the 'railway' key (Table 2) are used to assign the mode to each node, or to exclude nodes that are not relevant. The resulting nodes are then checked against other sources of information, mentioned in Sect. 2.2. It was found that the OSM data set could have inconsistent quality, namely missing stations, duplicate stations, inactive stations, and wrongly classified stations. In the present case, the rail network was mostly complete and correct; the metro network was mostly complete but whole segments were incorrectly classified as 'tram', and the tram network had 2 % of the stations missing and had classification issues such as missing names or name inconsistencies.

The 'match_name' attribute of the public transport network nodes table stores a lowercase and simplified version of the stop or station name, standardizing certain prefixes or suffixes that can occur differently in different modes. This attribute can be used to compare and merge data from different sources, and relate public transport nodes of different modes. When verifying station/stop names of local transport modes, such as metro, tram and bus, across a larger region, one must take care to include the municipality and/or district name in the query. Often there are general names, historical dates, or historical names, occurring in different cities or districts, raising false matches.

The public transport network links connecting the nodes are created next. The OSM data set includes ways representing the physical rail tracks, which have a railway key (see Table 2 for values). However, these are not necessarily complete and do not match the stations in the same way the street segments match intersection nodes. Therefore, these line segments can only be used for visual support when digitizing links between stations, or after generating links from sequences of node names. The first method was carried out in this case, as the number of links is not excessive, and it allows a careful verification of public transport nodes.

4.1.2 The Bus Network

While the semi-automated verification and manual digitization method can be adequate for rail track-based public transport modes, it is impractical to create the bus network, because it is quite extensive and is difficult to identify routes by visual inspection. The recommended approach for creating a network model of public transport modes is to use an automated and systematic procedure, and for that one requires route data and/or timetable data. The OSM data set can include the bus network, with bus stops in the nodes table identified with the key 'highway' = 'bus_stop', and routes connecting stops encoded as relations of ways with the keys 'type' = 'route' and 'route' = 'bus'. However, the relation data is often missing and only the nodes information is available. In such cases, one should look for public transit timetable data, the General Transit Feed Specification (GTFS) being a well-known standard used worldwide. In this case, the OpenOV timetable dataset was used.

Typically, public transport timetable datasets have a complex relational data structure, i.e. includes stops, links, services, cars, directions, and times. The first stage (B1, Fig. 5) is to understand the data structure and how to extract relevant information, namely the stop name, stop id and its geographic coordinates, and the links' mode, origin stop, and destination stop. Then we extract this data into the simplified node/link public transport network data model. It is important to visualize the automatically generated public transport network to have visual feedback on the progress, and to assess its completeness and overlap with the available OSM dataset. The second stage (B2, Fig. 5) is to clean and reduce the data, keeping only bus nodes and links, and eliminating links and stops that are exact duplicates in terms of the 'match_name' and geometry attributes. The third stage (B3, Fig. 5) is to complete missing data, if required. In this case, the data from one of the bus network operators was missing and one had to resort to a semi-automated process similar to the rail networks. The fourth stage (B4, Fig. 5) is to generalize the resulting bus network, merging stops around a location that correspond to different routes or different route directions.

Ideally this is done using a universal unique feature identifier, otherwise one has to use name matching, paying special attention to different naming conventions for stops used by different operators. When merging equivalent stops, using the 'match_name' together with the municipality and district attributes as recommended in Sect. 4.1.1, one must limit the operation to a given distance (in this case we used 200 m) because in the case of bus networks, there are routes crossing the same street a few kilometers apart but all the stops receive the same name. The merging operation involves assigning a unique identifier to stops and links that meet the above criteria, and removing features with duplicate unique identifiers, keeping one stop for each unique identifier, and one link for each start and end node stop pair. The link's geometry is then updated based on the new start and end stops location, giving a reduced set of new generalized links (Fig. 6).

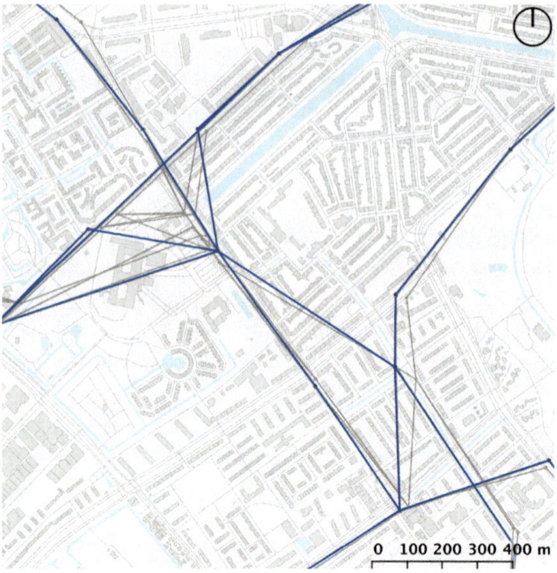

Fig. 6 Map with sample result of the bus network generalization procedure, comparing the original route data set with the simplified network of single bus stops and links

4.2 Public Transport Network Interfaces Generation

After creating the public transport network tables, we integrate the public transport modes together and connect them to the street network of the private transport system. The 'transit-mode interfaces' (Fig. 1c) are links connecting public transport networks between stations and stops that share the same name. Typical examples are the stops around central train stations where all modes converge, which have the same stop name. Using the 'match_name' attribute, in conjunction with the municipality and district names, we insert in the modal interfaces table pairs of nodes that share the name but belong to a different public transport mode. The 'transit-street interfaces' (Fig. 1d) are links connecting the public transport nodes to the street network. These links allow the measurement of availability of and proximity to public transport, and the measurement of multimodal trips, e.g. where walking is combined with a longer public transport journey. Each public transport stop is connected to several of the nearest street segments, making sure that they connect to each of the private transport modes (pedestrian, bicycle and car). The procedure followed to produce these interfaces is detailed in Fig. 7. This is a nested iterative procedure, with a sequence of steps gradually loosening restrictions to find connections, this same sequence of steps being repeated at increasing scales, and the full sequence being repeated to connect to each private transport mode. Without describing every step in detail, it is important to raise a couple of points. The relaxation of crossing buildings and water is introduced because these features come from different data sets and are not topologically consistent with the mobility networks. The restriction of connecting to endpoints of street segments aims to reduce the number of links created initially, and to obtain links perpendicular to street segments.

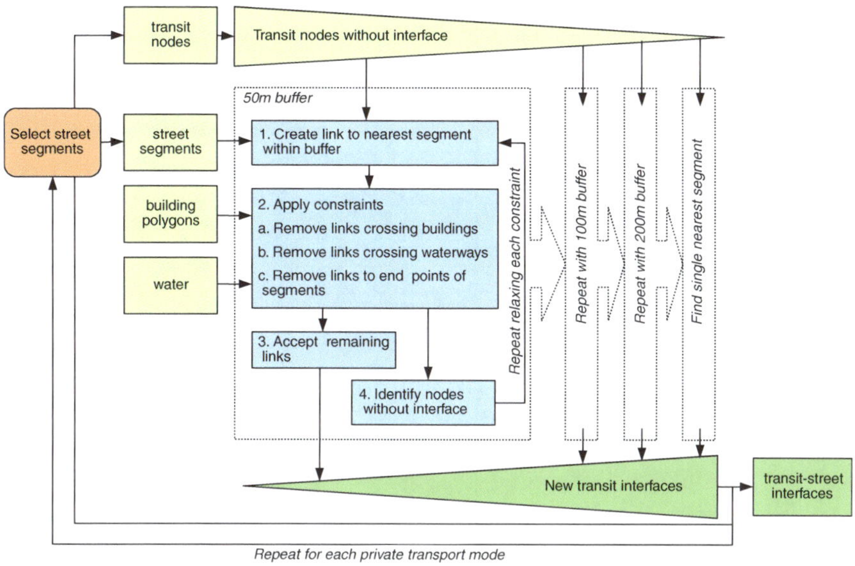

Fig. 7 Procedure to create the public transport system interface with the street network

5 Creating the Land Use System

The final stage in building the multimodal urban network model is the creation of the land use system. This system provides the origins and destinations for the journeys across the mobility infrastructure, and furthermore allows the calculation of urban form characteristics related to functional density and accessibility. The land use system is made of three components, namely the land use nodes, the building polygons and the land use–street interfaces. The procedure to produce this system is illustrated in Fig. 8.

The first stage is to verify the quality of the data, even when it comes from official data sources, namely the attribute values' statistical and spatial distribution. In this case, it was found that 'null' values, which should be empty fields as per specification, were in certain areas recorded using numbers, e.g. 999999, 99999, 949999; and in some areas the decimal values had lost the decimal sign. In order to obtain a smaller number of nodes in the network model, the next stage is to aggregate the land use nodes information on the corresponding building polygon using a spatial join, calculating the total number of land use units and total area, and the number of units and area of each land use category. Any buildings without land use points were discarded from land use related analyses. The final stage is to create interfaces between the buildings and street segments or intersection nodes. In some data sets, the land use address attribute includes postcode, street name and door number, and can be used to define these interfaces. However, geocoding is not always an option, as the complete address is rarely available on both the land use

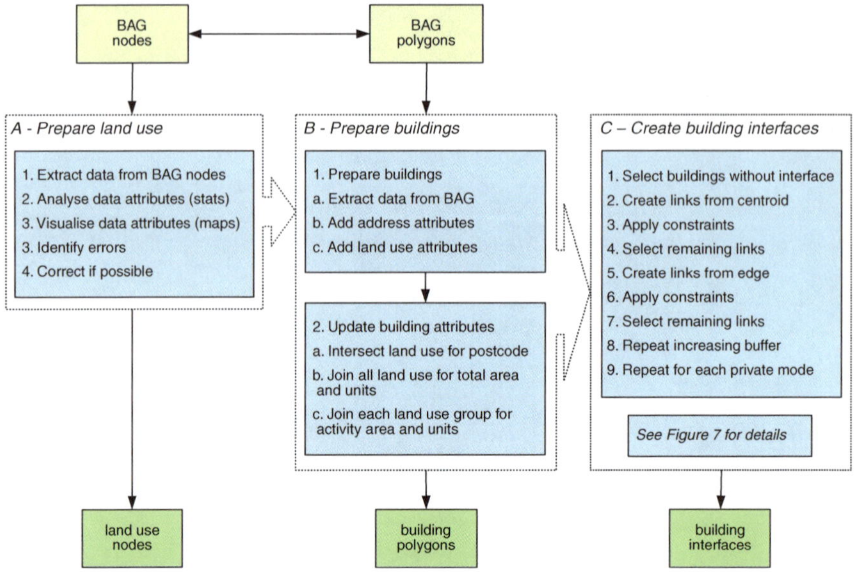

Fig. 8 Procedure to create the land use system data classes

and the OSM streets data set. Hence, one can define the interface as the (set of) line (s) from the building to the nearest the street segment(s) following a procedure largely similar to the one depicted in Fig. 7, with the difference that each iteration uses the building centroid first, and then repeats using the building perimeter.

6 Results

Following the procedures described in Sects. 3 to 5 we have produced a large-scale, detailed, multimodal transport and land use network model for the Randstad region of the Netherlands. It consists of 676,248 street segments, 462,384 street intersection nodes, 161 rail, 186 metro, 614 tram and 7,680 bus nodes, and 2,430,945 building polygons (Fig. 9). This model offers a rich representation of the region, supporting analysis of a regional scale but also goes down to the level of detail of the local neighborhoods and individual buildings.

By querying the network model to use relevant layers, e.g. pedestrian and local public transport, and to select specific sets of origins and destinations, e.g. residential buildings or education establishments, one can use network analysis and routing algorithms to calculate multimodal shortest paths and catchment areas. This is demonstrated next, with the results of shortest routes calculated around pedestrian areas, and catchment areas calculated for different transport modes.

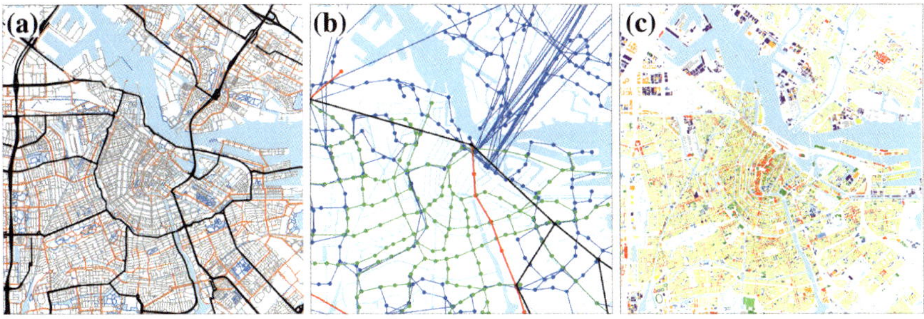

Fig. 9 Maps of the three main systems of the multimodal urban network model, for a data sample around central Amsterdam: **a** private transport system, **b** public transport system, **c** land use system

6.1 Multimodal Network Model Analysis

The shortest routes were calculated between buildings in a section of the network model in central Amsterdam, which has several pedestrian-only streets and pedestrian areas (Fig. 10). The calculation uses three different model filters: one that includes all pedestrian links (A), one that excludes the pedestrian area links described in Sect. 3.3 (B), and a third that does not include any exclusively pedestrian links and corresponds to the car network represented in official street network data sets (C). These results in Table 3 show the impact of adding pedestrian links on local shortest route calculations.

As found by other studies (Chin et al. 2008), the pedestrian routes can be considerably shorter in model A when compared to model B and in particular with model C. This impact is obviously dependent on the quantity and location of pedestrian areas and pedestrian streets. As there are only a few pedestrian areas, certain routes in model B are not affected at all (e.g. C–T and S–T), while others are considerable affected (e.g. E–F). However, pedestrian streets are common in this urban area, hence model C gives a significant increase in distance travelled in all

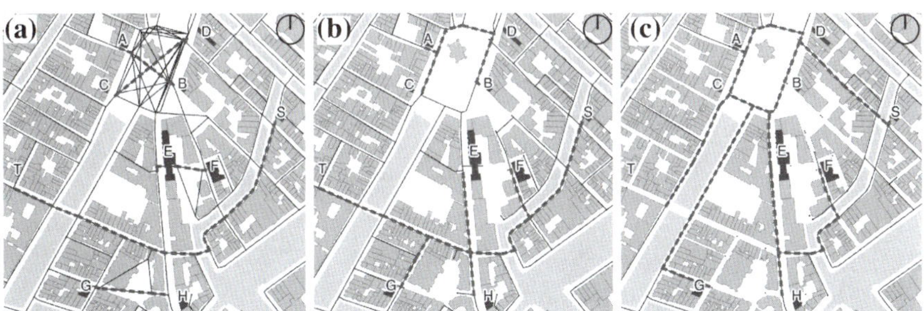

Fig. 10 Shortest routes between pairs of buildings (AB, CD, EF, GH, ST) on the private transport system, **a** including all pedestrian links, **b** including pedestrian paths without pedestrian area links, and **c** including only car accessible links

Table 3 Difference between model A and models B and C, in terms of shortest route distance between pairs of points (Fig. 10), and in terms of catchment area

Analysis	Model A (m)	Model B (m)	Model B (%)	Model C (m)	Model C (%)
A–B route	84	177	+109.1	177	+109.1
C–D route	121	145	+19.9	145	+19.9
E–F route	65	270	+313.2	270	+313.2
G–H route	90	174	+92.2	526	+481
C–T route	207	207	0	269	+29.9
D–T route	328	352	+7.4	379	+15.4
S–T route	407	407	0	470	+15.6
5 min catchment	5,636	5,374	−4.6	3107	−44.8
10 min catchment	36,884	36,058	−2.2	21411	−41.9

cases. The difference between models A, B and C is smaller in the results of the catchment area calculation for 5 and 10 min travel time, and this difference drops with a greater distance travelled. This seems to indicate that in aggregate calculations of many origin destination pairs including longer routes (e.g. S–T), the reduction in distance travelled from the presence of pedestrian paths and areas will have a small overall impact. From these results, it is clear that the location and concentration of pedestrian movement features will have a variable local effect on the results of network analysis. However, to draw definitive conclusions on the level of this impact on a given multimodal model, one would have to conduct a systematic study on the whole model using the full range of analyses that are required.

The results of a catchment area analysis for different travel modes are shown in Fig. 11, where one can observe the number of buildings accessed within a 10 min travel time: walking (Fig. 11a, b), comparing models A and C; using public transport (Fig. 11c), where pockets of buildings near public transport stops become accessible; and driving (Fig. 11d), with a more extensive coverage than public transport due to the fact that the central core of Amsterdam has very few public transport stops and uses a large part of the travel time budget walking to the nearest stop.

These results are mainly for demonstrating the use of the multimodal urban network model in basic network analysis calculations. From the calculation of shortest paths and catchment areas we can derive a wide range of analysis metrics of proximity, density and accessibility of the different mobility networks, which support the comparative assessment of sustainable accessibility of urban neighborhoods in the city-region. This is further described and developed in work by Gil and Read (2012, 2014), and Gil (2014).

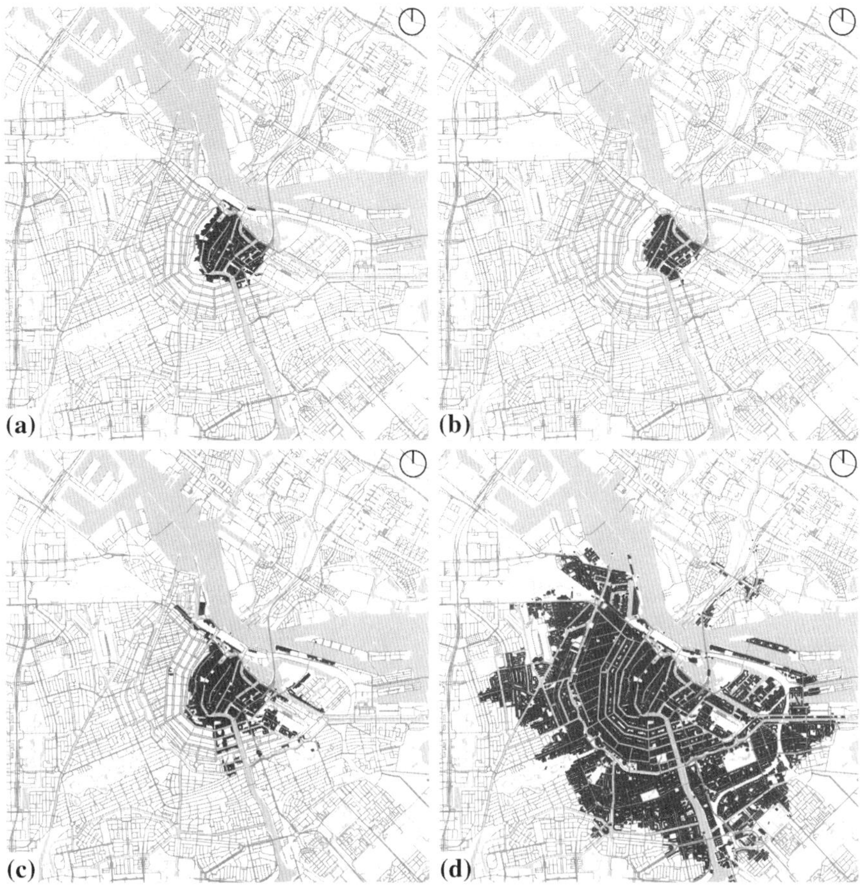

Fig. 11 Catchment area maps for a 10 min travel time **a** walking using model A, **b** walking using model C, **c** using public transport, and **d** driving

6.2 Reflections on the Multimodal Network Model

At this stage we would like to reflect on aspects of the multimodal urban network model's design and lessons learned from the process of producing it. The network model is intentionally simple for the purpose of analyzing the multimodal structure and accessibility of city-regions, and this is reflected in its data model. However, the geometry of the private transport system network could be generalized to address inconsistent representation of multiple lanes, cycle lanes, pedestrian pavements and crossings, complicated intersection layouts, and to allow a correct classification of crossings typology. We opted for a general classification of the street segments using mode attributes, because the geometric generalization using OSM data is not a trivial task, and the current procedure interferes less with the original data model and facilitates update and maintenance of the network model.

If the decision goes for a more detailed model, for example in a smaller study area, the private transport system procedure from OSM data remains the same but one should add parking space information to the street network, and complete the pedestrian network with correct representations of crossings (Ballester et al. 2011). On the other hand, one should produce a more detailed public transport system using the full timetable data, including multiple stops, platforms, lines and services, as seen in other models (Friedrich 1998; Butler and Dueker 2001; Li 2007; Liu 2011; Hadas 2013).

The network structure and level of detail of the model is also related to the type of analysis algorithm being used. In this case we use basic undirected shortest route and catchment area analysis suitable for a simpler network representation, but more detailed models would require advanced network analysis algorithms based on specific multimodal graph representations, as can be found in the works of Lozano and Storchi (2001), Bielli et al. (2006), or Ayed et al. (2011).

Data set selection and processing remains one of the biggest tasks in network model production. Some important lessons are: all data sets must be carefully verified and possibly corrected, whatever their source; data sets have strengths and weaknesses regarding the desired data model, which encourages the combination of the best available data sets from different sources; however, this can be problematic and requires workarounds, because data sets are not topologically consistent (e.g. network and building geometry), lack common unique feature identifiers, and use different naming or classification conventions (e.g. public transport attribute data).

Regarding the OSM data set, it has the potential to become a unified source for this type of urban analysis work: it is user-contributed and freely accepts updates, it is open to imports of large official datasets, it supports a wide range of features and attributes, and is based in a single geographically and topologically consistent platform. However, it still faces many of the challenges raised by Haklay and Weber (2008). For OSM to be further used in urban analytical studies, the challenge is to improve data quality (Mooney et al. 2010; Goodchild and Li 2012), especially of its attribute data (Neis and Zielstra 2014), supported by software applications for data editing that are user friendly and intuitive to attract many contributors—the crowd—but sophisticated enough in the background to provide quality control—automated. For example, in order to ensure a standard street network representation, OSM could include topology processing scripts on the server side to operate on features as they are edited on the database, similar to the procedures described in Sect. 3.1.2. Furthermore, in order to facilitate data analysis and modelling based on attributes, OSM could require a minimum level of attribute classification of features. This would be supported by tools including an ontology-based data specification (e.g. Fonseca et al. 2000; Scheider and Khun 2010; Scioscia et al. 2014) that would guide and reduce the efforts of individuals in complying with data standards.

7 Conclusions

This chapter proposes a set of procedures to build a multimodal urban network model, including transport infrastructure and land use, for the quantitative analysis of the urban regional environment in urban design and strategic planning studies. The general scope and reproducibility of this type of model is much greater than in the past, thanks to the availability of OSM data. In addition, the level of detail afforded by crowd-sourced data sets, with the capabilities of today's open GIS and statistical analysis platforms, increase the possibilities for the analysis of complex urban regions.

OSM has great potential: its specification offers 'on paper' all the features required for this type of network models; and its open nature and worldwide standard availability can turn it into a preferred data source. But there is still some way to go before its feature classification becomes more complete and consistent and before its coverage becomes comprehensive in most geographic regions, either through individual contribution, or through the merging of open and donated official data sets. One should expect this evolution to continue over time, and new urban network models to be built and further contribute to urban form and travel studies, but also that these models become operational and applied in urban design and planning practice.

Acknowledgments This research was generously funded by the Fundação para a Ciência e Tecnologia (FCT)—Portuguese Science and Technology Foundation—with grant SFRH/BD/46709/2008.

References

Ayed H, Galvez-Fernandez C, Habbas Z, Khadraoui D (2011) Solving time-dependent multimodal transport problems using a transfer graph model. Comput Ind Eng 61:391–401. doi:10.1016/j.cie.2010.05.018

Ballester MG, Pérez MR, Stuiver J (2011) Automatic pedestrian network generation. pp 1–13

Bertolini L, le Clercq F, Kapoen L (2005) Sustainable accessibility: a conceptual framework to integrate transport and land use plan-making. Two test-applications in the Netherlands and a reflection on the way forward. Transp Policy 12:207–220. doi:10.1016/j.tranpol.2005.01.006

Bielli M, Boulmakoul A, Mouncif H (2006) Object modeling and path computation for multimodal travel systems. Eur J Oper Res 175:1705–1730. doi:10.1016/j.ejor.2005.02.036

Butler JA, Dueker KJ (2001) Implementing the enterprise GIS in transportation database design. J Urban Reg Inf Syst Assoc 13:17

Chen S, Tan J, Claramunt C, Ray C (2011) Multi-scale and multi-modal GIS-T data model. J Transp Geogr 19:147–161. doi:10.1016/j.jtrangeo.2009.09.006

Chin GKW, Van Niel KP, Giles-Corti B, Knuiman M (2008) Accessibility and connectivity in physical activity studies: The impact of missing pedestrian data. Prev Med 46:41–45. doi:10.1016/j.ypmed.2007.08.004

Estima J, Painho M (2013) Exploratory analysis of OpenStreetMap for land use classification. In: Proceedings of the second ACM SIGSPATIAL international workshop on crowdsourced and volunteered geographic information, pp. 39–46. ACM, New York. doi:10.1145/2534732.2534734

Fonseca FT, Egenhofer MJ, Davis CA, Borges KAV (2000) Ontologies and knowledge sharing in urban GIS. Comput Environ Urban Syst 24:251–272. doi:10.1016/S0198-9715(00)00004-1

Friedrich M (1998) A multi-modal transport model for integrated planning. In: World conference on transport research society, pp 1–14. Antwerpen

Galvez-Fernandez C, Khadraoui D, Ayed H et al (2009) Distributed approach for solving time-dependent problems in multimodal transport networks. Adv Oper Res 2009:e512613. doi:10.1155/2009/512613

Gil J (2014) Analyzing the configuration of multi-modal urban networks. Geogr Anal 46:368–391. doi:10.1111/gean.12062

Gil J, Read S (2012) Measuring sustainable accessibility potential using the mobility infrastructure's network configuration. In: Proceedings: 8th international space syntax symposium, pp 8104:1–8104:19. Pontificia Universidad Catolica de Chile, Santiago

Gil J, Read S (2014) Patterns of sustainable mobility and the structure of modality in the Randstad city-region. A|Z ITU J Fac Architect 11

Girres J-F, Touya G (2010) Quality assessment of the French OpenStreetMap dataset. Trans GIS 14:435–459. doi:10.1111/j.1467-9671.2010.01203.x

Goodchild MF (2000) GIS and transportation: status and challenges. GeoInformatica 4:127–139–139

Goodchild MF, Li L (2012) Assuring the quality of volunteered geographic information. Spat Stat 1:110–120. doi:10.1016/j.spasta.2012.03.002

Graser A, Straub M, Dragaschnig M (2014) Towards an open source analysis toolbox for street network comparison: indicators, tools and results of a comparison of OSM and the official Austrian reference graph. Trans GIS 18:510–526. doi:10.1111/tgis.12061

Hadas Y (2013) Assessing public transport systems connectivity based on Google Transit data. J Transp Geogr 33:105–116. doi:10.1016/j.jtrangeo.2013.09.015

Haklay M (2010) How good is volunteered geographical information? A comparative study of OpenStreetMap and Ordnance Survey datasets. Environ Plan 37:682–703. doi:10.1068/b35097

Haklay M, Weber P (2008) OpenStreetMap: user-generated street maps. IEEE Pervasive Comput 7:12–18. doi:10.1109/MPRV.2008.80

Ismail MA, Said MN (2014) Integration of geospatial multi-mode transportation systems in Kuala Lumpur. IOP Conf Ser: Earth Environ Sci 20:012–027. doi:10.1088/1755-1315/20/1/012027

Jiang B (2011) A Short Note on Data-Intensive Geospatial Computing. In: Popovich VV, Claramunt C, Devogele T et al (eds) Information fusion and geographic information systems. Springer, Berlin, pp 13–17

Jiang B (2013) Volunteered geographic information and computational geography: new perspectives. In: Sui D, Elwood S, Goodchild M (eds) Crowdsourcing geographic knowledge. Springer, Netherlands, pp 125–138

Jokar Arsanjani J, Vaz E (2015) An assessment of a collaborative mapping approach for exploring land use patterns for several European metropolises. Int J Appl Earth Obs Geoinf (in press)

Jokar Arsanjani J, Barron C, Bakillah M, Helbich M (2013a) Assessing the quality of OpenStreetMap contributors together with their contributions. In: 16th AGILE international conference on geographic information science, Leuven, Belgium. 14–17 May 2013

Jokar Arsanjani J, Helbich M, Bakillah M et al. (2013b) Toward mapping land-use patterns from volunteered geographic information. Int J Geogr Inf Sci 27:2264–2278. doi:10.1080/13658816.2013.800871

Karimi HA, Kasemsuppakorn P (2012) Pedestrian network map generation approaches and recommendation. Int J Geogr Inf Sci 27:947–962. doi:10.1080/13658816.2012.730148

Kim J, yong Park S, Bang Y, Yu K (2009) Automatic derivation of a pedestrian network based on existing spatial data sets

Le Clercq F, Bertolini L (2003) Achieving sustainable accessibility: an evaluation of policy measures in the Amsterdam area. Built Environ 29:36–47. doi:10.2148/benv.29.1.36.53949

Li Z-C, Huang H-J, Lam WHK, Wong SC (2007) A Model for evaluation of transport policies in multimodal networks with road and parking capacity constraints. J Math Model Algor 6:239–257. doi:10.1007/s10852-006-9040-7

Liu L (2011) Data model and algorithms for multimodal route planning with transportation networks. Technische Universität München, PhD

Lozano A, Storchi G (2001) Shortest viable path algorithm in multimodal networks. Transp Res Part A Policy Pract 35:225–241. doi:10.1016/S0965-8564(99)00056-7

Mavoa S, Witten K, McCreanor T, O'Sullivan D (2012) GIS based destination accessibility via public transit and walking in Auckland, New Zealand. J Transp Geogr 20:15–22. doi:10.1016/j.jtrangeo.2011.10.001

Miller HJ, Shaw S-L (2001) Geographic information systems for transportation: principles and applications. Oxford University Press, Oxford

Mooney P, Corcoran P, Winstanley AC (2010) Towards Quality Metrics for OpenStreetMap. Proceedings of the 18th SIGSPATIAL international conference on advances in geographic information systems, pp 514–517. ACM, New York

Mouncif H, Boulmakoul A, Chala M (2006) Integrating GIS-Technology for Modelling Origin-Destination Trip in Multimodal Transportation Networks. Int Arab J Inf Technol 3:256–264

Neis P, Zielstra D (2014) Generation of a tailored routing network for disabled people based on collaboratively collected geodata. Appl Geogr 47:70–77. doi:10.1016/j.apgeog.2013.12.004

Neis P, Zielstra D, Zipf A (2011) The street network evolution of crowdsourced maps: OpenStreetMap in Germany 2007–2011. Future Internet 4:1–21. doi:10.3390/fi4010001

Scheider S, Kuhn W (2010) Affordance-based categorization of road network data using a grounded theory of channel networks. Int J Geogr Inf Sci 24:1249–1267. doi:10.1080/13658810903514198

Scheurer J, Curtis C (2008) Spatial network analysis of multimodal transport systems: developing a strategic planning tool to assess the congruence of movement and urban structure. Curtin University of Technology, Perth

Scioscia F, Binetti M, Ruta M et al (2014) A framework and a tool for semantic annotation of POIs in OpenStreetMap. Procedia—Soc Behav Sci 111:1092–1101. doi:10.1016/j.sbspro.2014.01.144

Sester M, Jokar Arsanjani J, Klammer R et al. (2014) Integrating and generalising volunteered geographic information. In: Burghardt D, Duchêne C, Mackaness W (eds) Abstracting geographic information in a data rich world. Springer International Publishing, New York, pp 119–155

Stead D, Marshall S (2001) The relationships between urban form and travel patterns. Int Rev Evaluat EJTIR 1:113–141

Van Nes R (2002) Design of multimodal transport networks. PhD, Civil Engineering, Delft Technical University

Van Wee B (2002) Land use and transport: research and policy challenges. J Transp Geogr 10:259–271. doi:10.1016/S0966-6923(02)00041-8

Yigitcanlar T, Sipe N, Evans R, Pitot M (2007) A GIS-based land use and public transport accessibility indexing model. Aust Plan 44:30–37. doi:10.1080/07293682.2007.9982586

Part IV
Land Management and Urban Form

Assessing OpenStreetMap as an Open Property Map

Mohsen Kalantari and Veha La

Abstract Approximately 5/6th of the world's land and property rights, restrictions and responsibilities (RRRs) are unrecorded. This applies to both the developing and developed world and suggests that a considerable gap exists in worldwide land tenure information. Crowdsourcing can be used as a methodology for addressing this information gap. This chapter investigates the potential of OpenStreetMap (OSM) as a crowdsourcing system in collecting and recording land tenure information including RRRs. The chapter uses a case study in Victoria, Australia, to investigate the quality of the current OSM data and its potential as a crowdsourced record of public properties. The chapter studies the completeness of the public property records in the OSM, and the location, shape, area and description of the existing records. The analysis concludes with a mixed result. While the results show a considerable gap in the OSM records, at same time they indicate an acceptable quality for the existing records. The paper finally discusses the potential of OSM as an Open Property Map (OPM).

Keywords Openstreetmap · Open property map · VGI · Cadastre · Land · Property · Crowdsourcing · Victoria · Australia

1 Setting the Scene

Land is an important economic asset and sustains the livelihoods of many. Community identity, history and culture also have their roots in land. Communities, therefore, can readily mobilize around land issues, making land a central object of

M. Kalantari (✉) · V. La
Centre for Spatial Data Infrastructure and Land Administration, Department of Infrastructure Engineering, Melbourne School of Engineering, The University of Melbourne, Melbourne, Australia
e-mail: mohsen.kalantari@unimelb.edu.au

V. La
e-mail: lav@student.unimelb.edu.au

J. Jokar Arsanjani et al. (eds.), *OpenStreetMap in GIScience*,
Lecture Notes in Geoinformation and Cartography,
DOI 10.1007/978-3-319-14280-7_13

conflict and divergence (Williamson et al. 2010). In a guideline to prevent land and natural resource conflicts, the United Nations (UN) argues that the challenges associated with managing land and natural resource issues may have significant impacts on defining global peace and security in the twenty first century. Global trends such as demographic changes, increasing consumption, environmental degradation and climate change are placing significant and potentially unsustainable pressures on the availability and usability of natural resources such as land, water and ecosystems (UN-FPA 2014).

Since 1990, at least 18 violent conflicts have been fuelled by the exploitation of land (UN-FPA 2014). Land conflicts commonly become violent when linked to wider processes of political exclusion, social discrimination, economic marginalization (Fig. 1), and the perception that peaceful action is no longer a viable strategy for change. For instance, a study in a dozen less developed countries suggests the existence of the land tenure risk that involves overlapping land claims. The risk decreases the viability of land development and investment and increases conflicts between local communities and potential investors. The analysis identifies the lack of micro-level data on the land status including boundaries of ownerships as a major factor fuelling the tenure risk (de Leon et al. 2013).

While this issue is largely relevant to developing countries, there are many land issues in developed nations too. For instance, in Australia, despite the introduction of the Native Title Act in 1993, Aboriginal native title claim groups are also required to participate in other processes outside of the native title domain. Under the Native Title Act (1993), an aboriginal interest can only be recognized if it arises from customs and has continuity from the first European contact. There are examples when aboriginals have engaged in the processes outside of the Native

Fig. 1 Developers risk losing billions if they fail to address land conflicts, companies who ignore the claims of local communities when buying up land face protests, financial damage and court action (Provost 2013)

Title Act to protect their interests in natural resources and land. The engagements in these processes do not 'count' as an occurrence of traditional interests (Weiner 2011). This issue suggests a gap as to what land rights exist and where they are, and if they can be managed under the Native Title Act. While the latter is a legal issue, the former is a gap in the tenure information of Australia.

In addition, in developed countries, many new land and property interests created by governments in response to concerns for sustainability are often poorly managed and understood (Bennett et al. 2008) causing problems in the urban planning and development process where public and private interests compete (Bishop and Jenkins 2011). Figure 2 depicts potential new interests in space as a result of high-rise development in a city. The owners of the new building have a responsibility to install and maintain a heliostat system. This system ensures enough sunlight is provided to the shorter building overshadowed by the new building. Developed societies will be facing more complex tenure arrangements such as this in the future.

Addressing land issues is fundamental in creating sustainability and stability of societies. Recording of land and property RRRs is essential for social stability, economic development and environmental sustainability of the world. To better understand the complexity and diversity of RRRs, they are often conceptualized as a continuum that provides different sets of RRRs (Payne 2002), degrees of security and responsibility, and varying degrees of enforcement (Fig. 3).

Considering both developed and less developed countries, only almost 1/6th of the world's RRRs are recognized and recorded (McLaren 2011). This issue indicates a gap in the worldwide tenure information including developed and developing countries. While methodologies for the recognition of the RRRs has been addressed previously in the literature (Kalantari et al. 2008), a critical gap remains in selecting an efficient and cost-effective approach to recording the remaining 5/6th of the world's land RRRs.

Fig. 2 Modern property rights, restriction and responsibilities: how do we ensure all are recorded (Perkins 2012)?

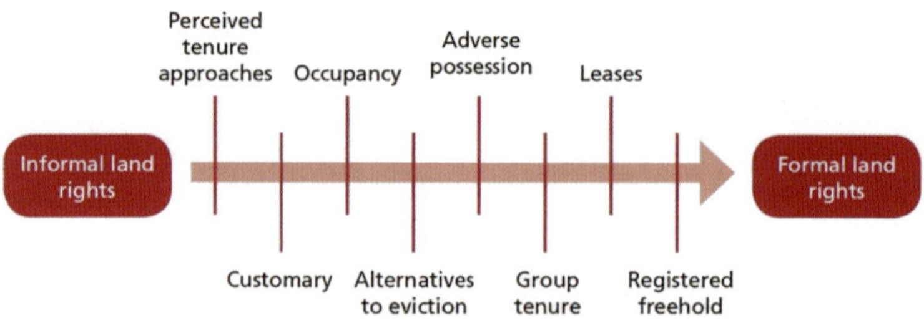

Fig. 3 Global land tool network's continuum of land rights (Augustinus 2010)

Cadastral surveying including ground and aerial methods is known as the primary method of identifying boundaries of RRRs. Establishing and maintaining a complete land and property record based on cadastral surveying is a costly and lengthy process that demands leadership of governments and their partnership with citizens and communities.

As an alternative approach, this chapter investigates the potential of crowdsourcing systems as citizen-centered and economical approaches in establishing and maintaining the record of RRRs with particular attention to information gaps about land and property RRRs. In doing so, the chapter studies spatial and attribute accuracies of OSM's public properties data in Victoria, Australia. The chapter compares OSM's park polygons with a reference data from the Victorian Government. This authoritative data is known as Features of Interest (FOI).

The remainder of the chapter is organized as follows: Sect. 2 provides an overview of the existing methods for assessing the data quality in the OSM. Section 3, with a combination of the methods identified in Sect. 2, develops the methodology of the study. Section 4 assesses the data quality of OSM's public properties in Victoria. Sections 5 and 6 discuss the results, draw conclusions and set directions for future research.

2 Assessing Spatial Data Quality

ISO 19157:2013 Geographic Information—Quality Principles sets the way by which one can measure the spatial data quality including completeness, logical consistency, positional accuracy, temporal accuracy, and thematic accuracy. Based on the ISO data quality elements, a number of investigations have been conducted to measure the OSM data quality. This section reviews a selected number of these investigations.

Haklay (2010) compared the OSM road network datasets in England against those of the Ordnance Survey (OS). The study focused on two aspects of spatial data quality including positional accuracy and completeness. Positional accuracy was measured by comparing buffered lines of the OS (in which buffer width varied according to the type of lines) with one-meter buffered lines of OSM. The completeness was measured by length comparison, coverage areas, and the number of features.

A similar study was conducted in Germany by Zielstra and Zipf (2010) where they compared the quality of OSM datasets against TeleAtlas in term of its completeness. The authors used total lengths of the road network as an indicator for completeness and the comparisons were done at different scales including the federal territory of Germany, five of the biggest German cities (Berlin, Hamburg, Munich, Cologne, and Frankfurt), five medium-sized cities, and its buffered zones (as an indicator for moving toward rural areas). The OSM datasets were also acquired at three different times (April, July, and December 2009) to evaluate its change growth.

Another study by Koukoletsos et al. (2012) assessed completeness in terms of omission (missing data) and commission (excessive data) using a multi-stage feature-based matching approach on the OSM road network dataset against the Integrated Transport Network layer of MasterMap from the Ordnance Survey (OS) in the UK. The automated matching process applied spatial constraints (including search distance and feature orientation) and attribute constraints (through text similarity ratio), and the completeness errors (omission and commission) were calculated based on matching errors. The study found that the matching technique was efficient with errors between 2.08 % in an urban area and 3.38 % in rural areas, and omission errors were between 0.54 % in rural areas and 1.70 % in urban areas.

Mooney et al. (2010) assessed the quality of polygon features (lakes and forests) in OSM in several European countries and regions including Ireland, Austria, Lower Saxony, Scotland, Wales, Latvia, Estonia, and Spain. The latter did the comparisons in terms of polygon representation, tagging and documents, and shape similarities.

In addition to the quality issues in OSM, (Jokar et al. 2013b) studied the evolution of OSM across space and over time to investigate collaborative contributing and to predict potential future states in OSM. This paper presents a methodology based on a cellular automata (CA) model for the period 2007–2012, in the city of Heidelberg in Germany. From a slightly different point of view, (Spielman 2014) suggests that crowdsourced maps such as the OSM are complex social systems and hence the quality should be assessed accordingly. The study argues that the structure of the crowd is more important than the ability of individuals in contributing to the system. So the quality of the outputs including credibility and accuracy is a function of the social structure of the crowd.

The methodology used in this chapter was adopted from existing research, which is summarized in Table 1.

Table 1 Comparison of methods used access quality of Volunteered Geographic Information (VGI)

Paper	Paper
Assessing data completeness of VGI through an automated matching procedure for linear data (Koukoletsos et al. 2012)	A comparative study of proprietary geodata and volunteered geographic information for Germany (Zielstra and Zipf 2010)
An automated method to assess data completeness and positional accuracy of OpenStreetMap (Koukoletsos et al. 2011)	How good is volunteered geographical information? A comparative study of OpenStreetMap and ordnance survey datasets (Haklay 2010)
Assessing completeness and spatial error of features in volunteered geographic information (Jackson et al. 2013)	Quality analysis of OpenStreetMap data based on application needs (Mondzech and Sester 2011)
Quality assessment of the French OpenStreetMap dataset (Girres and Touya 2010)	Towards quality metrics for OpenStreetMap (Mooney et al. 2010) Quality assessment for building footprints data on OpenStreetMap (Fan et al. 2014)

3 Assessment Methodology

To assess the quality of OSM as an OPM, this section compares the public properties in OSM with the public properties in the FOI. By adopting some of the methods introduced in the previous section, this section describes the steps and criteria for undertaking a quality assessment of the public properties data in OSM.

3.1 Area of Study

For undertaking the assessment, the OSM park polygons in Victoria, Australia (Fig. 4) with the tag name of "leisure" and value of "park" were extracted and downloaded from GEOFABRIK on 16 September 2013. Similarly, the FOI data was downloaded on 21 October 2013 (Department of Environment and Primary Industries 2013). At the time, FOI included 18,561 park polygons.

3.2 Feature Matching

Automatic feature matching was performed by introducing spatial and attributes constraints. The spatial constraint was in the form of search distances. In this study, the search distances ranged from 1 to 100 m. Several OSM features could be located within the specified distances, but they must also have a textual name similarity ratio of 51 % to be considered as matches (with reference features).

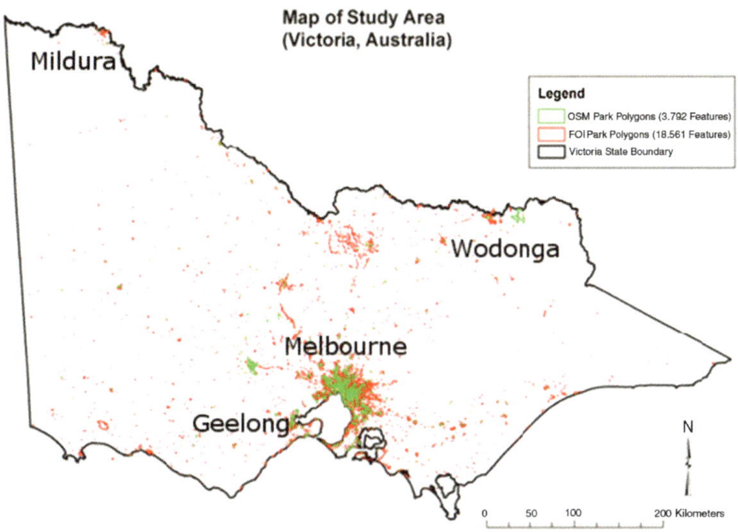

Fig. 4 An overview of the study area

An increase of 5 m of search distance was incorporated each time when there was no matched feature. The feature matching outcomes include final search distance where features were considered as matches, number of match features at a minimum search distance, attribute similarity ratio when features were considered as matches, unique match IDs for linking reference features with their OSM match features, and final conclusion of whether or not a feature has its match feature.

3.3 Quality of Feature Representations

A polygon can be represented by a list of vertices in which each vertex is a pair of coordinates or it can be represented by turning functions which are tangent angles measured counter clockwise and are the functions of arc lengths from a chosen reference point on the polygon's boundary (Arkin et al. 1991). The number of vertices could slightly or greatly impact the polygon's properties, such as its shape, perimeter, and area. The polygon could be under-representation with fewer vertices, and it could be over-representation with more than necessary vertices. A greater number of vertices will not guarantee better representation but may unnecessarily increase the size of the data.

This study assessed quality of feature representations of OSM polygons by comparing the number of vertices, mean vertex spacing distances, feature areas, and shape similarity ratio with the FOI polygons. The shape similarity ratio was computed by normalizing results from the algorithm developed by Arkin et al. (1991) as explained by Mooney et al. (2010).

3.4 Positional Uncertainty

The search distances used to automatically match the OSM and FOI park features were ranging between 1 to 100 m and were treated as approximations of the positional accuracy. However, search distance is not always a good representation of positional accuracy because any parts of the OSM polygons could fall within the buffered zone of the FOI polygons as shown in Fig. 5. Therefore, to better understand the positional accuracy of the OSM park polygons, distances between centroids of the OSM and FOI features were calculated and analyzed.

3.5 Attribute Accuracy

Attribute accuracy is an important aspect of the study because in the context of land administration, property data may contain descriptive information such as the rightful claimants of the property, different types of RRRs, or other types of descriptions. This study compared the names of OSM and FOI public properties. Being an authoritative data source, the names in the FOI comply with the Register of Geographic Names, VICNAMES in Victoria. VICNAMES is a public register, so the names in the FOI are the official and legitimate description of the features. The study compared sequences of characters from the FOI park name and OSM park name. It then computed the comparison result in the form of similarity ratio with values ranging from 0 to 1 where 1 represents identical texts, and 0 is when they have nothing in common (Python 2013).

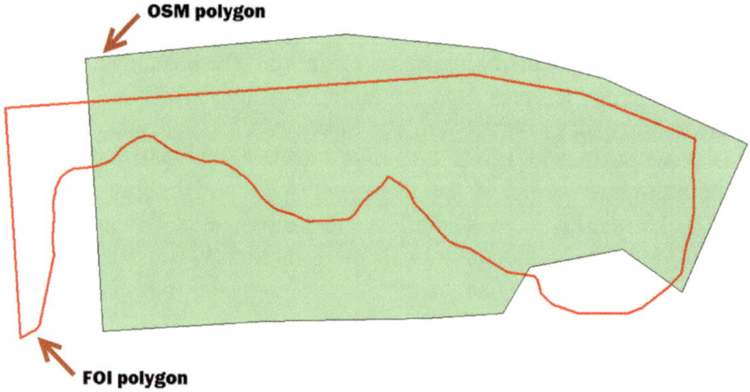

Fig. 5 Partly overlapping area of the OSM polygon on the FOI polygon

4 Results of the Assessment

4.1 Data Completeness

During the data matching process, 810 features of the FOI parks were initially matched to 729 OSM parks. Manual checks were performed to find and fix non-unique matches. Also, the study excluded multi-part features of the FOI polygons with single-part features of the OSM polygons as the feature representations of multi-part features (composed of two or more polygons) are different from feature representation of single-part features. While all of the OSM polygons were single-part features, only a small proportion of the FOI polygons, 7 % (48 features), were multi-part features with the number of parts ranging from 2 to 5. Given the small number of multi-part features, they were removed from further analysis to prevent exaggerated comparison results. Based on feature counts, the OSM park polygons were 20 % complete comparing to the FOI park polygons. Table 2 shows actual data completeness and feature matching.

4.2 Feature Vertex and Part Differences

The differences in number of vertices of the OSM and FOI features were ranging from 2,937 (over-representation) to −1,237 (under-representation) but after visual checks, any difference values over 100 (24 polygon pairs) were considered as outliers and removed from the analysis because of the unusual number of vertices used to represent the features as shown in Fig. 6.

"Unusual use of vertex numbers" means that the volunteers have used too many vertices to represent the polygons (GPS receivers have probably been used to trace the polygons rather than marking each turning point of the polygons). Additional visual checks were performed on 100 randomly selected polygon pairs to determine a reasonable threshold for good use of vertices, and it was found that the threshold values should be between −10 and 10 vertices difference. Figure 7 shows the histogram distribution of differences in vertex numbers in which 43 % of the OSM polygons had the same number of vertices as the FOI polygons, 30 % had differences between −10 and 10 vertices, 24 % was over-representation, and only less than 1 % was under-representation. Therefore, more than 73 % of the OSM polygons have made good use of vertex numbers to represent the polygons.

Table 2 Summary of data completeness and feature matching

Feature count/feature type	OSM park polygons	FOI park polygons
All features	3,792 (20 %)	18,561
Final matched features distance within 1–100 m name similarity ratio ≥51 %	729 (729 initially)	729 (810 initially)

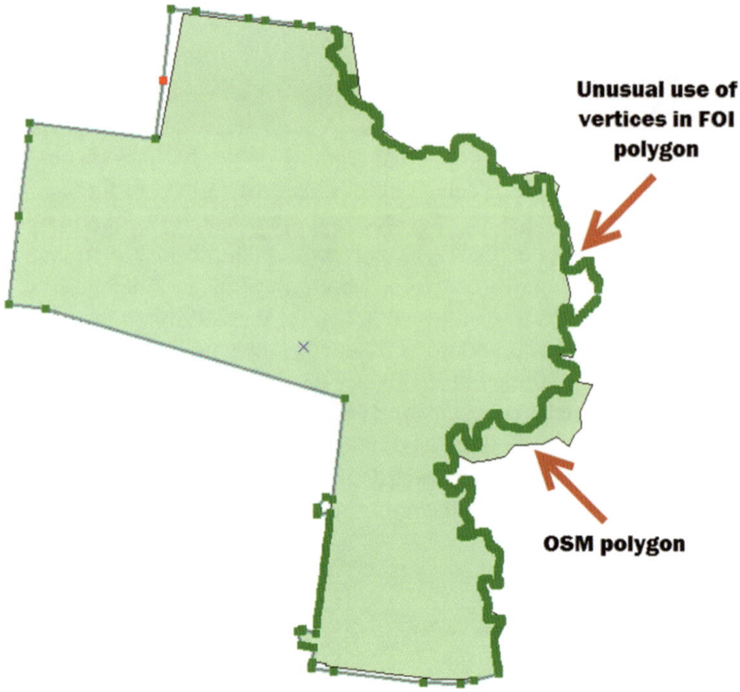

Fig. 6 Unusual use of vertex numbers

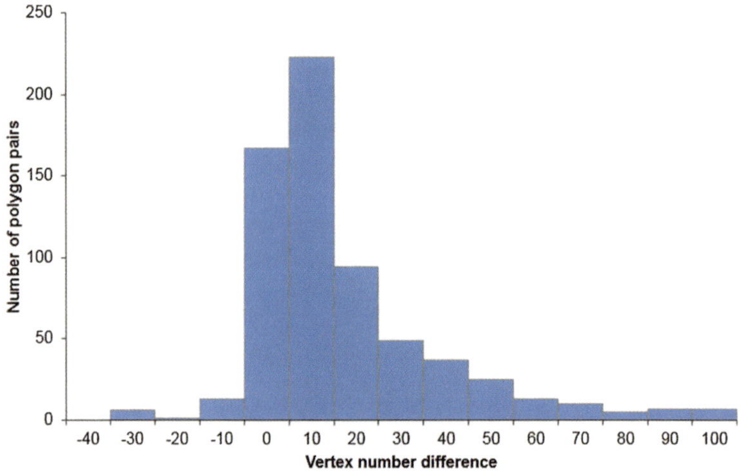

Fig. 7 Histogram of vertex number differences

4.3 Mean Vertex Spacing Differences

The study looked at the average distance (spacing) that the OSM volunteers and FOI data generation process have used for marking/defining a new point/vertex to represent a polygon. The mean spacing between vertices of OSM ranged from 7 to 800 m with an average of 86 m while they were from 2.5 to 605 m with an average of 56 m for the FOI. Only 15 % of polygon pairs had the same mean vertex spacing while 76 % of them had mean vertex spacing differences ranging from −50 to 50 m. Figure 8 shows the actual distribution of the mean vertex spacing differences.

Moreover, there was a weak negative correlation with a correlation coefficient of −0.39 between differences in the number of vertices and mean vertex spacing. This means if the difference in the number of vertices increases, the mean spacing difference would slightly decrease.

4.4 Feature Area Differences

Differences in feature areas between the OSM and FOI park polygons were ranging from −171 ha (larger than FOI) to 53 ha (smaller than FOI) with an average difference of −0.92 ha and standard deviation of 11.63. However, 42, 35, and 10 % of them had the same areas, differences between −0.5 and 0.5 ha, and differences between −1 and 1 ha, respectively. Figure 9 shows the actual distribution in the form of a histogram.

Correlation between differences in vertex numbers and feature areas was very weak with a correlation coefficient of 0.08.

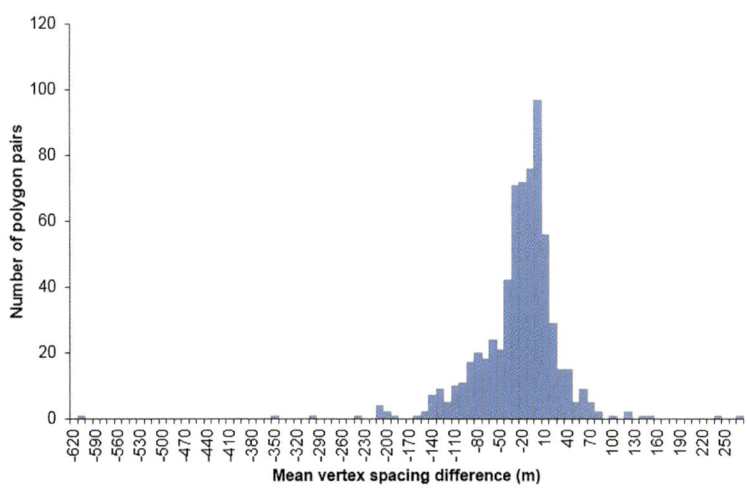

Fig. 8 Histogram of mean vertex spacing differences

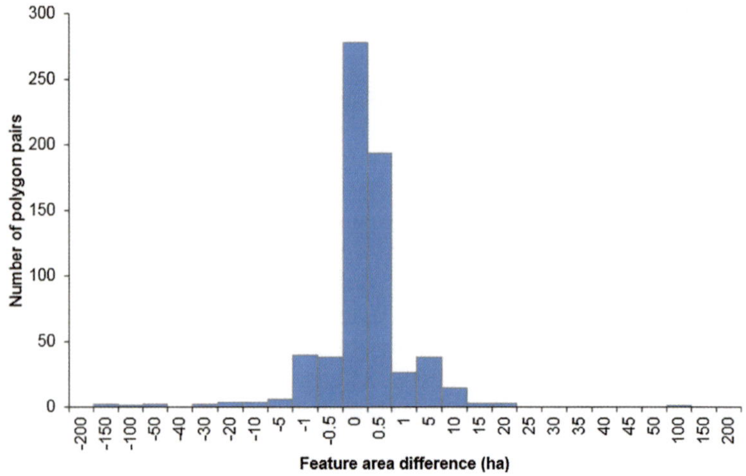

Fig. 9 Histogram of area differences

4.5 Shape Similarity

Mooney et al. (2010) found that a pair of polygons had very similar shape when its similarity ratio was greater than or equal to 0.8 while a value of 0.5 or less represented very dissimilarity of the shape by performing visual checks on 100 randomly selected polygon pairs. The same observations were made on 50 polygon pairs with a shape similarity ratio value greater than 0.8 and 50 polygon pairs with a value less than 0.5, and the same result was confirmed in this research.

Figure 10 shows the histogram distribution of the shape similarity ratio of the 681 polygon pairs in which 56, 40, and 4 % had similarity ratio values of 0.8 or greater, between 0.6 and 0.7, and 0.5 or less, respectively. It is important to note again that the shape similarity algorithm by Arkin et al. (1991) is scale independent and number of vertices could slightly or significantly affect polygon shapes. Therefore, it was interesting to find out about correlations between shape similarity values and differences in the number of vertices and feature areas between the OSM and FOI features.

Correlation coefficient values of 0.09 and −0.28 were derived for relationships between shape similarity ratio and feature area differences and between shape similarity ratio and vertex number differences, respectively. The values were very small suggesting that there was no (or very small) correlation between those variables and it could not be expected that higher values of shape similarity ratio would correspond to smaller differences in shape areas and number of vertices.

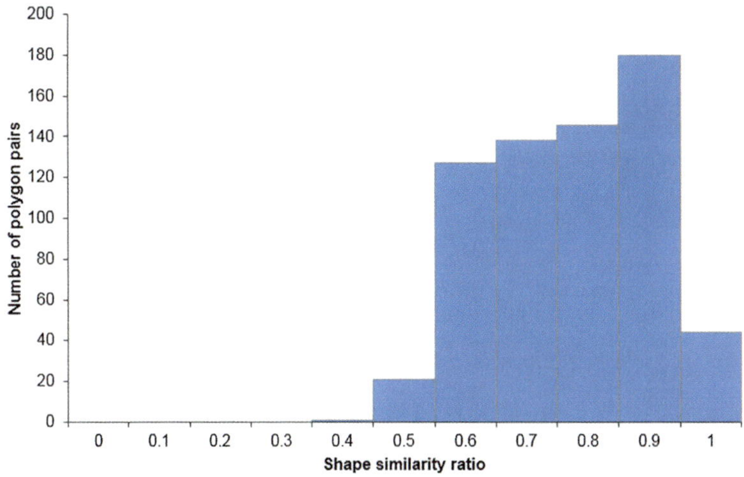

Fig. 10 Histogram of shape similarity ratio

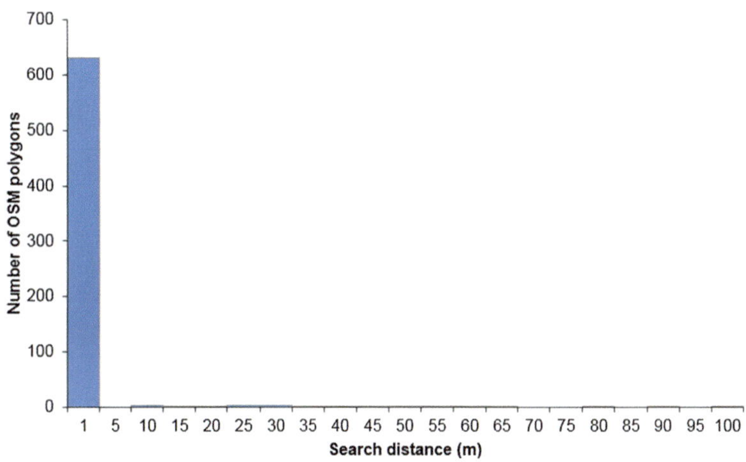

Fig. 11 Histogram of the distribution of search distances

4.6 Position Accuracy

Figure 11 shows the histogram distribution of the number of the OSM features within particular search distances where they were successfully matched with the corresponding FOI features in which 96 % of the OSM polygons were found within

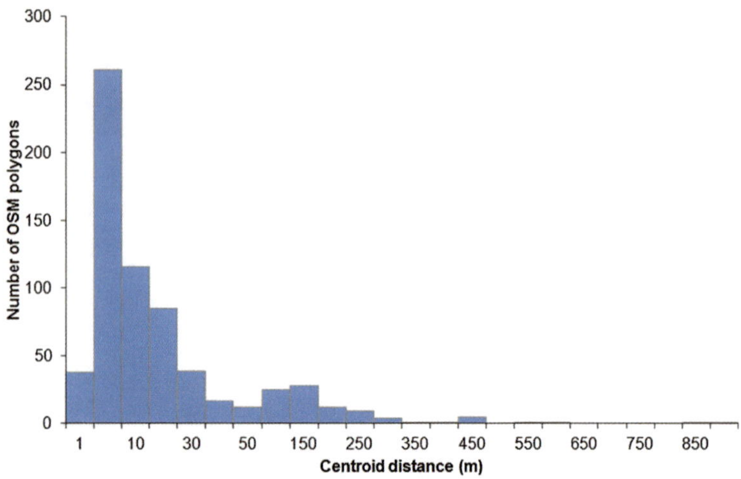

Fig. 12 Histogram of the distribution of centroid distances

a search distance of 1 m from the FOI polygons. On the other hand, Fig. 12 shows distributions of centroid distances between the OSM and FOI polygons in which 6, 40, 18, 23, and 14 % of the OSM park centroids were within 1, 5, 10, 50, and more than 100 m from the corresponding FOI park centroids, respectively. Therefore, more than 63 % of the OSM polygon centroids were within 10 m from the FOI polygon centroids.

Because positions of polygon centroids are dependent on polygon shapes, it is interesting to see relationships between centroid distances with shape similarity ratios. A low negative correlation between those two testing variables with coefficient value of −0.38 was found, and it means that the polygon pairs with higher values of shape similarity ratio tend to have closer centroids.

4.7 Attribute Accuracy

The attribute accuracy of the 657 OSM park polygons ranged from 52 to 100 % in which 77 % of them had the same exact attribute names as the FOI park polygons and 13 % of them had attribute accuracy of over 80 %. Figure 13 shows the histogram distribution of the attribute accuracy of OSM.

The correlation between shape similarity ratio and attribute accuracy was small with coefficient value of 0.12 and therefore higher values of shape similarity ratio do not guarantee higher values of attribute accuracy.

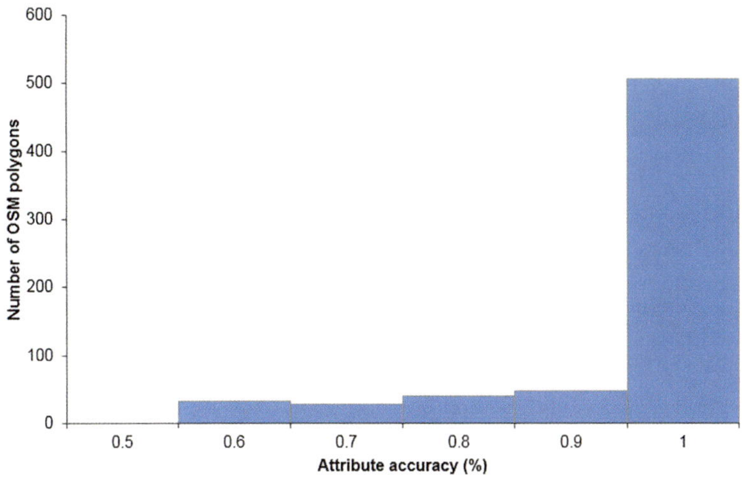

Fig. 13 Histogram of attribute accuracy

5 Discussion of the Results

There are studies which highlight the utility of OSM in fine-scale land use mapping (Jokar et al. 2013a; Arsanjani et al. 2014). Besides this, land records, being the heart of land administration, should unambiguously identify entitlements in land and assist in resolving disputes that arise from land issues. Land records are used to confirm the existence of RRRs in land, to specify the location of the RRRs, the spatial extent of the RRRs and the rightful claimants of the RRRs (Basiouka and Potsiou 2012).

This study revealed that only 20 % of the public properties in the FOI are recorded by volunteers in OSM. This was expected as the contribution of volunteers in OSM tends to focus more on road and street features. This result shows that OSM, in its current state, does not represent a complete record of the public properties in Victoria. In other words, only 20 % of the public properties are recorded in OSM and the remaining public properties are not collected and recorded by volunteers. Considering an ideal inventory of properties, such a state of OSM suggests an information gap in terms of the public properties in OSM.

However, by analyzing the records of OSM that matched with those of FOI, the results demonstrate that 78 % of the OSM public properties had similar shapes with ratio of greater than or equal to 0.7. In addition, 64 % of the polygons in OSM had centroid distances less than 10 m from those in the FOI. This implies that even though OSM does not have all the public properties recorded, for a high percentage of the public properties that exist in OSM, the shape and location of representative polygons are comparable with those of the FOI. In other words, OSM represents the shape and location of the boundaries of a high percentage of the existing public properties in the same way as the FOI. For land administration, this is considered as the first step in creating an inventory of properties. If the land administration can verify the approximate location and boundaries of properties so it can protect the interests of rightful claimants. For Victoria, this means that the majority of the public properties in OSM that are found in the FOI possess reasonable quality as they confirm the location of the properties and to a large degree they depict their shape in accordance with the data from authoritative source.

In many land records, the area is considered as an important attribute of properties. A similar quality assessment result can be observed for the areas of the public properties. The results of the analysis indicate that 77 % of the OSM public property polygons had area differences less than 0.5 ha with those of the FOI. This implies volunteers have recorded the spatial extent of the majority public properties close to what has been recorded in the FOI.

The study also demonstrates that the contribution of the volunteers in terms of describing and naming the public properties is close to the description of the geographic name register. 90 % of the recorded descriptions are alike with a similarity ratio of 80 % or over. This implies that the volunteers have taken good care when choosing a name and describing the properties, so the names are in accordance with the authoritative sources.

Overall, the study observes a considerable information gap in terms of the public property data in OSM. This gap undermines the utility of OSM data as a representative of existing public properties. However, the study reveals that those properties that are found in OSM possess a reasonable quality. The location, size, shape and description of the majority of the properties are close to those of the authoritative source. It is also important to emphasize that, as illustrated in the previous sections, there are small correlations between assessment criteria meaning that they were most likely independent from each other.

6 Conclusions and Future Research

This chapter provided another contribution to the research area of crowdsourcing geographic information by studying OSM's potential as an OPM. The chapter studied spatial and attribute accuracies of OSM's data in Victoria, Australia. More specifically, it compared the OSM park polygons as public properties with the reference data from Vicmap known as FOI. The analysis suggests a significant gap in the public property data where the data was clearly not complete. At the same time, it discovered a reasonable quality for the existing public property data suggesting that OSM has the potential to become an OPM.

The study focused on the geometrical aspects of the data quality and only investigated names as a non-geometrical property. The reason for this limitation is that other property attributes such as different types of RRRs, which may exist on properties, were absent from the OSM data. The volunteers were able to collect the name of the public properties to describe them. However, an important question to ask is what types of land RRRs can potentially be collected by volunteers. Moreover, future research can investigate the addition of new features in the existing OSM data model so that land RRRs can be recorded.

References

Arkin EM, Chew LP, Huttenlocher DP, Kedem K, Mitchell JSB (1991) An efficiently computable metric for comparing polygonal shapes. IEEE Trans Pattern Anal Mach Intell 13(3):209–216

Augustinus C (2010). Social tenure domain model: what it can mean for the land industry and for the poor. XXIV FIG International Congress

Basiouka S, Potsiou C (2012) VGI in cadastre: a Greek experiment to investigate the potential of crowd sourcing techniques in cadastral mapping. Surv Rev 44(325):153–161

Bennett R, Wallace J, Williamson I (2008) Organising land information for sustainable land administration. Land Use Policy 25(1):126–138

Bishop P, Jenkins V (2011) Planning and nuisance: revisiting the balance of public and private interests in land-use development. J Environ Law 23(2):285–310

De Leon R, Garcia T, Kummel G, Munden L, Murday S, Pradela L (2013) Global capital, local concessions: a data-driven examination of land tenure risk and industrial concessions in emerging market economies. The Munden Proj

Department of Environment and Primary Industries (2013) Vicmap features of interest, from http://www.dse.vic.gov.au/property-titles-and-maps/maps-imagery-and-data/data/vicmap/vicmap-products/vicmap-features-of-interest. 21 Oct 2013

Fan H, Zipf A, Fu Q, Neis P (2014) Quality assessment for building footprints data on OpenStreetMap. Int J Geogr Inf Sci 28(4):700–719

Girres JF, Touya G (2010) Quality assessment of the French OpenStreetMap dataset. Trans GIS 14 (4):435–459

Haklay M (2010) How good is volunteered geographical information? A comparative study of OpenStreetMap and ordnance survey datasets. Environ Plan 37(4):682

Jackson SP, Mullen W, Agouris P, Crooks A, Croitoru A, Stefanidis A (2013) Assessing completeness and spatial error of features in volunteered geographic information. ISPRS Int J Geoinf 2(2):507–530

Jokar Arsanjani J, Helbich M, Bakillah M, Hagenauer J, Zipf A (2013a) Toward mapping land-use patterns from volunteered geographic information. Int J Geogr Inf Sci 27(12):2264–2278

Jokar Arsanjani J, Helbich M, Bakillah M, Loos L (2013b) The emergence and evolution of OpenStreetMap: a cellular automata approach. Int J Digit Earth 1–15

Jokar Arsanjani J, Vaz E, Bakillah M, Mooney P (2014) Towards initiating OpenLandMap founded on citizens' science: the current status of land use features of OpenStreetMap in Europe. Proceedings of the 17th AGILE Conference on Geographic Information Science, Castellon, Spain, AGILE Digital Editions

Kalantari M, Rajabifard A, Wallace J, Williamson I (2008) Spatially referenced legal property objects. Land Use Policy 25(2008):173–181

Koukoletsos T, Haklay M, Ellul C (2011) An automated method to assess data completeness and positional accuracy of OpenStreetMap. In: 11th international conference on geocomputation, London, UK

Koukoletsos T, Haklay M, Ellul C (2012) Assessing data completeness of VGI through an automated matching procedure for linear data. Trans GIS 16(4):477–498

McLaren R (2011) Crowdsourcing support of land administration: a new, collaborative partnership between citizens and land professionals. Royal Institution of Chartered Surveyors (RICS) Report November

Mondzech J, Sester M (2011) Quality analysis of OpenStreetMap data based on application needs. Cartographica: Int J Geogr Inf Geovis 46(2):115–125

Mooney P, Corcoran P, Winstanley AC (2010) Towards quality metrics for openstreetmap. In: Proceedings of the 18th SIGSPATIAL international conference on advances in geographic information systems, ACM

Payne GK (2002) Land, rights and innovation: improving tenure security for the urban poor. ITDG publishing, London

Perkins M (2012) Developer wants to let the sunshine in and it will all be done with mirrors. The Age, from http://theage.domain.com.au/developer-wants-to-let-the-sunshine-in-and-it-will-all-be-done-with-mirrors-20121031-28kc0.html. 11 Feb 2014

Provost C (2013) Developers risk losing billions if they fail to address land conflicts, from http://www.theguardian.com/global-development/2013/sep/19/developers-land-rights-conflicts. 11 Feb 2014

Python (2013) difflib—helpers for computing deltas, from http://docs.python.org/2/library/difflib.html. Retrieved 10 Oct 2013

Spielman SE (2014) Spatial collective intelligence? Credibility, accuracy, and volunteered geographic information. Cartography Geogr Inf Sci 41(2):115–124

UN-FPA (2014) Land, natural resources and conflict: from curse to opportunity. An UN-EU Partnership in action, from http://www.un.org/en/land-natural-resources-conflict/. 11 Feb 2014

Weiner JF (2011) Conflict in the statutory elicitation of aboriginal culture in Australia. Anthropol Forum 21(3):257–267

Williamson I, Enemark S, Wallace J, Rajabifard A (2010) Land administration for sustainable development. ESRI Press Academic Redlands, CA

Zielstra D, Zipf A (2010) A comparative study of proprietary geodata and volunteered geographic information for Germany. In: 13th AGILE international conference on geographic information science

Investigating the Potential of OpenStreetMap for Land Use/Land Cover Production: A Case Study for Continental Portugal

Jacinto Estima and Marco Painho

Abstract In the last decade, volunteers have been contributing massively to what we know nowadays as Volunteered Geographic Information (VGI). Through the research that has been conducted recently, it has become clear that this huge amount of information might hide interesting and rich geographical information. The OpenStreetMap (OSM) project is one of the most well-known and studied VGI initiatives. It has been studied to identify its potential for different applications. In the field of Land Use/Cover, an earlier study by the authors explored the use of OSM for Land Use/Cover (LULC) validation. Using the COoRdination of INformation on the Environment (CORINE) Land Cover (CLC) database as the Land Use reference data, they analyzed the OSM coverage and classification accuracy, finding an interesting global accuracy value of 76.7 % for level 1 land classes, for the study area of continental Portugal, despite a very small coverage value of approximately 3.27 %. In this chapter we review the existing literature on using OSM data for LULC database production and move this research forwards by exploring the suitability of the OSM Points of Interest dataset. We conclude that OSM can give very interesting contributions and that the OSM Points of Interest dataset is suitable for those classified as CLC class 1 which represents artificial surfaces.

Keywords Volunteered geographic information (VGI) · OpenStreetMap (OSM) · Land use · Land cover

J. Estima (✉) · M. Painho
ISEGI, Universidade Nova de Lisboa, Lisbon, Portugal
e-mail: jacinto.estima@gmail.com

M. Painho
e-mail: painho@novaims.unl.pt

J. Jokar Arsanjani et al. (eds.), *OpenStreetMap in GIScience*,
Lecture Notes in Geoinformation and Cartography,
DOI 10.1007/978-3-319-14280-7_14

1 Introduction

1.1 Land Use and Land Cover Production

Land Use/Land Cover (LULC) databases, as they characterize land, are fundamental inputs for a variety of applications such as LULC change monitoring, climate change and biodiversity monitoring, among others (Caetano et al. 2006; Ellis 2013; Fritz et al. 2009). While Land Cover (LC) is more related to natural environments characterizing biophysical features, Land Use (LU) represents human-related environments attempting to describe the human interaction with these natural features (Baulies and Szejwach 1997). The production of LC and LU maps is usually undertaken by highly trained and skilled people interpreting and classifying remote sensing data, and involves a complex and long process of four main steps: acquisition of remote sensing data used as the basis for the classification process, pre-processing data into a proper format to extract information, analysis/classification including quality assessment, and product generation and documentation (Cihlar 2000). Aerial photography or in situ data are required in some of these phases, either to clarify the interpretation in areas of uncertainty or for quality assessment and validation (Caetano et al. 2006), thus increasing the time and cost of production. These constraints mean that the focus of LULC mapping is on themes and areas that are considered more important and for use within multiple applications.

This also has a negative impact on the update strategies, which therefore happen less frequently. Consequently, these databases become outdated very quickly in some areas (Goodchild 2008b).

According to Cihlar (2000, p 1108), "*The research agenda needs to address the best ways of taking advantage of the new capabilities and, importantly, the ways of resolving problems identified during the production of the land cover maps over large areas*". This statement drives us to think that VGI data, and in this particular case OSM data, have to be investigated and exploited in order to be used for LULC database production.

1.2 Volunteered Geographic Information and OpenStreetMap

The term Volunteered Geographic Information (VGI) was coined by Michael Goodchild in 2007 to describe "*the widespread engagement of large numbers of private citizens, often with little in the way of formal qualifications, in the creation of geographic information, a function that for centuries has been reserved to official agencies*" (Goodchild 2007, p 212). Other related terms were also introduced, in different years by different authors, such as Neogeography (Turner 2006) or Crowdsourcing geospatial data (Hudson-Smith et al. 2009).

There are a range of initiatives to which volunteers might participate and contribute and both the initiatives and the volunteers have been growing over the years according to the inventory made by Elwood et al. in 2009 where 99 VGI initiatives were identified (Elwood et al. 2012). Initiatives such as Wikimapia (2014), Google MyMaps (2014), GMapCreator (2014), London Profiler (2014), Map Tube (2014) and Flickr (2014) are just a few examples from this comprehensive list. According to the same authors, OpenStreetMap (OSM), also part of the list, is one of the most important and studied VGI initiatives. Started in August 2004 by Steve Coast and developed by the OpenStreetMap Foundation since 2006, this initiative is a worldwide mapping effort that includes more than a million volunteers around the globe, the number reached during 2013, and aims at providing free geographic data to anyone. Users collect data, including topographic data, mostly with GPS or GPS-enabled equipment. The collected data are then uploaded to the OSM database, along with descriptions, names and other attributes, through the OSM web page or established editors. The data then become available to anyone in the form of rendered maps and other services, including the possibility to download the data in vector format or embed it in websites using their Application Programming Interface (API).

The interest in exploring some of these initiatives has been increasing for areas such as navigation (Holone et al. 2007), emergency response (Goodchild and Glennon 2010; Zook et al. 2010), vernacular geography (Hollenstein and Purves 2010) and LULC production (Estima and Painho 2013b; Fritz et al. 2009). In this chapter we provide the current status on the use of data from OSM for LULC database production. We then move forward in our continuing research in this matter to explore a point-based dataset from the OSM database. We also discuss possible and interesting contributions OSM can give for LU database production, also considering some of the main issues related with its use and approaches to overcome them.

The chapter is structured as follows: After a brief introduction some related work is presented with an emphasis on studies using OSM data for LULC database production. We then describe the data and methods used for the practical part followed by the results and discussion. The paper ends with some conclusions, summarizing possible contributions of OSM for LULC database production purposes, and future research directions.

2 Volunteered Geographic Information and OpenStreetMap for Land Use/Land Cover Production

In this chapter we review studies already conducted on using VGI, with particular emphasis on OSM, for LULC database production. We start by summarizing the research on using VGI followed by the development of a more extended overview on the particular case of OSM.

2.1 Volunteered Geographic Information for Land Use/Land Cover Production

As stated before, VGI has been increasingly used to research novel applications for different areas, including LULC database production. In this particular domain two different approaches have been used so far: (1) asking volunteers to actively contribute to a specific project such as the validation of global land cover datasets (Fritz et al. 2009; Perger et al. 2012), and (2) using data contributed for other purposes/projects to extract valuable information and develop new ways to use it in this domain (Estima and Painho 2013a, b, 2014). Geo-Wiki.Org (Fritz et al. 2009) is a project that fits in the first approach, described as a global network of volunteers who wish to help improve the quality of global land cover maps. “GLC-2000”, “MODIS”, and “GlobCover” global land cover databases are overlaid on a platform based on Google Earth (GE) and their areas of divergence highlighted. Then, a network of registered volunteers helps to solve these discrepancies using their local knowledge along with available GE satellite imagery and other ancillary data coming from other VGI projects such as pictures from Panoramico (http://www.panoramio.com/) and Degrees of Confluence Project (http://www.confluence.org/). Another example is the Virtual Interpretation of Earth Web-Interface Tool (VIEW-IT) initiative based on GE high-resolution imagery to collect LULC reference data (Clark and Aide 2011). It was tested with a small group of selected users acting as volunteers and not yet in a real crowdsourcing environment. Nevertheless they found important issues with using GE and its satellite imagery, e.g. the legal restrictions in the free use of the Google Maps/Earth APIs and that some classes that cannot be discriminated with the available imagery (e.g. different annual crops). In these examples, volunteers need to be available to contribute to these specific projects and they also need to have some familiarity with these tools, which might be discouraging for some groups of participants. To overcome this difficulty, some projects occasionally use contests and a mechanism of rewards to increase contributions and participation (Fritz et al. 2012; Perger et al. 2012).

Using the second aforementioned approach, some experiments were conducted by Leung and Newsam (2010) to derive maps of what-is-where from large collections of georeferenced photos in an automated way. In this initial work the authors derived LC classifications from georeferenced image collections for locations where ground-truth was available. The aim was to evaluate the quality of the results obtained from the automatic classification by comparing them with the available ground truth. They achieved a classification accuracy of approximately 75 %. Another interesting work was conducted by Estima and Painho (2013b) to explore the possibility of using Flickr photos as a source of ground-truth data to help in the accuracy assessment phase of LULC production. Using continental Portugal as the study area and COoRdination of INformation on the Environment (CORINE) Land Cover (CLC) as a reference LULC database, the authors explored all the publically available and geotagged Flickr photos in terms of their temporal and spatial distributions and their distribution over the different CLC classes.

The number of photos and their temporal resolution were the most positive aspects whereas their asymmetry and irregular distribution over different CLC classes the most negative. They concluded stating that this could be a valuable source of ground truth data if combined with other sources but could not be used alone. Foody and Boyd (2013) used two sources of volunteered data to illustrate the potential of amateur or neogeographical activity in map validation. They used photographs acquired from an internet-based collaborative project and interpreted by other volunteers to evaluate the Globcover map's representation of tropical forests in West Africa. They confirmed the potential value of VGI projects, such as the Degrees of Confluence project, for the provision of useful, spatially extensive, data to support map evaluation.

2.2 OpenStreetMap for Land Use/Land Cover Production

The exploration of data from the OpenStreetMap (OSM) project is also recent. In 2013, Estima and Painho (2013a) explored the use of OSM data for LULC database production, particularly for validation purposes. The authors explored three OSM datasets—buildings, land use, and natural areas—using continental Portugal as the study area and CLC as the reference LULC database. They analyzed the spatial coverage and distribution, established a correspondence between OSM and CLC nomenclatures, and explored the coverage and accuracy when compared with CLC level 1 classes (CLC nomenclature can be found in Table 1). They found that the coverage area is not homogeneous among all the CLC level 1 classes, with values of approximately 75, 20, 2, 0.8, and 0.2 % for classes water bodies, artificial surfaces, forest and semi natural areas, agricultural areas, and wetlands, respectively. A table summarizing discrepancies between OSM and CLC classification nomenclatures was also reported, showing that multiple correspondences increase when we move from CLC level 1 to CLC level 3, given the increasing level of detail. To give just a brief example, the class "Farm" from the OSM Landuse dataset corresponds to the class 2 from the CLC level 1, to classes permanent crops, pastures, and heterogeneous agricultural areas from the CLC level 2, and to classes fruit trees and berry plantations, pastures, annual crops associated with permanent crops, or complex cultivation patterns from the CLC level 3. In terms of classification accuracy, the values reported were very promising, showing around 99.5, 84.3, 83.5, 46.6, and 1.2 % for classes water bodies, artificial surfaces, forest and semi-natural areas, agricultural areas, and wetlands, respectively, of the CLC level 1 nomenclature, and a global value of 76.7 %. They conclude that OSM might be very useful for LULC classification for classes with good coverage and accuracy such as classes' artificial surfaces and water bodies.

The possibility of using VGI data to replace training data acquired from in-site visits in the process of LULC classification was also investigated by Jokar Arsanjani et al. (2013a). Using the city of Koblenz, Germany, as the study area, they applied a supervised classification approach to classify data from the RapidEye

Table 1 Corine land cover nomenclature

Level 1	Level 2	Level 3
1 Artificial surfaces	11 Urban fabric	111 Continuous urban fabric
		112 Discontinuous urban fabric
	12 Industrial, commercial and transport units	121 Industrial or commercial units
		122 Road and rail networks and associated land
		123 Port areas
		124 Airports
	13 Mine, dump and construction sites	131 Mineral extraction sites
		132 Dump sites
		133 Construction sites
	14 Artificial, non-agricultural vegetated areas	141 Green urban areas
		142 Sport and leisure facilities
2 Agricultural areas	21 Arable land	211 Non-irrigated arable land
		212 Permanently irrigated land
		213 Rice fields
	22 Permanent crops	221 Vineyards
		222 Fruit trees and berry plantations
		223 Olive groves
	23 Pastures	231 Pastures
	24 Heterogeneous agricultural areas	241 Annual crops associated with permanent crops
		242 Complex cultivation patterns
		243 Land principally occupied by agriculture
		244 Agro-forestry areas
3 Forest and semi natural areas	31 Forests	311 Broad-leaved forest
		312 Coniferous forest
		313 Mixed forest
	32 Scrub and/or herbaceous vegetation associations	321 Natural grasslands
		322 Moors and heathland
		323 Sclerophyllous vegetation
		324 Transitional woodland-shrub
	33 Open spaces with little or no vegetation	331 Beaches, dunes, sands
		332 Bare rocks
		333 Sparsely vegetated areas
		334 Burnt areas
		335 Glaciers and perpetual snow

(continued)

Table 1 (continued)

Level 1	Level 2	Level 3
4 Wetlands	41 Inland wetlands	411 Inland marshes
		412 Peat bogs
	42 Maritime wetlands	421 Salt marshes
		422 Salines
		423 Intertidal flats
5 Water bodies	51 Inland waters	511 Water courses
		512 Water bodies
	52 Marine waters	521 Coastal lagoons
		522 Estuaries
		523 Sea and ocean

Source Corine Land Cover Nomenclature 2011

sensor, and they used data downloaded from the OSM project as field measurements to select the most optimal training sites. They performed a comparison of the resultant LU map with the Global Monitoring for Environment and Security Urban Atlas (GMESUA) map achieving a Kappa index of 89 %, which proves that OSM data is suitable to use as a source for training site definition. They also stress that the quality of VGI is heterogeneous and location-dependent, and they recommend checking the amount of contributions and also considering other VGI data, such as Flickr photos.

Another study investigated a new approach to generating land-use patterns from VGI without applying remote-sensing techniques and/or engaging official data (Jokar Arsanjani et al. b). Using OSM datasets and Vienna, Austria, as the study area, the authors applied a Hierarchical GIS-based decision tree approach to classify and segment parcels. The results were evaluated by conducting a texture-variability analysis of the LU maps generated using each dataset, and producing a confusion matrix to compare each LU class in the two datasets. Results of the texture analysis showed that the LU patterns derived from OSM data are richer than those derived from GMESUA. The confusion matrix showed a high level of agreement between the two classifications but this decreased when we move from level 1 towards the more detailed level 3. Although they conclude that VGI can be a potential data source for mapping LU patterns, they only used one source of VGI, OSM, and they did not test any other sources. Nevertheless, they pointed out as advantages of such an approach that no inputs from remote-sensing or any other administrative data were used, no financial cost exists as the OSM data is freely available and no field work was required, a number of incorrectly labeled features in the GMESUA were identified when OSM was incorporated, and the process of updating LU maps is facilitated due to the updating rate of OSM while GMESUA requires time and high financial costs to be updated by authorities.

A different approach was previously proposed by Hagenauer and Helbich (2012). They applied Artificial Neural Networks (ANNs) and Genetic Algorithms

(GAs) as a machine learning methodology to delineate continuous urban areas using all the information diversity of OSM, where a large set of potential OSM attributes was derived for inductive learning. Using OSM and GMESUA data, they applied this methodology to 42 randomly selected GMESUA urban regions and analyzed the significance of the attributes used and the performance of the model. The model performed comparatively well for most regions, with a few remarkable exceptions. The study shows that if enough OSM data for reasoning is present, urban patterns can be predicted to a large extent. This approach could be very useful to help map continuous fabric classes, from OSM data, for LULC databases.

The representation of natural features in OSM was also explored by Mooney et al. (2010), who examined the level of detail present in the representation of such polygon features. They tried to verify if there was enough detail in the representation of those features to provide a high-quality spatial representation. They used data for Austria, Estonia, Switzerland, Bretagne, Lower Saxony, Iceland, Ireland, and Scotland to calculate the statistical distribution of the mean distance between connected vertices of polygons. They found that many of the features are under-represented, with a small number of vertices used to delineate them, while some of them might be considered over-represented (e.g. small urban green spaces and golf courses). Some OSM data collection characteristics, such as the different GIS skill levels of OSM volunteers or the differences in accuracy of equipment and methods used, influence the under-representation of some features. These under-represented features have a serious impact on using OSM data in certain Earth science applications, mainly those that use OSM as ground-truth data. They recommend that the quality of the OSM representation of "natural" polygons and other features should be established against a recognized ground-truth dataset.

In this sense, other authors have been exploring the quality of OSM data that are of interest for LULC database production. Barron et al. (2014) developed a comprehensive framework for intrinsic OSM quality analysis that included the logical consistency of "natural" and "landuse" polygons. They developed a tool to generate information about OSM data quality for a selectable area without a reference dataset but using only OSM's data history. This tool intends to help users to assess the OSM data quality of a given area for a specific application. As an example, for map applications such as LULC database production, the tool automatically identifies erroneously overlapping land use polygons and analyzes not only the equidistance between the polygons' adjacent vertices, which is a good way to determine the quality of those polygons, but also the evolution of their equidistance over time.

Methods to analyze the completeness of building footprints over space and time were described and analyzed by Hecht et al. (2013) for the German states of North Rhine-Westphalia and Saxony. They used unit-based and object-based methods to analyze the level of completeness of building footprints contained within OSM by always comparing them with a reference dataset regarded as complete. They conclude that unit-based methods require less computation but have limitations in their level of detail when compared with object-based methods. Their results in applying these methods to the mentioned areas of Germany showed that OSM building footprints, as of November 2012, are characterized by a low degree of completeness, below 30 %, and a strong geometrical heterogeneity, and the level of completeness is higher in urban than in rural areas.

A similar study for the German city of Munich was developed by Fan et al. (2014). In this study the authors developed a quality assessment of building footprint data, after they found that the number of buildings in OSM was over 77 million on 5 May 2013. Building footprints were assessed using four criteria: (1) completeness, (2) semantic accuracy, (3) position accuracy, and (4) shape accuracy, where OSM data were compared with the reference data from the German Amtliches Topographisch-kartographisches Informatiosystem—Authorative Topographic-Cartographic Information System (ATKIS) to perform a quantitative assessment. They concluded that, for the case study of Munich (Germany), a high level of completeness was found but OSM building footprints still lack important attributes such as name, type, and height, among others. They found, however, more than 1,200 newly constructed buildings which were not documented in the ATKIS data. On the other side, although OSM building footprints are very similar in terms of shape, they have on average a 4 m offset to their corresponding ones in ATKIS in terms of position accuracy. Building footprints might be an important source of information to help in the classification or validation of urban areas, and these results are a very good indicator. Jokar Arsanjani and Vaz (2015) analyzed the completeness and thematic accuracy of seven European metropolises and thanks to the promising accuracy values concluded that these parameters greatly vary from location to location, which confirms the heterogeneity of contributions.

3 Materials and Methods

In this chapter we introduce the practical part of this study, which is an advance on our previous studies in exploring the suitability of OSM data for LULC purposes (Estima and Painho 2013a). In this case, the Points of Interest dataset was explored. We start by presenting and describing the study area and data used for this study, and explaining the methodology used to accomplish our objective.

3.1 Study Area and Data

The defined study site is continental Portugal, located on the southwestern side of Europe covering a total area of 8,908,220.16 Ha. The land cover is mainly composed of agricultural and forest areas covering around 95 % of the country.

The OSM database under analysis covers the area of continental Portugal and was downloaded from the Geofabrik website (Geofabrik 2014). This database is current as of 23 July 2013, and is divided into six datasets: places, points, railways, roads, waterways, buildings, landuse, and natural areas. Places and points are represented by point geometries; railways, roads and waterways by line geometries;

and buildings, landuse and natural areas by polygon geometries. As already mentioned, moving further in our research, the points dataset was used in this study.

The CLC database is composed of version 16 (04/2012) of Corine Land Cover (CLC) for the CLC2006 inventory, downloaded from the European Environment Agency (EEA 2014). This dataset, in vector format, was developed using the European Terrestrial Reference System 1989 (ETRS89) with the Lambert Azimuthal Equal Area, also known as ETRS89-LAEA. Using a Minimum Mapping Unit (MMU) of 25 Ha, the land cover is classified according to the CLC nomenclature, which is hierarchically divided into three levels of classes, as shown in Table 1. For the purpose of this investigation we used the five classes from level one: (1) artificial surfaces (AS), (2) agricultural areas (AA), (3) forests and semi-natural areas (F), (4) wetlands (W), (5) water bodies (WB).

3.2 Methods

The methodology adopted to conduct this analysis was as follows:

1. We explored the point dataset defined in the previous section in terms of content and coverage;
2. We established a relationship between each point type and the CLC classes, based on their description documented on the OSM Map Features website (OpenStreetMap Map Features 2014);
3. For each point location, we compared the classification given in the previous step with the respective class extracted from the CLC database, using a confusion matrix approach. We also analyzed the classification accuracy for each OSM point type.

4 Results and Discussion

In this chapter we present and discuss the results obtained by applying the methodology described in the previous section.

4.1 Analysis of the OSM Dataset

In this first step we explored the point dataset in terms of content and coverage. This data are composed of a collection of 49,861 Points of Interest (PoI) within the study area, classified according to type of PoI. A list of predefined types is available for use when a new point is registered (OpenStreetMap Map Features 2014), but each user can also define new types. Although this possibility gives a lot of flexibility in the mapping and classification process, it creates additional difficulties to perform further analysis, mainly related to the lack of proper descriptions but also to the possibility of introducing spelling errors.

Table 2 shows a list of POI types found within the collection of points. A closer look shows some types that are not of interest for the purpose of our study, mainly because they do not represent any type of LULC or related feature, or the relation is not clear (e.g. "attraction", "heritage", "no", "yes"). Different spelling for the same type were also found (e.g. "community_centre", "comunity centre", and "Comunity_centre"), a typical error related to the possibility of users creating their own types. Taking into account the description available for each feature type, and only for those types available in the wiki list, the types marked with asterisk (*) in Table 2 were considered attributable to a CLC class and selected for further analysis. This represents a total of 26,290, corresponding to around 52 % of the total number of initial points.

Figure 1 shows the spatial distribution of the selected PoIs over the study area. It is possible to observe the concentration of points over the coast, where touristic places and larger cities are represented, as well as along some of the main roads.

4.2 Correspondence Between OSM Point Types and CLC Classes

After selecting the types of PoI to use in the previous task, a CLC equivalent class was attributed to each type according to their description in the wiki website. Only two CLC classes were used: classes 1 and 5, representing AS and WB, respectively. This was already expected due to the higher probability of more volunteers visiting places fitting in these classes. There were some special cases where we also took into account our knowledge of the feature type class versus their surroundings. The case of the "bridge" feature type, which would apparently be classified as Artificial Surfaces, was classified as Water Bodies since bridges are usually over water bodies and are not represented in LULC databases due to their size. Table 3 shows the list of PoI types for each given CLC level 1 class.

4.3 Classification Accuracy Analysis

After assigning a CLC level 1 class to each PoI type, the evaluation of the classification was the next step. In this task we first filled the PoI dataset with the CLC class, based on the correspondence defined in the previous step. We then intersected it with the CLC database to have, for each point location, the classification defined by the PoI description and the classification taken from the CLC database. A new attribute was created to identify agreements/disagreements between the two

Table 2 List of types of OSM PoIs

arts_centre*	charging station*	flagpole	marketplace*	reservoir*	tertiary
adit	charging_station*	food_court*	mast	reservoir_covere*	tertiary_link
alpine_hut	chimney	footway	measurement_stat	residential*	theatre*
animal_shelter	cinema*	ford*	megalith	resort	theme_park*
antenna	city_gate*	forester's lodge	memorial	rest_area*	toilets
archaeological_s	clinic*	fort	milestone	restaurant*	tower
artwork	clinica fisiote	forte de sao jo	mineshaft	road*	townhall*
ashtray	clock	fountain*	mini_roundabout*	ruins	track
atm*	college*	fuel*	moinho do cuco	satellite_centre	traffic_signals
attraction	communications_t	gasometer*	monument*	school*	traffic-signs
baby_hatch	community_centre*	gate	motel*	scout_hut	trail_riding_sta
bank*	comunity centre*	give_way	motorcycle_parki	secondary	tram_stop*
bar*	comunity_centre*	grave_yard	motorway_junctio*	seguranca socia	trunk_junction
battlefield	conference_centr	guest_house	museum*	service	turning_circle*
bbq	construction	halt	newspaper*	services*	turntable*
beacon	convent	health	newstand	shelter	undefined
beauty	courthouse*	health_centre	nightclub*	shop*	university*
bed & breakfast	coutada	healthcare	no	shower*	user defined
bench	crane*	heritage	nursing_home*	silo*	vending_machine
biblias e casa	critpy	horses	oil_tank	snack_bar	veterinary*
bicycle_parking	cross	hospital*	old_cafe	social_centre*	viewpoint
bicycle_rental	crossing	hostel*	optical	social_facility*	waste_basket
biergarten	dentist*	hotel*	park*	solicitor	waste_deposal
boundary_stone	disused	hunting_stand	parking*	souvenirshop	waste_disposal
bridge*	diving_center	ice_cream*	parking_entrance*	spa	waste_dispostal

(continued)

Table 2 (continued)

brothel	doctor*	icon	parking_space*	spa	wastewater_plant*
buffer_stop	doctors*	incline	passing_place	speed_camera*	water_tank
buoy	drinking_water	incline_steep	path	sport clube leir	water_tower*
buoy	driving school	incline_up	pharmacy*	station*	water_well
bus_station*	driving_school	info	picnic_site	steps	water_works*
bus_stop	elevator	information	pier*	stop	waterfall
café*	embassy*	junction	pillar buoy	storage_tank	watering_place
cairn	emergency_access	kindergarten*	place_of_worship*	street_lamp	watermill
caixa geral de d	emergency_phone	laboratory	police*	studio*	wayside_cross
camp_site	escola superior	landmark	post_box	subway entrance*	wayside_shrine
camping park	ev_charging*	lavoir	post_office*	subway_entrance*	wifi
capela	farmacia	lawyer*	posto abastecime	survey point	wind_turbine
car_rental*	fast_food*	leisure_centre	primary_link	survey_pillar	windmill
car_wash*	ferry_terminal	level_crossing*	prison*	survey_point	works*
chalet	fitness_center	lookout_tower	register_office*	telephone	
caravan_site	fire_hydrant*	library*	pub*	swimming_pool*	yes
castle*	fire_station*	lift	public_building*	taxi	zoo*
cemiterio	first_aid	lighthouse	recycling	teahouse	

Legend types marked with asterisk (*) were considered attributable to a CLC class and selected for further analysis

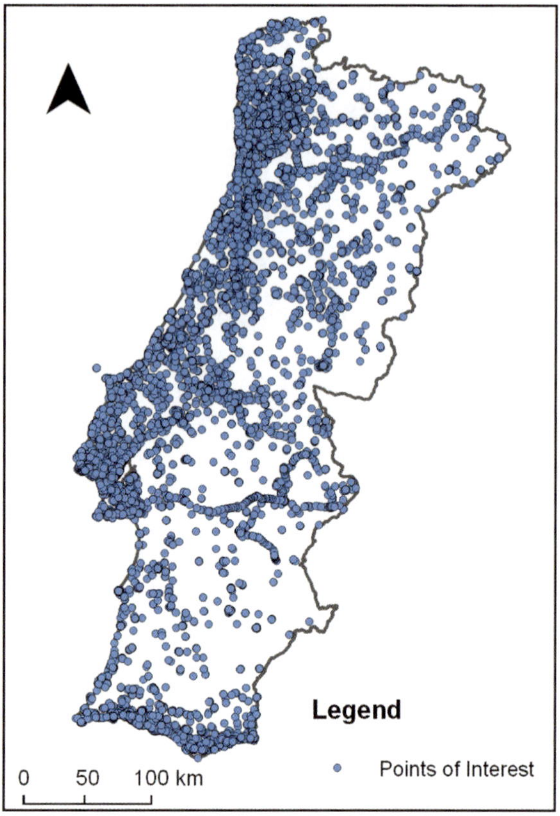

Fig. 1 Spatial distribution of the points of interest over the study area

classifications. This agreement/disagreement is depicted, along with their spatial distribution, in Fig. 2. Red points represent locations where both classifications are not matching and green points represent locations where both classifications are equal.

Table 4 summarizes the classification of the OSM point accuracy. Points classified as and WB classes obtained 77.96 and 1.47 % correct classification, respectively, when compared with the CLC classification for the same locations. One of the reasons for the poor result of the WB class might be related with the MMU of 25 Ha of the CLC database. It is natural that body areas of small dimension do not represent the predominant class when using such a MMU value.

Finally we analyzed the classification accuracy for each OSM point type. In Table 5, each PoI type is classified according to its range of accuracy. This is important to understand the suitability of each OSM PoI type to use in LULC databases. The lower accuracy of some OSM point type might be also related with the MMU. A "rest_area", for instance, might be located within a forest crossed by a motor way. In the same way, a "water_tower" might be located within an area where another class is predominant.

Table 3 CLC classes given to each PoI type

Class agricultural areas (AA)					Class water bodies (WB)
arts_centre	crane	lawyer	post_office	subway entrance	bridge
atm	dentist	level_crossing	prison	subway_entrance	ford
bank	doctor	library	pub	swimming_pool	pier
bar	doctors	marketplace	public_building	theatre	reservoir
beauty	embassy	mini_roundabout	register_office	theme_park	
bus_station	ev_charging	monument	residential	townhall	
cafe	fast_food	motel	rest_area	tram_stop	
car_rental	fire_hydrant	motorway_junctio	restaurant	turning_circle	
car_wash	fire_station	museum	road	turntable	
castle	food_court	newspaper	school	university	
charging station	fort	nightclub	services	veterinary	
charging_station	fountain	nursing_home	shop	wastewater_plant	
cinema	fuel	park	shower	water_tower	
city_gate	gasometer	parking	silo	water_works	
clinic	hospital	parking_entrance	social_centre	works	
college	hostel	parking_space	social_facility	zoo	
community_centre	hotel	pharmacy	speed_camera		
comunity centre	ice_cream	place_of_worship	station		
courthouse	kindergarten	police	studio		

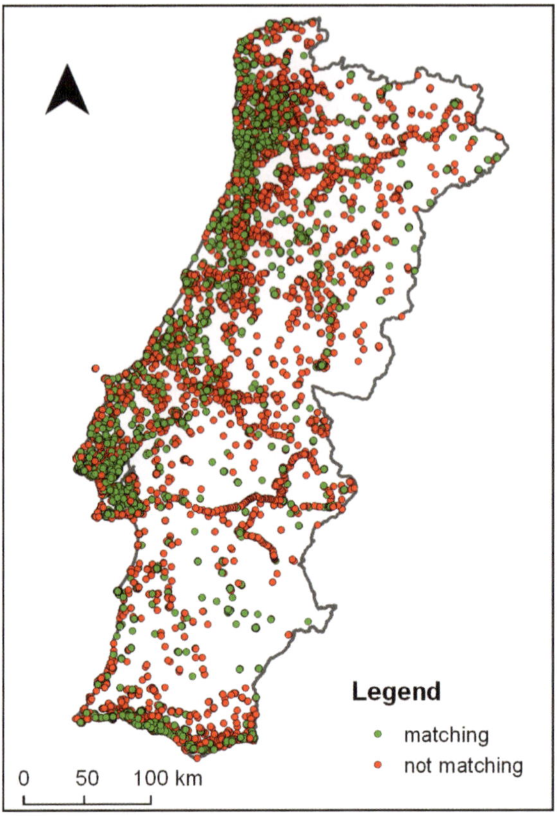

Fig. 2 PoI type class versus CLC class

Table 4 Classification of OSM points

		Classification based on OSM points	
		AS	WB
CLC classes containing the point locations	AS	20,421	1
	AA	4,110	46
	F	1,556	20
	W	19	0
	WB	85	1
Total		26,191	68
Correct (%)		77.96	1.47
Wrong (%)		22.03	98.53

Table 5 Classification accuracy by PoI type

Accuracy classes (%)						
0–50	50–60	60–70	70–80	80–90	90–100	100
water_tower	place_of_worship	works	clinic	fire_station	cinema	charging station
castle	social_facility	station	townhall	parking_space	bank	charging_station
rest_area	speed_camera	motel	hotel	car_wash	courthouse	comunity centre
motorway_junctio	water_works	city_gate	museum	hospital	university	doctor
zoo	silo	food_court	parking	bus_station	pharmacy	embassy
level_crossing	monument		mini_roundabout	nightclub	veterinary	ev_charging
pier	fire_hydrant		swimming_pool	arts_centre	theatre	fort
theme_park	residential		turning_circle	kindergarten	police	ice_cream
gasometer	studio		nursing_home	crane	car_rental	lawyer
services			fuel	public_building	post_office	newspaper
wastewater_plant			fountain	cafe	library	park
beauty			hostel	pub	dentist	parking_entrance
bridge				fast_food	doctors	prison
ford				school	marketplace	register_office
reservoir				restaurant	atm	road
shower				tram_stop	college	shop
turntable				bar		social_centre
				community_centre		subway entrance
						subway_entrance

5 Conclusions and Future Research Directions

In this chapter we looked into what has been done, in terms of research, to use VGI, with particular emphasis on OSM data, for LULC database production and we extended our previous investigations in this area. The explored studies have shown that we are still at the beginning, and more research is needed to address the identified issues. Based on the literature, OSM data have been investigated in terms of their suitability to be used as a source of ground-truth or ground measurements for validation purposes, with some studies focusing on urban areas, one of the CLC level 1 classes. Other studies have tried to derive information from the set of unstructured attributes of OSM data.

In the practical part of this study, we explored the OSM PoI dataset and demonstrated its suitability for the purpose of helping in the LULC production process. From the total of 26,191 points classified as class 1, almost 78 % of them matched the classification of the CLC level 1 at the same location. For some point types, the classification accuracy was actually 100 %. In contrast, those points falling into the WB class did not correspond well to the Corine LC databases and therefore more research is needed to compare these results with others at different locations.

All the reviewed literature, along with the practical part, demonstrated the suitability of OSM data to be used in the process of LULC database production. The main contributions in this activity would be in validation or helping the classification process when needed (e.g. when the remote sensing images or aerial photographs are not clear in a particular location). This suitability is different for different classes and for different locations. Urban and touristic areas, for instance, are more likely to have more data available in comparison to other places. This phenomenon is related to many factors including the amount of people living in larger cities, the amount of people visiting touristic places, and the Digital Divide already reported by Goodchild (2008a), among others.

Future research needs to be oriented towards a higher level of detail regarding the nomenclature of CLC. More research is needed to find ways to use these data, not only from OSM but also from other VGI sources, within CLC levels 2 and 3. The integration of different sources would increase the quantity of available data, and in some cases the coverage, proving a way to have a certain level of uncertainty, at least among the available resources. It is also important to compare results among different countries and continents to understand if they are location-dependent. This would help understand whether the potential of using OSM data for LULC production can be generalized or not.

References

Barron C, Neis P, Zipf A (2014) A comprehensive framework for intrinsic OpenStreetMap quality analysis. Trans GIS. doi:10.1111/tgis.12073

Baulies X, Szejwach G (1997) Survey of needs, gaps and priorities on data for land use/land cover change research. Report presented at LUCC data requirements workshop. Barcelona, Spain, 11–14 Nov 1997

Caetano M, Mata F, Freire S (2006) Accuracy assessment of the Portuguese CORINE Land Cover map. Glob Dev Environ Earth Obs Space:459–467

Cihlar J (2000) Land cover mapping of large areas from satellites: status and research priorities. Int J Remote Sens 21(6–7):1093–1114. doi:10.1080/014311600210092

Clark ML, Aide TM (2011) Virtual interpretation of earth web-interface tool (VIEW-IT) for collecting land-use/land-cover reference data. Remote Sens 3(3):601–620. doi:10.3390/rs3030601

Corine Land Cover Nomenclature (2011) Corine land cover nomenclature illustrated guide. In: Joint meeting Geoland2—EAGLE. Málaga, Spain, 23–24 June. http://sia.eionet.europa.eu/EAGLE/EAGLE_6thMeeting_g2_Malaga/04d_Nomenclature_CLC.pdf. Accessed 5 Jun 2014

Ellis E (2013) Land-use and land-cover change. In: The encyclopedia of earth. http://www.eoearth.org/view/article/51cbee4f7896bb431f696e92. Accessed 10 May 2014

Elwood S, Goodchild MF, Sui DZ (2012) Researching volunteered geographic information: spatial data, geographic research, and new social practice. Ann Assoc Am Geogr 102(3):571–590. doi:10.1080/00045608.2011.595657

Estima J, Painho M (2013a) Exploratory analysis of OpenStreetMap for land use classification. In: Proceedings of the second ACM SIGSPATIAL international workshop on crowdsourced and volunteered geographic information—GEOCROWD '13. ACM Press, pp 39–46. doi:10.1145/2534732.2534734

Estima J, Painho M (2013b) Flickr geotagged and publicly available photos: preliminary study of its adequacy for helping quality control of corine land cover. In: Murgante B, Misra S, Carlini M, Torre CM, Nguyen HQ, Taniar D, Gervasi O (eds) ICCSA 2013: computational science and its applications. The 13th international conference on computational science and its Applications, Ho Chi Minh City, Vietnam, 24-27 June 2013 . Lecture notes in computer science, vol 7974. Springer, Heidelberg, pp 205–220. doi:10.1007/978-3-642-39649-6_15

Estima J, Painho M (2014) photo based volunteered geographic information initiatives: a comparative study of their suitability for helping quality control of corine land cover. Int J Agric Environ Inf Syst 5(3):75–92. doi:10.4018/ijaeis.2014070105

European Environment Agency (2014) http://www.eea.europa.eu/data-and-maps/data/clc-2006-vector-data-version-2. Accessed 5 Jun 2014

Fan H, Zipf A, Fu Q, Neis P (2014) Quality assessment for building footprints data on OpenStreetMap. Int J Geogr Inf Sci 28(4):700–719. doi:10.1080/13658816.2013.867495

Flickr (2014) https://www.flickr.com/. Accessed 5 Jun 2014

Foody GM, Boyd DS (2013) Using volunteered data in land cover map validation: mapping West African forests. IEEE J Sel Top Appl Earth Obs Remote Sens 6(3):1305–1312. doi:10.1109/JSTARS.2013.2250257

Fritz S, McCallum I, Schill C, Perger C, Grillmayer R, Achard F, Obersteiner M (2009) Geo-Wiki.Org: The use of crowdsourcing to improve global land cover. Remote Sens 1(3):345–354. doi:10.3390/rs1030345

Fritz S, McCallum I, Schill C, Perger C, See L, Schepaschenko D, Obersteiner M (2012) Geo-Wiki: an online platform for improving global land cover. Environ Model Softw 31:110–123. doi:10.1016/j.envsoft.2011.11.015

Geofabrik (2014) http://www.geofabrik.de/. Accessed 5 Jun 2014
GMapCreator (2014) http://www.bartlett.ucl.ac.uk/casa/latest/software/gmap_creator. Accessed 5 Jun 2014
Goodchild M (2007) Citizens as sensors: the world of volunteered geography. GeoJournal 69 (4):211–221. doi:10.1007/s10708-007-9111-y
Goodchild M (2008a) Assertion and authority: the science of user-generated geographic content. In: Proceedings of the Colloquium for Andrew U. Frank's 60th birthday. Department of Geoinformation and Cartography, Vienna, Austria
Goodchild M (2008b) Commentary: whither VGI? GeoJournal 72(3–4):239–244. doi:10.1007/s10708-008-9190-4
Goodchild M, Glennon JA (2010) Crowdsourcing geographic information for disaster response: a research frontier. Int J Digit Earth 3(3):231–241. doi:10.1080/17538941003759255
Google MyMaps (2014) https://www.google.com/maps/d/. Accessed 5 Jun 2014
Hagenauer J, Helbich M (2012) Mining urban land-use patterns from volunteered geographic information by means of genetic algorithms and artificial neural networks. Int J Geogr Inf Sci 26(6):963–982. doi:10.1080/13658816.2011.619501
Hecht R, Kunze C, Hahmann S (2013) Measuring completeness of building footprints in OpenStreetMap over space and time. ISPRS Int J Geo-Inf 2(4):1066–1091. doi:10.3390/ijgi2041066
Hollenstein L, Purves R (2010) Exploring place through user-generated content: using Flickr to describe city cores. J Spat Inf Sci 1(1):21–48. doi:10.5311/JOSIS.2010.1.3
Holone H, Misund G, Holmstedt H (2007) Users are doing it for themselves: pedestrian navigation with user generated content. In: International conference on next generation mobile applications, services and technologies. IEEE, pp 91–99. doi: 10.1109/NGMAST.2007.4343406
Hudson-Smith A, Batty M, Crooks A, Milton R (2009) Mapping for the masses: accessing web 2.0 through crowdsourcing. Soc Sci Comput Rev 27(4):524–538. doi:10.1177/0894439309332299
Jokar Arsanjani J, Helbich M, Bakillah M (2013a) Exploiting volunteered geographic information to ease land use mapping of an urban landscape. In: International archives of the photogrammetry, remote sensing and spatial information sciences. 29th Urban data management symposium, vol XL-4/W1. London, United Kingdom, 29–31 May 2013
Jokar Arsanjani JJ, Helbich M, Bakillah M, Hagenauer J, Zipf A (2013b) Toward mapping land-use patterns from volunteered geographic information. Int J Geogr Inf Sci 27(12):2264–2278. doi:10.1080/13658816.2013.800871
Jokar Arsanjani J, Vaz E (2015: in-press) An assessment of a collaborative mapping approach for exploring land use patterns for several European metropolises. Int J Appl Earth Obs Geoinf
Leung D, Newsam S (2010) Proximate sensing: Inferring what-is-where from georeferenced photo collections. In: Conference on computer vision and pattern recognition CVPR. IEEE, San Francisco, CA, pp 2955–2962, 13–18 June 2010. doi:10.1109/CVPR.2010.5540040
London Profiler (2014) http://128.40.111.250/casa/websites/profiler.asp. Accessed 5 Jun 2014
MapTube (2014) http://www.maptube.org/. Accessed 5 Jun 2014
Mooney P, Corcoran P, Winstanley A (2010) A study of data representation of natural features in OpenStreetMap. In Proceedings of the 6th GIScience international conference on geographic information science, vol 150. Zurich, Switzerland, 14–17 Sept 2010
OpenStreetMap Map Features (2014) http://wiki.openstreetmap.org/wiki/Map_Features. Accessed 5 Jun 2014
Perger C, Fritz S, See L, Schill C, Van Der Velde M, Mccallum I, Obersteiner M (2012) A campaign to collect volunteered geographic information on land cover and human impact. In GI Forum 2012: Geovizualisation, Society and Learning. pp 83–91

Turner AJ (2006) Introduction to neogeography. Sebastopol, CA
Wikimapia (2014) http://wikimapia.org/. Accessed 5 Jun 2014
Zook M, Graham M, Shelton T, Gorman S (2010) Volunteered geographic information and crowdsourcing disaster relief: a case study of the Haitian Earthquake. World Med Health Policy 2(2):6–32. doi:10.2202/1948-4682.1069

Using Crowd-Sourced Data to Quantify the Complex Urban Fabric—OpenStreetMap and the Urban–Rural Index

Johannes Schlesinger

Abstract To date, hardly any classification of the urban–rural continuum exists that is based on objective and reproducible criteria. This particularly applies to regions of the world where accurate and up-to-date geodata is scarce Therefore, an Urban–Rural Index (URI) was developed as a contribution to the theoretical debate about the spatiality of urban–rural gradients as well as to make use of the increasing amount of crowd-sourced data especially in traditionally data-scarce regions of the developing world. The URI was calculated based on two subindexes representing: (1) the kernel density of existing buildings derived from high-resolution satellite imagery and (2) the travel times from the city center calculated based on OpenStreetMap data. The advantage of this index over common categorizations of urban, periurban, and rural areas lies in its ability to quantify the spatial implications of urban morphology. This paper draws on the analysis of three study sites: Bamenda in Cameroon, Moshi in Tanzania, and Bangalore in India. The URI as a reproducible representation of the spatial complexity of the urban landscape and its surrounding areas has the potential to contribute to the understanding of urban development patterns. Furthermore, it is a time- and cost-effective way for municipal town planning institutions to increase their knowledge of past, current, and future urbanization trends in their respective areas of responsibility.

Keywords Urban–Rural Index (URI) · Urbanization · Building density · Travel time · Urban–rural continuum

1 Introduction

For the first time in human history, more people live in urban agglomerations than in the countryside (UN-HABITAT 2013). In the past, the global population distribution was dominated by the overwhelming majority of rural dwellers. However,

J. Schlesinger (✉)
Institute of Environmental Social Sciences and Geography,
University of Freiburg, Werthmannstr. 4, 79085 Freiburg, Germany
e-mail: johannes.schlesinger@geographie.uni-freiburg.de

J. Jokar Arsanjani et al. (eds.), *OpenStreetMap in GIScience*,
Lecture Notes in Geoinformation and Cartography,
DOI 10.1007/978-3-319-14280-7_15

urbanization processes lead to the reversal of those patterns. According to the UN (2014), more than half of the approximately seven billion people living on the planet reside in urban areas. The rapid increase in this population is thereby likely to continue for the decades to come (UN 2014; White et al. 2008). Urban areas in most countries in the western world are largely consolidated typically showing small urban growth rates (UN 2014). In the global south, however, the urbanization process is characterized by a dynamic increase in urban population with growth rates of up to 20 % (UN-HABITAT 2010, 2013). Settlement patterns are changing rapidly, even though the rural population is still in the majority. The African continent alone already has an urban population as large as that of North America (UN 2014). With a high natural increase in the urban population and ongoing rural–urban migration in many regions of the global south, a rapid transformation from rural to urban societies can be observed (UN DESA 2011). This trend is fuelled by economic development, an improved health care system, especially in the cities, but also rather soft factors, such as the often-desired modern lifestyle of the urban areas.

In most of the developing countries, urbanization tends to be rapid as a consequence of the high growth rates in most cities. According to UN-HABITAT (2010), some of these cities have quadrupled the area they occupy within just two decades. The changes in urban areas and the resulting settlement patterns, however, are largely context-specific and can differ significantly between different cities.

Urbanization research oftentimes is hampered by the lack of standardized, objective, reproducible, and spatially consistent definitions of urban, periurban, as well as rural areas.[1] Furthermore, census campaigns conducted to count the population are time-consuming and costly. In highly dynamic urban contexts of the developing world, census data tends to be outdated once it is published. Thus, statistics on population numbers and their distribution can only give an estimate, but hardly ever an appropriate reflection of the actual situation. In most cases, only a total population for a respective administrative area is provided, making more detailed intra-urban differentiation difficult. The aggregated data thereby impedes the analysis of spatial changes along the whole urban–rural gradient.

Administrative bodies in the developing world tend to be overwhelmed by the rapid and unregulated growth of urban agglomerations. One of the reasons for that is the lack of resources in terms of finances, personnel, education, and equipment. In many cases, however, municipal administration branches responsible for the strategic planning of urban development simply lack data on the current state of urban development and the underlying dynamics. Often, outdated survey data—sometimes decades old—derived from aerial imagery or expensive field campaigns is used to plan urban development. Unfortunately, urban administrations more commonly—if at all—react to current unplanned and unintended urban development than proactively planning the future layout of the cities.

[1] In Senegal and Malaysia, for example, settlements with more than 10,000 inhabitants are categorised as urban (UN 2001), while in Ethiopia, Liberia, or Cuba, a threshold of 2,000 people is applied (UN 2001; IFPRI and EDRI 2009).

OpenStreetMap (OSM) thereby has the potential to increase municipalities' capacities and data basis for more adjusted spatial planning. The completeness—hence its suitability for urban planners—of urban datasets, of course, depends on the number, activities, and reliability of OSM contributors. No general statement can be made regarding the overall reliability and completeness of OSM data in the global south due to the nature of crowd-sourced data. Some positive examples, however, illustrate the potential of volunteered geographic information. In Ouagadougou (Burkina Faso), Kathmandu (Nepal), Kibera Slum (Kenya), or Chennai (India), OSM data matches or even overtrumps proprietary platforms such as GoogleMaps or Yahoo Maps. In many cases, the collaboration between the local population (as contributors of raw data), local public agencies (as contributors of e.g. survey data and facilitators), and development institutions (as facilitators and funders) have been working together to increase the coverage of OSM data. In many of the smaller towns and cities in the global south lacking active OSM communities, capable public administrations, and external funding and knowledge transfer, however, OSM coverage remains sparse and is likely to remain sparse in the future.

There have been approaches in the past to efficiently analyze, quantify, and visualize the urban morphology and land use dynamics in urban and periurban contexts, some of them making use of OSM data. Most of these studies, however, were based on a remote sensing based approach, focusing on the (semi-)automatic classification of satellite images. Forkuor and Cofie (2011), for example, quantified the impact of urbanization on agricultural land as well as other land uses in and around Freetown, the capital of Sierra Leone, based on multi-temporal Landsat data. They unveiled a 140 % increase in the built-up area within about 25 years, showing that much of this increase took place in the periurban fringes. Their study has in common with most other remote sensing based analyses the focus on medium- to large-scale patterns and dynamics, a restriction that is inherent in the spatial resolution of coarse satellite imagery. It is the heterogeneous, small-scale character, however, that makes urban growth patterns so complex. Therefore, capturing them requires more detailed raw data. The analysis of the urban fabric based on a landscape metrics approach, as propagated by Eiden et al. (2000) and other studies, has also proven successful to a certain degree. However, there are significant limitations with regard to up- and down-scaling of the analyses, as pointed out by Uuemaa et al. (2009) and Mander et al. (2005). Thapa and Murayama (2010), for example, extended a remote sensing based approach with a Geographic Information System (GIS) component, adding significant value to their quantification of urban growth in Kathmandu. Similar studies were conducted by Alsharif and Pradhan (2014), Jokar Arsanjani et al. (2013), or Vermeiren et al. (2012). Yet, some studies remain at the case study level without drawing more general conclusions or providing a standardized and reproducible approach of measuring the urban fabric.

It is the holistic and comprehensive understanding of the current state of urban development that is lacking in many cases. However, it is a fundamental prerequisite for any appropriate spatial planning and implementation measures. In order to plan urban development under oftentimes unfavorable conditions, administrative bodies need to be informed with detailed and up-to-date information to support sustainable decision-making. Clinging to arbitrary and rough classifications of urban areas that hardly reflect the situation *on the ground* hinders that process. Therefore, there is a need for low-cost and inclusive approaches to generate current data. Furthermore, town planners need to be aware of *hot spots* of urban growth in order to appropriately target them and plan the future development accordingly. Even though information about these areas can be derived from coarse-resolution satellite data, a high-resolution dataset is desirable in this context to allow for locally adjusted planning.

In order to overcome the abovementioned limitations of previous studies, a comprehensive methodological approach to analyze and—similarly important—to visualize the urban fabric was developed in the course of this study. Based on high-resolution remote sensing data and OSM data, a URI was developed. The standardized approach was then applied to three case study sites in order to evaluate the potentials and shortcomings of the approach.

The objective of this paper is to contribute to the understanding of the complex urban fabric, especially in the context of rapidly growing cities in the global south. By providing a reproducible method to assess the current state of the urbanization process as well as the historical dimension, the aim is to replace commonly used arbitrary classifications of spatial entities along the urban–rural context with a more appropriate reflection of the actual situation. Furthermore, this study aimed at developing an easy-to-understand and transparent analysis and visualization of urban growth to allow for more informed and adjusted spatial planning in areas lacking any other data on the urban morphology. The final objective was to identify selected spatial characteristics of three case study sites based on the abovementioned analyses in order to draw more general conclusions on the shaping of urban agglomerations.

2 Materials and Methods

2.1 Study Sites

A total of three cities were selected for this study in order to allow for the evaluation of the methodology in different spatial contexts as well as to compare the metrics of urban morphology of multiple sites. Accordingly, site selection aimed at covering a wide spectrum of cities with regard to: total population, annual growth rate of the urban population, state of economic development and technology, settlement history, and administrative context. Furthermore, the availability of up-to-date and high-resolution remote sensing data and at least a rudimentary coverage of OSM data were necessary. Based on these selection criteria three cities were selected for the study: Bamenda, located in the Northwest Region of Cameroon; Moshi, situated on the foothills of Mt. Kilimanjaro in northern Tanzania; and Bangalore, the capital of the Indian state of Karnataka. Figure 1 and Table 1 provide an overview of the selected sites and their respective characteristics.

While all selected cities have the rapid nature of urban growth in common, they were intentionally chosen due to differing substantially in their population sizes and other parameters. This allows for the application of the developed methodology in different contexts and for the comparison of various study sites.

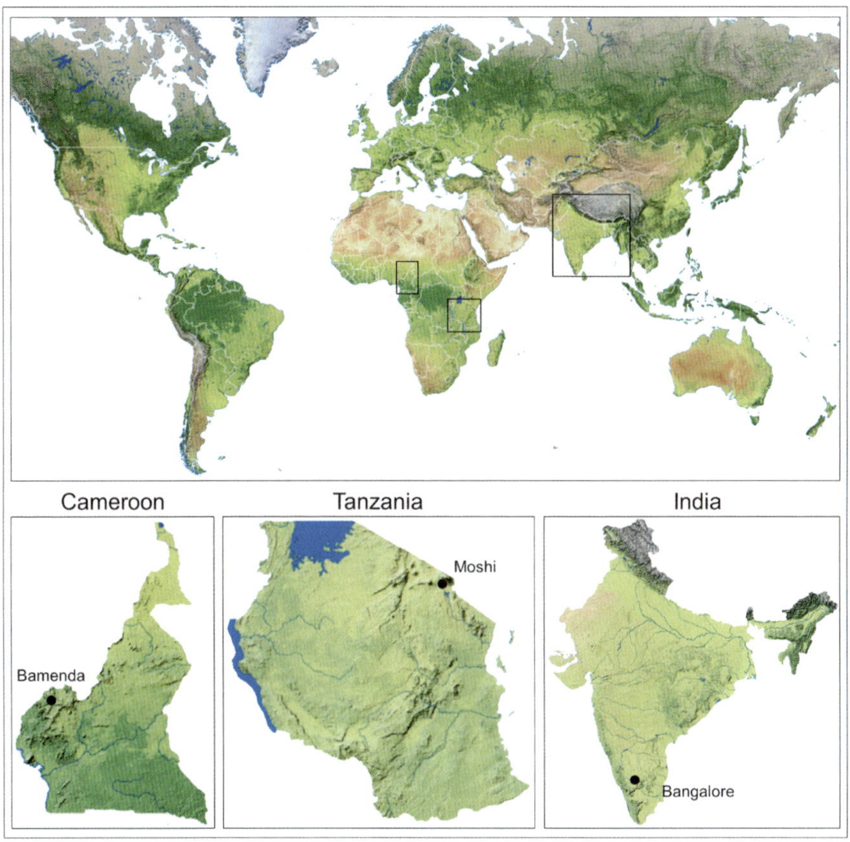

Fig. 1 Study site locations (*Data* CC BY-SA by OpenStreetMap.org, map layout: R. Hologa)

Table 1 Study sites and selected parameters

	Bamenda, Cameroon	Moshi, Tanzania	Bangalore, India
Total population	380,000 (2010)	183,000 (2009)	9,589,000 (2011)
Growth rate (in %)	8.0 (1987–2005)	2.8 (1998–2008)	4.7 (2001–2011)
Context	Important trading hub and service center for northwestern Cameroon	Transformation from coffee industry to tourism and regional service hub	Rapid development due to IT industry development from about the year 2000 onwards
Administrative steering capacities	Medium	Rather weak	Rather strong

Bureau Central des Recensements et des Etudes de Population (2010), Moshi Municipal Council (2010), Schlesinger (2013), Ramachandra et al. (2012), Mahendra et al. (2010)

2.2 Methods

Integrated and interdisciplinary approaches are necessary to capture the complexity of urban fabric shaped by a wide range of ecological, political, as well as societal processes (McIntyre et al. 2008; Knox and Pinch 2007; Schlesinger 2013). According to Coquery-Vidrovitch (1991), urbanization is a result of the interplay of these processes with numerous interlinkages and reciprocities with a strong temporal component. Geography with its strong focus on spatial processes and their manifestations on the earth's surface is "inherently interdisciplinary" (Baerwald 2010: 493) providing "a wide range of methods suitable to unravel the urban fabric and its fuzzy spatial and socio-economic patterns" (Schlesinger 2013: 81). This study concentrates on the spatiality of urbanization rather than the social implications of this process. Therefore, the URI developed in the course of this study is solely based on the integrative analysis of remotely sensed data as well as OSM data sets. The URI was thereby calculated based on two subindexes derived from the two different data sources. While the former allowed for an object-based classification of buildings and the calculation of building densities, the latter was used to calculate the travel times from the city center to any given point within the research areas. Eventually, both subindexes were combined and the resulting general index was used to quantify the complex morphology of the selected cities.

Various studies have applied remote sensing-based approaches in building extraction and building density calculation. The use of satellite imagery is therefore quite common in the classification of impervious surface or built-up areas. Several studies have applied respective methods to quantify the growth of urban areas during the last decades, such as Vermeiren et al. (2012) in the case of Kampala/ Uganda, Wakode et al. (2014) concentrating on Hyderabad/India, or a study by

Sarvestani et al. (2011) on Shiraz/Iran. In most of these case studies, free of cost Landsat MSS, TM, and ETM+ sensor data was used as the foundation for image classification. However, spatial resolutions of panchromatic data do not exceed 15 m, in the case of multispectral data not even 30 m. Such a spatial resolution, however, is not sufficient to appropriately detect small-scale urban structures. Yet, these structures, including single houses and huts, are of particular interest in the context of urban growth in the developing world that is in most cases dominated by urban sprawl. With regard to building density calculation, airborne laser-scanning techniques allowing for the collection of high-resolution got more attention in recent years. Yu et al. (2010), González-Aguilera (2013), and Brenner (2010) provided studies on the extraction of building densities based on the analysis of light detection and ranging (LiDAR) data. The acquisition of LiDAR data, however, is costly and time-consuming and therefore difficult in the case of a multi-city comparison. Jensen and Cowen (1999) suggest that building density analysis should be based on datasets with a spatial resolution of ≤0.25 to 5 m. Analysts should thereby aim at high spatial resolution rather than high spectral resolution, as long as a sufficient spectral contrast between the objects of interest (e.g. buildings) and the background (e.g. vegetation around buildings) is ensured.

A quantitative index solely based on building density, however, does not appropriately reflect the complex urban fabric in urban agglomerations of the developing world. By only concentrating on this—certainly important and easily quantifiable—aspect of urban morphology, the location of urban settlements is underrepresented. A periurban village or leapfrog-development projects, for example, can have the same building densities as the highly developed urban center. Yet, due to their specific location that is spatially separated from the high-density urban center with its entire central infrastructure (e.g. banks, hospitals, administration), these areas have a different level of *urbanity*. Partly due to the difficulties associated with the quantification of these levels, some quantitative urbanization studies have restricted themselves to building densities as the only indicator (e.g. Wu et al. 2011; Salvati et al. 2012). Therefore, a second subindex for accessibility represented by travel times from the city center was developed. It needs to be understood as an indirect proxy for urban influence at any given location. Especially in the developing world, accessibility of urban infrastructure in general and urban markets for agricultural products from the periurban and rural areas in particular is an important component of urban development. At least to a certain degree, von Thünen's (1910) concept of regional land use that was later adapted to an urban context (Park et al. 1925), is still valid in present development patterns. Ye and Van Nes (2013), Wang et al. (2013) showed that building density and street morphology are strongly correlated. Causalities do exist, yet are often disregarded. Nevertheless, substituting one indicator with the other would lead to a loss of information. Therefore, in this study both subindexes were combined.

OSM data has been used in several studies on the topic of urbanization. Moghadam and Helbich (2013), for example, used traffic infrastructure derived from the project's data as one of four variables to model urban growth of the city of Mumbai. Hagenauer and Helbich (2012) estimated urban land use patterns based on

OSM data, while Wurm et al. (2010), for example, used it to increase accuracy of their quantification of urban structures based on remote sensing data. Zollweg et al. (2012) simulated urban nightlight distribution based on OSM data as an indirect indicator for urban activities. Furthermore, Jokar Arsanjani et al. (2013) as well as Bakillah et al. (2014) introduced innovative solutions for land use and population mapping based on volunteered geographic information showing the potential and wide range of applications of OSM data in the field of human geography.

In this study, OSM data was used to calculate theoretical travel times. Unfortunately, the availability of comprehensive datasets on road infrastructure is rather limited, especially in the global south. In cases where those datasets exist, they often lack spatial reference, metadata, comprehensiveness, or a decent road classification scheme. In some cases, access to respective datasets is limited due to administrative restrictions. Automatically detecting road infrastructure in high-resolution remote sensing data is an option that has proven successful in past studies (e.g. Kavzoglu et al. 2009; Liu et al. 2013). However, the automatic derivation of the respective road class is difficult, especially in an urban context where the types of roads range from well-planned paved multi-lane trunk roads to small roads with loose gravel in informal settlements. Therefore, OSM data was used in the course of this study as it has significant advantages over the abovementioned datasets. OSM data is free of charge, easily accessible through the internet, highly standardized, and downloadable in several formats. Furthermore, it includes a road classification scheme that—despite the sometimes lacking attribution—provides a good foundation for road network analyses. The increasing amount of crowd-sourced data especially in traditionally data-scarce regions is an additional advantage. In some municipalities, OSM data has a better coverage of the roads than the official datasets used by the respective local road planning institutions. Some cities, such as Kathmandu for example, even establish partnerships with OSM in order to benefit from the increasing body of crowd-sourced data (Open Cities Project 2014).

2.2.1 Datasets and Data Preparation

Remote Sensing Data

As a compromise between cost-efficiency and spatial as well as spectral accuracy, high-resolution satellite imagery was used in this study as summarized in Table 2.

Furthermore, multi-temporal satellite images as well as aerial imagery were obtained for subsets of the study areas in order to allow for the analysis of the temporal dynamics of urban growth. The multispectral channels were combined in layer stacks before pan-sharpening algorithms were applied in order to obtain images with the spectral resolution of the multispectral images and a spatial resolution of the panchromatic channel.

Table 2 Remote sensing data used in this study

		Bamenda, Cameroon	Moshi, Tanzania	Bangalore, India
Sensor		Ikonos-2	GeoEye-1	SPOT-6
Acquisition date		2008-02-16	2010-10-18	2013-03-13
Area covered by the scene (km^2)		138	290	3,550
Spatial resolution	Panchromatic (m)	1.0	0.5	1.5
	Multispectral (m)	4.2	1.8	6.0
Spectral resolution	Blue (μm)	0.445–0.516	0.450–0.520	0.455–0.525
	Green (μm)	0.506–0.595	0.520–0.600	0.530–0.590
	Red (μm)	0.632–0.698	0.625–0.695	0.625–0.695
	Near IR (μm)	0.757–0.853	0.760–0.900	0.760–0.890
Radiometric resolution (bits)		11	11	12

OpenStreetMap Data

OSM data for this study was downloaded from Geofabrik GmbH (www.geofabrik.de) and the website OSM2GIS (www.osm974.re/osm2gis/). Datasets were downloaded for perusal in GIS software. As streets are represented by lines in OSM, only the respective line vector data was used for further analysis. While data from Geofabrik is provided in seven separate files based on content,[2] OSM2GIS data can be downloaded as three aggregated files based on the data geometry. Therefore, additional data cleaning was necessary in the latter case in order to extract road data from the line dataset. All datasets were assessed in terms of spatial accuracy and coverage of the respective area of interest. As coverage is still rather low in some developing countries, missing transport infrastructure was manually digitized based on recent satellite imagery and field observations. Then, all road segments were categorized based on their type. Accordingly, trunk roads were assigned to the highest category, other paved roads to the second, and all smaller (usually unpaved) roads to the third category. Hypothetical average velocities (Table 3) were defined for each class based on local expertise and measurements by the author.

As the second input variable for travel time calculation, the length of each road segment in m was calculated. Afterwards, the travel time per road segment in minutes was calculated as an additional attribute based on the following equation:

[2] The data is provided in separate files with different geometries: building footprints, land use categorisations, and natural features as polygons; populated places, points of particular interest as points; and railways, waterways, and roads as lines.

Table 3 Average velocities used for travel time calculation (based on Schlesinger 2013, edited)

Road category		Average velocity	
		km/h	m/min
1	Main roads (paved)	60	1,000
2	Other paved roads	40	667
3	Unpaved roads	10	167

$$t_x = \frac{d_x}{v_x}$$

where t_x is the time needed to pass road segment x in minutes, d_X represents the length of the road segment x, and v_x is the average velocity on the road segment x as defined above.

2.2.2 Building Density Quantification Based on Satellite Imagery

Building densities were calculated based on an object-based extraction of building footprints in the high-resolution scenes. An object-based approach was chosen as conventional pixel-based methods, including unsupervised classification and maximum likelihood supervised classification, have severe limitations when applied to high-resolution remotely sensed images. These methods have been developed for the analysis of coarse resolution satellite images, however, should not be applied to high-resolution data. One reason being that these classification methods fail to incorporate the high spatial content and associated information in the classification scheme (Blaschke and Strobl 2001). While pixel-based approaches only take into account the spectral characteristics of single pixels, object-based methods consider pixel regions as objects or features. As they evaluate pixels within their context, other variables such as shape and texture of the objects are more important than the mere spectral characteristics (Blaschke 2010; De Maeyer et al. 2010; Qian et al. 2007). Object-based classifications have been applied in a wide range of studies so far (e.g. Mathieu et al. 2007; Phinn et al. 2012; Benz et al. 2004). Some studies comparing this approach with conventional pixel-based methods showed significant advantages of the former with regard to the accuracy of the resulting classifications, particularly in the field of (semi-)automatic building extraction (e.g., De Maeyer et al. 2010; Freire et al. 2010; Myint et al. 2011).

In this study, Trimble eCognition 8.8 software package was used for the object-based segmentation and classification of the satellite images. Several (semi-)automatic extraction and conversion steps were necessary for the calculation of a standardized raster-based index value for each location within the study areas (Fig. 2).

As a first step, a Multiresolution Segmentation algorithm was applied to the original satellite image. The segmentation of buildings has proven particularly successful, as this method is especially capable of detecting features based on shape homogeneity in addition to their spectral characteristics. Secondly, training areas were defined as an input for a Nearest Neighbor Classification. Buildings were

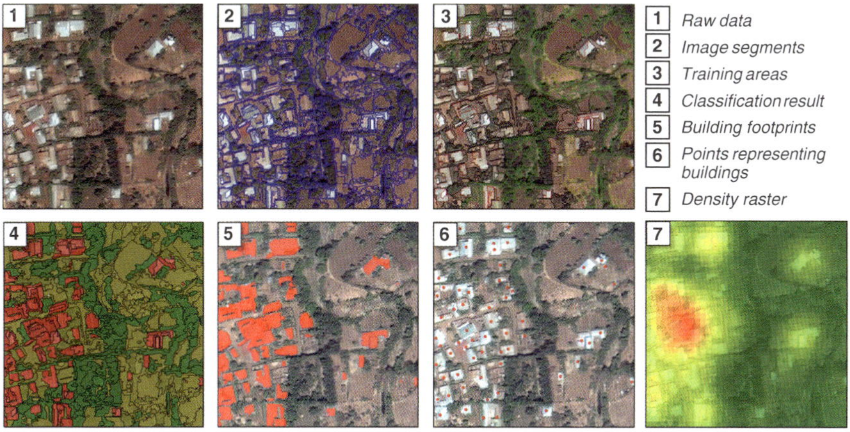

Fig. 2 Schematic representation of workflow of semi-automatic building extraction and density calculation (based on Schlesinger 2013, extended; DigitalGlobe/GeoEye-1 © 2010; OpenStreetMap.org)

selected as the positive output class, while the negative output class consisted of all other land cover classes within the areas of interest. It was ensured that all different kinds of buildings were covered, despite the wide range of shapes, textures, and colors appearing in the heterogeneous urban landscape (e.g. big houses with flat concrete roofs versus huts with corrugated metal roofs). Subsequently, all segments classified as buildings were exported as polygon features. For each building footprint, the total area was calculated before converting the polygons into point data. The resulting points were visually evaluated in terms of spatial accuracy in order to avoid errors in the successive processing steps.

As a comprehensive raster dataset covering all parts of the study sites was necessary for further analysis, a Weighted Kernel Density analysis was conducted based on the point data providing only spatially selective as well as unnormalized information on a limited number of locations. It describes a statistical method used to calculate smoothened estimates of probability densities based on an actual sample of observations (Gatrell et al. 1996). In spatial analysis, it is applied to transform georeferenced point data—in this case, the points representing the buildings—into a continuous and smoothly curved surface, considering a weight variable—here the area calculated for each building footprint (Kloog et al. 2009). In this case, a quadratic kernel function as described in Silverman (1986) was applied. A fixed bandwidth—or search radius—of 250 m was chosen based on expert knowledge as a compromise between smoothness and complexity (Van Kerm 2003). In the course of the analysis, a curved surface was fitted over each point, theoretically reaching zero at the search radius. The higher the value for each point representing the area of a building, the steeper the resulting curve. An output raster with a spatial resolution of 10 m was created, reflecting not only the density of buildings, but also the respective surface areas. As a result, higher values were calculated for larger buildings, such as big office complexes, and lower values for smaller buildings, such as individual huts or sheds. Highest values were reached in the city centers characterized by high population densities and large shares of built-up areas. Accordingly, lowest values were calculated for areas lacking buildings, such as the rural hinterland that is dominated by agricultural land use and unpopulated forests. Eventually, the raster dataset was normalized on a scale of 0–1, with the former indicating low and the latter high building densities.

2.2.3 Travel Time Calculation Based on OSM Data

For comprehensive travel time calculation for all areas of interest a transportation network was created using the Network Analyst. The files containing the information outlined above were used as source features and converted into a network dataset. The travel time per segment in minutes t_x was thereby used as the network cost/impedance. Based on building densities derived from the abovementioned subindex, expert knowledge, and field observations the city center was defined as the origin for travel time calculation. Trimmed service area polygons at 5-min intervals were calculated, representing the area that can be reached from the origin within a given time. The resulting polygon layers were assessed in terms of completeness and absence of unintended artefacts before they were converted into isolines. Eventually, an interpolation was executed based on these isolines in order to create a raster dataset of the whole study area with a spatial resolution of 10 m. Pixel values thereby represented the absolute travel times from the city center in minutes. The raster dataset was inverted and normalized on a scale of 0–1, similar to building densities to allow for smooth combination of both subindexes (Fig. 3).

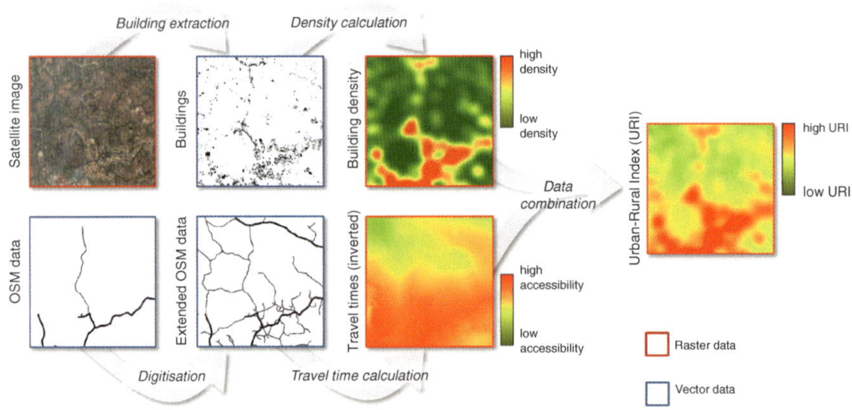

Fig. 3 Calculation of the URI based on building density and travel times (based on Schlesinger 2013, edited; data: DigitalGlobe/GeoEye-1 © 2010; OpenStreetMap.org)

Finally, the building density index raster and the travel time index raster were combined forming the URI. The output raster had a spatial resolution of 10 m and contained a single value between 0 (least urban) and 1 (most urban) for any given location within the study area. Even though URI values provided information about the relative level of urbanity rather than an absolute measurement, characteristics of the urban morphology of the study sites could be visualized and interpreted. Therefore, in a next step, the three cities were compared based on the histograms of the respective URI values. Furthermore, a multitemporal URI analysis was carried out for selected sites to quantify and visualize the urban growth dynamics of rapidly growing cities in the developing world. Building density values as well as road infrastructure information was derived from several satellite images covering the same spatial extent at different points in time. Eventually, differences in URI values were calculated.

3 Results and Discussion

Building densities, travel times, and the resulting URI were calculated for all study sites, as illustrated in Fig. 4. Highest building densities were typically reached in the cities' urban centers. However, in some cases, informal settlements showed similar or even higher building density values. Lowest values, in contrast, were calculated for intra-urban open spaces and—more commonly—for the rural hinterland. Travel times were obviously low in the urban centers due to the relative spatial proximity to the origin of travel time accumulation. Accordingly, more remote areas showed longer travel times. Areas along well-developed roads, however, appeared as linear *islands of accessibility*.

While the subindexes offered a detailed view of the individual variables, the URI provided a more comprehensive approximation of the actual state of urbanity by eradicating extreme values. The result was a smooth raster dataset, reflecting the morphology of the study sites. As expected, highest URI values were calculated for the urban core with a high building density and proximity to the city center. Adjacent to these urban areas, vast areas showing medium URI values could be identified. These areas could be interpreted as the heterogeneous periurban areas that did not belong to the urban core and were not yet purely rural. Three possible scenarios led to the visualization of transitional areas and their respective medium URI: medium values in both subindexes, high building densities, and low accessibility, or low building densities with good accessibility. Even though the resulting URI values were similar, the character of the areas could significantly differ, showing the heterogeneity of the periurban areas as described in previous studies (e.g. Drescher and Iaquinta 2002; Marshall et al. 2009). Consequently, lowest values were computed for the rural hinterland of the three cities, where building densities were very low and travel times from the city center long.

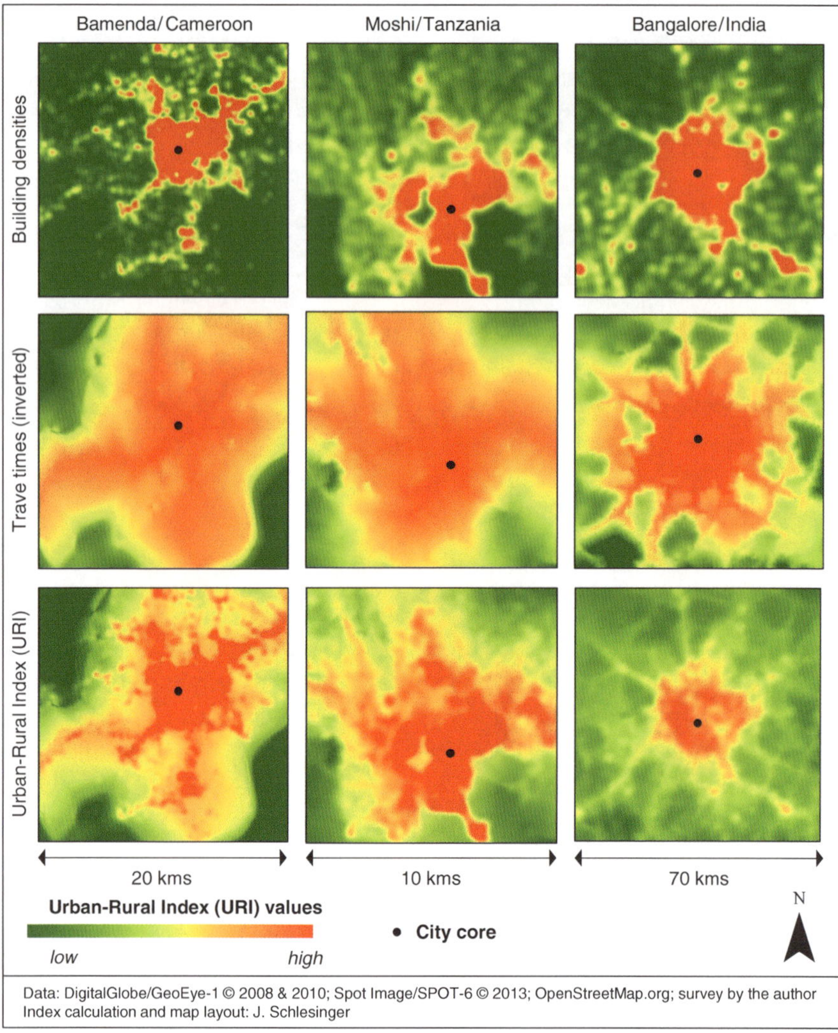

Fig. 4 URI and subindexes as calculated for all study sites based on remote sensing data and OSM street network

The URI visualization unveiled certain patterns in urban development. Especially in the case of Bangalore, roads could be clearly identified as development axes. The well-developed transportation system thereby often came with an above-average building density compared to the surrounding areas. Furthermore, isolated *local peaks* could be identified in some periurban and even rural areas, representing villages, leapfrog development projects, or sub-centers. Also, the influence of natural barriers for urban growth became visible. The south-western part of the Moshi study site, for example, was dominated by low URI values due to an unbridged river hindering urban development in the area. Simultaneously, the south-eastern area showed low values due to the existence of a protected forest that functions as a water reservoir. These examples show how the understanding of urban development patterns can be facilitated by such an analysis and the respective visualization.

The URI also allows for a distinction between more compact urban agglomerations and those dominated by scattered development. While steep declines in URI values were an indication for urban planning and urban zoning measures, a smooth transition from high to low values could be interpreted as a lack of planning resulting in heterogeneous land use patterns (Fig. 5). The analysis of these patterns could be operationalized based on the comparison of histograms.

While mean values did not differ significantly between the three datasets, the shape of the histograms was an indication for the differences in the urban morphology of the selected research sites. While the histogram of Bamenda data showed a rapid decrease in the frequency of URI values above the mean value—as indicated by the red line—the transition in the respective data from the other case study sites was less steep. Hence, there was a stark contrast between highly urbanized areas and those showing more rural characteristics in Bamenda. Moshi's surroundings, however, were dominated by scattered developments represented by a smooth transition between medium and high URI values. This can be seen as a quantification of Satterthwaite et al.'s (2010) hypothesis of scattered development. Bangalore's histogram showed a medium slope. Yet, higher URI values were more frequent there reflecting the bigger size of the urban area in terms of space and population.

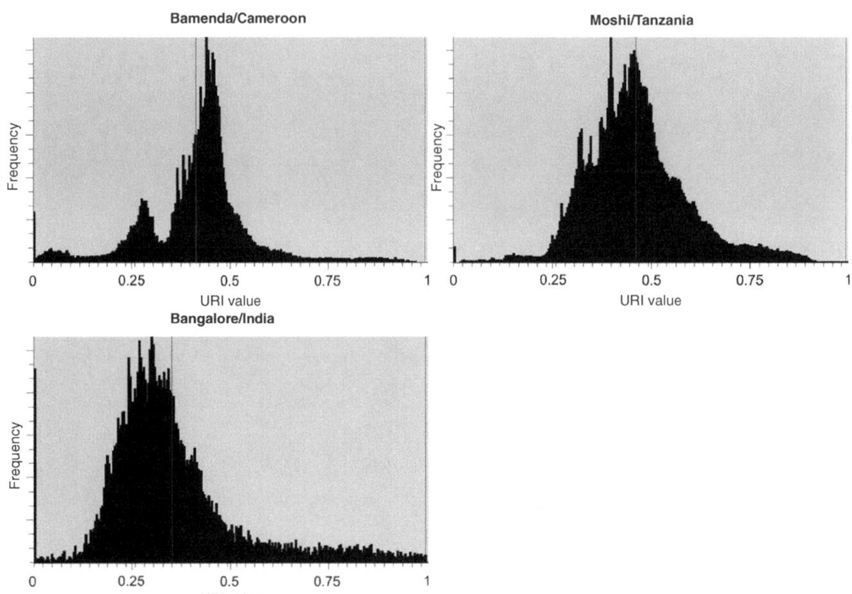

Fig. 5 Histograms of URI raster data of the three study sites

The characteristics of the histogram must be understood as a consequence of the complex interactions shaping the urbanization patterns. Pham et al. (2011: 230) state that "instruments related to land use in urban areas have a significant effect on the patterns and nature of urbanisation," meaning that there are strong reciprocal effects "between certain changes of spatial metric parameters and a particular type of city planning." However, it is not only the planning, but—probably even more importantly—the implementation of planning and the compliance with master plans and land zoning policies. Their absence typically leads to uncontrolled urban development in the form of scattered housing and informal settlements.

Due to a lack of resources of the respective administrative bodies, these patterns could be detected in the case of Moshi. Bamenda's planning institutions, in contrast, are rather assertive. Combined with natural features, such as an escarpment bordering the city in the south-west and a river hindering development in the north, this is a crucial factor for Bamenda's rather compact development despite the high population growth rates. In Bangalore, planned leapfrog development with isolated

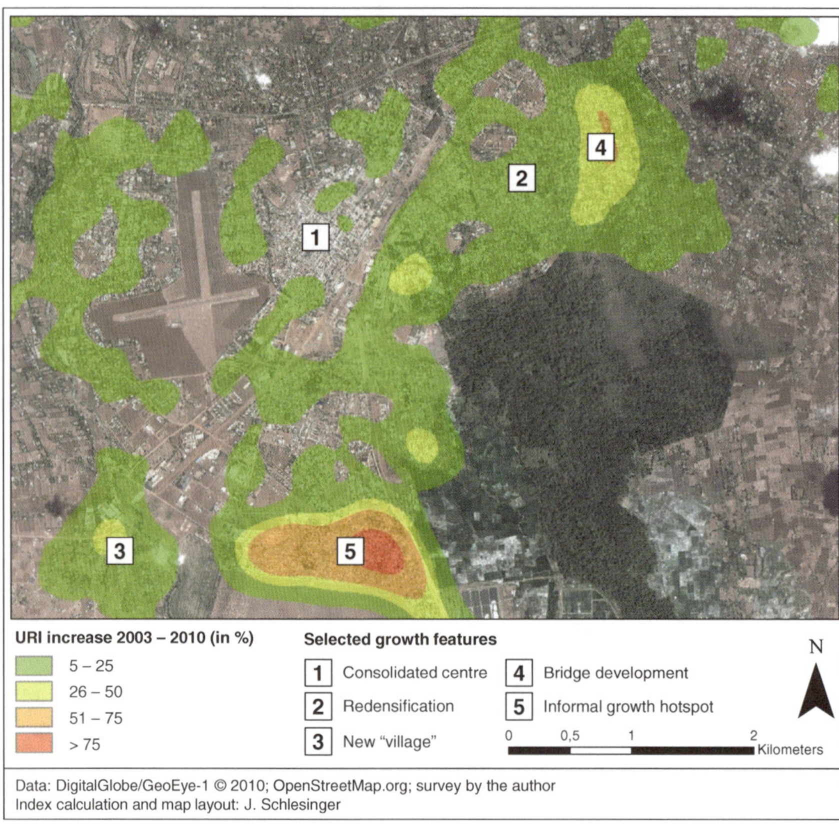

Fig. 6 Urban growth dynamics in selected parts of Moshi, Tanzania

clusters for residential, commercial, or industrial purposes mainly along the bigger roads dominates the scene.

The URI not only allows for the comparison of different cities, but also for the same areas at different points of time. Based on multi-temporal URI calculations for the Moshi study site, urban growth patterns were analyzed in detail. Figure 6 visualizes the quantification of urban growth in the area between the year 2003 and 2010. Several distinct patterns of urban growth could be identified based on this quantification.

Negligible changes in URI values were detected in the urban center (1) where there was little potential for further urban development. A slight increase due to redensification of residential areas could be identified just east of the city center. (2) The construction of a factory on the outskirts of the city (3) lead to the establishment of a new settlement showing village characteristics. One area (4) was experiencing significant linear growth between 2003 and 2010 due to the development of bridges over a river previously holding back urban development. The URI change analysis revealed a *hotspot of urban growth* (5) just south of the city center. Due to its proximity to the urban core, this area is highly attractive for housing development. However, the absence of appropriate planning measures lead to unregulated and uncontrolled informal development activities in this area, represented by a significant increase in the URI value.

4 Conclusions and Outlook

The URI calculation based on recent satellite imagery and OpenStreetMap data proved to be an appropriate approach of quantifying the complex urban fabric. By providing a reproducible method to assess the current state of the urbanization process as well as the historical dimension the URI could inform decision makers efficiently about the state of current urban spatial development as well as past dynamics. URI analyses can assist in predicting future growth patterns allowing for timely countermeasures in case of unintended developments. This particularly applies to municipalities in developing countries where administrative bodies including town planning departments lack resources in terms of data collection and implementation of planning measures. Even though data availability in the three case studies included in this research varied (e.g. different satellite sensors, spatial resolutions, coverage of OSM data), satisfactory results could be generated in a cost- and time-effective way. The availability of free of cost OSM data, especially in traditionally data scarce regions, needs to be emphasized as an important factor facilitating the analysis.

The study showed that cities' morphologies differ from one another and are largely context-specific. Furthermore, it was shown that arbitrary classifications of spatial entities along urban–rural gradients—such as urban, periurban, rural—come with substantial limitations in reflecting the complex urban patterns. Nevertheless, classifications based on absolute thresholds in the URI data might have to be developed in the future for the sake of simplification and operability. The definition of global URI maxima and minima could further increase its strengths in comparing urban agglomerations based on absolute values rather than relative data. More case study sites should be investigated in the future in order to operationalize the classification and differentiation of cities based on selected morphological metrics derived from a standardized approach. The analysis and visualization methods for urban growth quantification presented in this paper, however, are a starting point for further improvement of the understanding of past, current, and future urban development.

References

Alsharif A, Pradhan B (2014) Urban sprawl analysis of Tripoli Metropolitan city (Libya) using remote sensing data and multivariate logistic regression model. J Indian Soc Remote Sens 42 (1):149–163. doi:10.1007/s12524-013-0299-7

Baerwald TJ (2010) Prospects for geography as an interdisciplinary discipline. Ann Assoc Am Geogr 100(3):493–501. doi:10.1080/00045608.2010.485443

Bakillah M, Liang SHL, Mobasheri A, Jokar Arsanjani J, Zipf A (2014): Fine resolution population mapping using OpenStreetMap points-of-interest. Int J Geogr Inf Sci 28(9):1940–1963. doi:10.1080/13658816.2014.909045

Benz UC, Hofmann P, Willhauck G, Lingenfelder G, Heynen M (2004) Multi-resolution, object-oriented fuzzy analysis of remote sensing data for GIS-ready information. ISPRS J Photogram Remote Sens 58:239–258. doi:10.1016/j.isprsjprs.2003.10.002

Blaschke T (2010) Object based image analysis for remote sensing. ISPRS J Photogram Remote Sens 65(1):2–16. doi:10.1016/j.isprsjprs.2009.06.004

Blaschke T, Strobl J (2001) Angewandte Geographische Informationsverarbeitung XIII. In: Beiträge zum AGIT-symposium 2001, Salzburg

Brenner C (2010) Building extraction. In: Vosselman G, Maas H-G (eds) Airborne and terrestrial laser scanning. Whittles Publishing, Dunbeath

Bureau Central des Recensements et des Etudes de Population (2010) Rapport de présentation des résultats définitifs. Yaoundé

Coquery-Vidrovitch C (1991) The process of urbanization in Africa (from the origins to the beginning of independence). Afr Stud Rev 34(1):1–98. doi:10.2307/524256

De Maeyer M, Sotiaux A, Wolff E (2010) Comparison of standardized methods (object-oriented vs. pixel) to extract the urban built-up area: example of Lubumbashi (DRC). In: Proceedings of the GEOBIA 2010 conference, Ghent

Drescher AW, Iaquinta DL (2002) Urbanization—linking development across the changing landscape, Rome

Eiden G, Kayadjanian M, Vidal C (2000) Capturing landscape structures: tools. In: European Commission (ed) From land cover to landscape diversity in the European Union, pp 7–15. Brussels

Forkuor G, Cofie O (2011) Dynamics of land-use and land-cover change in Freetown, Sierra Leone and its effects on urban and peri-urban agriculture—a remote sensing approach. Int J Remote Sens 32(4):1017–1037. doi:10.1080/01431160903505302

Freire S, Santos T, Navarro A, Soares F, Dinis J, Afonso N, Fonseca A, Tenedório J (2010) Extraction of buildings from Quickbird imagery for municipal planning purposes. Quality assessment considering existing mapping standards. In: International Society for Photogrammetry and Remote Sensing (ISPRS) (ed.) GEOBIA 2010-geographic object-based image analysis. Ghent University, Ghent, Belgium, 29 June–2 July

Gatrell AC, Bailey TC, Diggle PJ, Rowlingson BS (1996) Spatial point pattern analysis and its implication in geographical epidemiology. Trans Inst Br Geogr 21(1):256–274. doi:10.2307/622936

Geofabrik GmbH (2011) OpenStreetMap-Daten download, Karlsruhe. Available via http://www.geofabrik.de. Accessed 20 Sept 2014

González-Aguilera D, Crespo-Matellán E, Hernández-López D, Rodríguez-Gonzálvez P (2013) Automated urban analysis based on LiDAR-derived building models. IEEE Trans Geosci Remote Sens 51(3):1844–1851

Hagenauer J, Helbich M (2012) Mining urban land-use patterns from volunteered geographic information by means of genetic algorithms and artificial neural networks. Int J Geogr Inf Sci 26(6):963–982. doi:10.1080/13658816.2011.619501

IFPRI, EDRI (2009) Urbanization and spatial connectivity in Ethiopia, summary of ESSP-II discussion paper no. 3: urbanization and spatial connectivity in Ethiopia: urban growth analysis using GIS (Dec 2009), Ethiopia strategy support program II (ESSP—II) research notes, No. 5, Washington, DC

Jensen JR, Cowen DJ (1999) Remote sensing of urban/suburban infrastructure and socio-economic attributes. Photogram Eng Remote Sens 65:611–622

Jokar Arsanjani J, Helbich M, Bakillah M, Hagenauer J, Zipf A (2013a) Toward mapping land-use patterns from volunteered geographic information. Int J Geogr Inf Sci 27(12):2264–2278. doi:10.1080/13658816.2013.800871

Jokar Arsanjani J, Helbich M, de Noronha Vaz E (2013b) Spatiotemporal simulation of urban growth patterns using agent-based modeling: the case of Tehran. Cities 32:33-42. doi:10.1016/j.cities.2013.01.005

Kavzoglu T, Sen YE, Cetin M (2009) Mapping urban road infrastructure using remotely sensed images. Int J Remote Sens 30(7):1759–1769. doi:10.1080/01431160802639582

Kloog I, Haim A, Portnov BA (2009) Using kernel density function as an urban analysis tool: Investigating the association between nightlight exposure and the incidence of breast cancer in Haifa, Israel. Comput Environ Urban Syst 33(1):55–63. doi:10.1016/j.compenvurbsys.2008.09.006

Knox PL, Pinch S (2007) Urban social geography: an introduction. 5th edn. Harlow

Liu X, Li X, Li J, Wang Q (2013) Object-oriented remote sensing image classification and road damage adaptive extraction. In: Proceedings of the international conference on remote sensing, environment and transportation engineering (RSETE 2013). pp 140–143

Mahendra B, Harikrishnan K, Krishne G (2010) Urban governance and master plan of Bangalore City. India J 7–2:1–18

Mander Ü, Müller F, Wrbka T (2005) Functional and structural landscape indicators: Upscaling and downscaling problems. Ecol Ind 5(4):267–272. doi:10.1016/j.ecolind.2005.04.001

Marshall F, Waldman L, MacGregor H, Meht L, Randhawa P (2009) On the edge of sustainability: Perspectives on peri-urban dynamics. STEPS working paper 35, Brighton: STEPS Centre

Mathieu R, Aryal J, Chong AK (2007) Object-based classification of Ikonos imagery for mapping large-scale vegetation communities in urban areas. Sensors 7(11):2860–2880. doi:10.3390/s7112860

McIntyre NE, Knowles-Yánez K, Hope D (2008) Urban ecology as an interdisciplinary field: differences in the use of "urban" between the social and natural sciences. In: Marzluff JM, Shulenberger E, Endlicher W (eds) Urban ecology, Boston, pp 49–65

Moghadam HS, Helbich M (2013) Spatiotemporal urbanization processes in the megacity of Mumbai, India: a Markov chains-cellular automata urban growth model. Appl Geogr 40:140–149. doi:10.1016/j.apgeog.2013.01.009

Moshi Municipal Council (2010) District agricultural development plans 2010/2011, Moshi

Myint SW, Gober P, Brazel A, Grossman-Clarke S, Weng Q (2011) Per-pixel versus object-based classification of urban land cover extraction using high spatial resolution imagery. Remote Sens Environ 115(5):1145–1161. doi:10.1016/j.rse.2010.12.017

Open Cities Project (2014) Kathmandu, Nepal. Available via http://opencitiesproject.com/cities/kathmandu/. Accessed 02 May 2013

OSM2GIS (2011) Data download section. Available via http://www.osm974.re/osm2gis/. Accessed 16 Aug 2014

Park R, Burgess EW, McKenzie RD (1925) The city. Chicago

Pham HM, Yamaguchi Y, Bui TQ (2011) A case study on the relation between city planning and urban growth using remote sensing and spatial metrics. Landscape Urban Plann 100:223–230. doi:10.1016/j.landurbplan.2010.12.009

Phinn SR, Roelfsem CM, Mumby PJ (2012) Multi-scale, object-based image analysis for mapping geomorphic and ecological zones on coral reefs. Int J Remote Sens 33(12):3768–3797. doi:10.1080/01431161.2011.633122

Qian J, Zhou Q, Hou Q (2007) Comparison of pixel-based and object-oriented classification methods for extracting built-up areas in aridzone. In: International Society for Photogrammetry and Remote Sensing (ISPRS) (ed) Proceedings of the ISPRS workshop on updating geo-spatial databases with imagery and the 5th isprs workshop on dynamic and multi-dimensional GIS, pp 163–171

Ramachandra TV, Bharath HA, Durgappa DS (2012) Insights to urban dynamics through landscape spatial pattern analysis. Int J Appl Earth Obs Geoinf 18:329–343. doi:10.1016/j.jag.2012.03.005

Salvati L, Munafo M, Morelli VG, Sabbi A (2012) Low-density settlements and land use changes in a Mediterranean urban region. Landscape Urban Plann 105:43–52. doi:10.1016/j.landurbplan.2011.11.020

Sarvestani MS, Ibrahim AL, Kanaroglou P (2011) Three decades of urban growth in the city of Shiraz, Iran: a remote sensing and geographic information systems application. Cities 28:320–329. doi:10.1016/j.cities.2011.03.002

Satterthwaite D, McGranahan G, Tacoli C (2010) Urbanization and its implications for food and farming. Philos Trans Roy Soc B Biol Sci 365(1554):2809–2820. doi:10.1098/rstb.2010.0136

Schlesinger J (2013) Agriculture along the urban–rural continuum: a GIS-based analysis of spatio-temporal dynamics in two medium-sized African cities. Freiburger Geographische Hefte, Freiburg

Silverman BW (1986) Density estimation for statistics and data analysis. Monographs on statistics and applied probability, vol 26. London, New York

Thapa RB, Murayama Y (2010) Drivers of urban growth in the Kathmandu valley, Nepal: Examining the efficacy of the analytic hierarchy process. Appl Geogr 30:70–83. doi:10.1016/j.apgeog.2009.10.002

UN (2001) Compendium of human settlements statistics 2001. Sixth Issue. New York

UN (2014) World urbanization prospects. The 2014 revision: highlights. New York

UN DESA (2011) Population distribution, urbanization, internal migration and development: an international perspective. New York

UN-HABITAT (2010) The state of African cities 2010: governance, inequality and urban land markets. Nairobi

UN-HABITAT (2013) State of the world's cities 2012/2013. Prosperity of cities. Nairobi

Uuemaa E, Antrop M, Roosaare J, Marja R, Mander Ü (2009) Landscape metrics and indices: an overview of their use in landscape research. Living Rev Landscape Res 3(1):5–28. doi:10.12942/lrlr-2009-1

Van Kerm P (2003) Adaptive kernel density estimation. The Stata J 3(2):148–156

Vermeiren K, Van Rompaey A, Loopmans M, Serwajja E, Mukwaya P (2012) Urban growth of Kampala, Uganda: Pattern analysis and scenario development. Landscape Urban Plann 106:199–206. doi:10.1016/j.landurbplan.2012.03.006

von Thünen JH (1910) Der isolierte Staat in Beziehung auf Landwirtschaft und Nationalökonomie. Neudruck nach der Ausgabe letzter Hand (2. bzw. 1. Auflage, 1842 bzw. 1850), Jena

Wakode HB, Baier K, Jha R (2014) Analysis of urban growth using landsat TM/ETM data and GIS—a case study of Hyderabad, India. Arab J Geosci 7:109–121. doi:10.1007/s12517-013-0843-3

Wang H, Shi S, Rao X (2013) A study of urban density in Shenzhen: the relationship between street morphology, building density and land use. In: Kim YO, Park HT, Seo KW (eds) Proceedings of the ninth international space syntax symposium, Seoul

White MJ, Mberu BU, Collinson MA (2008) African urbanization: recent trends and implications. In Martine G (ed) The new global frontier. Urbanization, poverty and environment in the 21st century. London, pp 301–16

Wu Q, Chen R, Sun H, Cao Y (2011) Urban building density detection using high resolution SAR imagery. In: Stilla U, Gamba P, Juergens C, Maktav D (eds) JURSE 2011—joint urban remote sensing event. Munich, 11–13 Apr 2011

Wurm M, Taubenböck H, Dech S (2010) Quantification of urban structure on building block level utilizing multisensoral remote sensing data. In: Michel U, Civco DL (eds) Proceedings of SPIE. Earth resources and environmental remote sensing/GIS applications, vol 7831

Ye Y, Van Nes A (2013) Measuring urban maturation processes in Dutch and Chinese new towns: combining street network configuration with building density and degree of land use diversification through GIS. J Space Syntax 4(1):18–37

Yu B, Liu H, Wu J, Hu Y, Zhang L (2010) Automated derivation of urban building density information using airborne LiDAR data and object-based method. Landscape Urban Plann 98:210–219. doi:10.1016/j.landurbplan.2010.08.004

Zollweg JD, Gartley MG, Roskovensky J, Mercier J (2012) Using GIS databases for simulated nightlight imagery. In: Pellechia MF, Sorensen RJ, Dockstader SL, Palaniappan K (eds) Proceedings of SPIE. Geospatial InfoFusion II, vol 8396

Part V
Outlook

An Outlook for OpenStreetMap

Peter Mooney

This volume has presented "*OpenStreetMap in GIScience: experiences, research and applications*" with a collection of experiences and research carried out with OpenStreetMap as the central and core theme. The volume has sought to build a firm foundation to highlight research work focused on OpenStreetMap. This was one of our original goals when we set out at the beginning of the editorial process. This is, to the best of our current knowledge, the first academically produced volume of its kind which focuses exclusively on OpenStreetMap. Approximately one decade on from the birth of OpenStreetMap in 2004 this volume appears at the most opportune of times. OpenStreetMap has emerged from one of the most tumultuous decades in Information and Communication Technologies (ICT) and possibly in the history of human communication. In the decade where ICT, social media, ubiquitous computing and the Internet of Things emerged OpenStreetMap arguably now proudly stands as one of the best examples of crowd and volunteered-based innovation of this time. Its past has been remarkable and the future for OpenStreetMap is bright.

The chapters in the volume have responded to various issues and criticisms aimed at crowdsourced mapping by demonstrating OpenStreetMap's current value and future potential. The cartographic data generation and map making industry has been fundamentally altered by the ICT advancements of the last decade or so. The mapping experience is being transformed (Dodge and Kitchin 2013). This foundation can be used to support a platform from which future research can be launched. These are exciting times in Geographic Information Science (GIScience). Our edited volume demonstrates that OpenStreetMap is now firmly established on the GIScience research agenda. At the time of writing OpenStreetMap data is increasingly being used as a source of geospatial data for researchers in a wide range of topic areas. These include diverse applications such as the use of OpenStreetMap data for hydraulic modelling in data scarce floodplain areas (Schellekens et al. 2014), to estimating associations between proximity to green spaces and surrounding

P. Mooney (✉)
Department of Computer Science, Maynooth University, Maynooth, Co. Kildare, Ireland
e-mail: peter.mooney@nuim.ie

J. Jokar Arsanjani et al. (eds.), *OpenStreetMap in GIScience*,
Lecture Notes in Geoinformation and Cartography,
DOI 10.1007/978-3-319-14280-7_16

greenness and pregnancy outcomes (Agay-Shay et al. 2014), to developing models for increased efficiency in road-network warning message dissemination (Fogue et al. 2013).

Many of the key research questions in volunteered Geographic Information (VGI) as outlined by authors such as (Mooney and Corcoran 2013) are prevalent and relevant to OpenStreetMap. The purpose of this short outlook chapter is to present a picture of where we, as the academic and research community, go from here with OpenStreetMap research. Our edited volume demonstrates that GIScience research on OpenStreetMap has matured. However we urge caution. It is important that the potential generated by this research is maintained and developed going forward. To remain consistent in how we have presented the chapters we will discuss the outlook for OpenStreetMap under the same headings as we have used in the volume.

1 Data Management and Quality

Amongst the greatest concerns in using VGI to geographers and GIScientists emerge from issues related to quality assurance of VGI. Is VGI of sufficient quality that it is fit for a specific purpose or use? Many geographers and GIScientists today begun their professional academic training amongst the practices of centralized, scientific, commercial and industrialized geographic information production. These processes were carried out by recognized professionals. Quality assurance procedures were implemented through all of the organizational or professional structures where the geographic information and data were produced. VGI and projects such as OpenStreetMap generate geographic information and data outside these types of organizational and professional structures. Because projects like OpenStreetMap generate geographic information and data in a different setting to established professional structures there are challenges in building quality assurance mechanisms capable of addressing the perceived weaknesses and lack of quality and accuracy in VGI.

There is an abundance of literature where academics and researchers have tackled the quality assurance problem in OpenStreetMap. Studies have shown that in comparison with professionally collected geographical data, such as that from National Mapping Agencies, OpenStreetMap is of equivalent accuracy and in some cases displays better geographical, temporal and attribute accuracy. One of the limitations of these studies surrounds the heterogeneity of OpenStreetMap. Urban areas and agglomerations are usually subject to greater mapping efforts and coverage in OpenStreetMap than rural areas. Some authors have demonstrated heterogeneity in OpenStreetMap related to the socio-economic characteristics of regions within larger areas of population (Jokar Arsanjani and Bakillah 2015).

However, issues remain. Authors such as Dodge and Kitchin (2013) see VGI projects like OpenStreetMap as perpetually unfinished mapping projects. There are always latent concerns amongst academics and professional users that “something untoward will happen" and data within OpenStreetMap will be damaged, deleted or updated incorrectly immediately rendering it unfit for most application purposes. The greatest challenge faced by OpenStreetMap as it moves forward into its second decade is conquering negative perceptions built up from the influence of decades of the established professional production of geographic information and recent commercial interests generating geographic information and services.

The academic and research community must continue to investigate methods for demonstrating the quality of OpenStreetMap data and information and assessing its fitness for purpose for various applications. This is of course a very extensive research mandate. Research efforts will need to be channeled into priority areas such as:

- semantics and interpretation of contributor supplied metadata on primitive OpenStreetMap objects
- geographic accuracy of OpenStreetMap data originating from a heterogeneous set of input sources (smartphones, bulk import, on-screen digitizing, etc.)
- update and refresh rates of attribute metadata and vector data
- conflation of OpenStreetMap data with other VGI datasets

2 The Social Context

Many studies have begun to appear which try to understand why, how, where and when citizens contribute to OpenStreetMap. This is not limited to contribution in the form of map edits but contribution in a wider form: software development, website and wiki development, etc. Contributors to OpenStreetMap share what Lin (2014) calls a rather strong shared identity of being an 'OpenStreetMapper'. A sense of community has played an important role in motivating these contributors. OSM can empower individuals and communities. As we have mentioned in this chapter, and have seen in the chapters contained in this volume, there are continuous efforts to understand the emerging practices and methodologies of crowd-sourced mapping and geodata collection. This leads us to ask social and critical geographers to consider how we can understand and reveal the social construction of OSM over time.

What are the social mechanisms which have connected thousands of people who have the appropriate interests and resources so that they can collectively contribute to OpenStreetMap? Collective action (Poteete and Ostrom 2004), as witnessed in OpenStreetMap, has arisen where groups of especially interested, motivated and resourceful individuals are in some way socially connected to each other. However little is really known about the effects of the social ties within the OpenStreetMap community (Mooney and Corcoran 2013b). Even less is known about the thousands of contributors to the project who *only stay for a while* and leave after contributing maybe only a small handful of edits. Sustaining contributor involvement is also crucial for the long-term future of OpenStreetMap. What are the best ways OpenStreetMap can recruit new contributors whilst ensuring the existing 'OpenStreetMappers' remain actively involved and engaged in the project? Gender divisions which are appearing on the GeoWeb and within VGI must be addressed as a matter of urgency (Stephens 2013).

3 Network Modelling and Routing

With the smartphone or smartdevice in our pocket people are using location-based services in ever increasing numbers throughout their everyday life. The concepts of directions, navigation and routing are becoming embedded in our social networking applications, web-searches and general smartdevice usage. We have seen examples in this volume where OpenStreetMap is being used as the data source for the development of advanced network models and route finding algorithms. References within these chapters further demonstrate the influence the availability of OpenStreetMap data is having on the development of network models and routing algorithms for Location-based Services and other Internet-based and smart device-based applications. There is still some way to go in regards to addressing and location-based identification in OpenStreetMap. There are millions of building objects in the OpenStreetMap database without any usable addressing information attributed to them. This is something that needs to be considered by the OpenStreetMap community. But perhaps there are opportunities for the academic and research community to use data conflation techniques to extract accurate addresses for these building objects from other openly available geographic datasets or from social media such as Twitter, Foursquare and Flickr (Quercia and Saez 2014). Addressing and improved location-based identification must be built into OpenStreetMap. The experience and expertise of the academic community from Computer Science, Networking, Mathematics and GeoComputation can assist in driving OpenStreetMap forward in this domain.

This type of collaboration and coordination provides an opportunity for better coordination or cooperation between traditional GIScience processes and VGI. Areas of mutual benefit can be found such as networking modelling and routing. At this stage in our technological life in GIScience it is no longer satisfactory to consider VGI part of a growing cohort of 'disruptive technologies'. As work presented in this volume has demonstrated OpenStreetMap and other VGI data sources are being considered as an integrated part of the GIScience technological landscape.

4 Land Management and Urban Form

Very often the research questions which academics and researchers can answer are constrained by the availability of data. Land management data and data from the urban environment can often be difficult to access for academics and researchers.

OpenStreetMap offers ready access to a crowdsourced global geodatabase not necessarily limited to the traditionally popular thematic areas of infrastructure, networks and natural features such as rivers and lakes. Should OpenStreetMap make concerted efforts to develop its landcover and landuse capabilities for example as indicated by authors such as (Jokar Arsanjani et al. 2013), Estima and Painho (2013, this volume)? If enhanced attribute information about buildings and structures such as height, facade type etc. were added to building objects in OpenStreetMap then opportunities for use in urban planning, urban modelling, sustainability and environmental modelling could begin to appear. Are these types of datasets and thematic areas suitable for OpenStreetMap? OpenStreetMap must prevent itself from becoming a collection area for suitably licensed open geographic data just because it fits into OpenStreetMap. This enlargement of the database opens the question of OpenStreetMap as a form of geographic Big Data. Big Data is commonly characterized as being large in volume, produced continuously, and varied in nature. Boyd and Crawford (2012: 663) argue that "*there is little doubt that the quantities of data now available are often quite large, but that is not the defining characteristic of this new data ecosystem*". OpenStreetMap contains billions of primitive nodes and ways along with billions of pieces of associated metadata in the form of object attributes, changeset documentation and edit histories. As data and metadata from other thematic areas are added to the OpenStreetMap database the maintenance and long-term sustainability of these additional data must be carefully considered. Extracting knowledge and information from OpenStreetMap as it grows in size may require approaches currently being used or developed from the Big Data domain.

Finally, in our outlook we consider how the academic and research community can collaborate with the OpenStreetMap community. The OpenStreetMap community should not be expected to shoulder the entire responsibility for the future of the project. There are roles which the academic community can play in the future development of OpenStreetMap. Presently there are few direct connections between these two communities. There are some academics and researchers who are active 'OpenStreetMappers' but with only empirical evidence we believe these numbers are small. We urge academics and researchers to become actively involved and engaged with the OpenStreetMap community. When academics use data or software purchased from commercial companies we are not slow to report our successes while at the same time reporting bugs, errors and other issues. This model has also worked incredibly well for the Open Source Software community within which the academic and research community are very heavily involved particularly in areas such as operating system design, software engineering and scientific computing. Why should this be any different for OpenStreetMap? It is crucial that the academic community feedback results, observations and recommendations to the OpenStreetMap community from their research. There are numerous ways of doing this. Academic papers and edited volumes such as this one are one such channel. However we urge academics to also interact with the OpenStreetMap community through mailing lists, wikis, social media, etc. Together, in this fashion, everyone will benefit. In a wider sense tantalizingly Graham and Shelton (2013: 259) remark that "the futures of geography and big data are still to be made". OpenStreetMap and VGI will have a central role to play in shaping these futures. The future begins here.

References

Agay-Shay K, Peled A, Crespo AV, Peretz C, Amitai Y, Linn S, Nieuwenhuijsen MJ (2014) Green spaces and adverse pregnancy outcomes. Occup Environ Med 71(8):562–569

Jokar Arsanjani J et al (2013) Toward mapping land-use patterns from volunteered geographic information. Int J Geogr Inf Sci 27(12):2264–2278

Jokar Arsanjani J, Bakillah M (2015) Understanding the potential relationship between the socio-economic variables and contributions to OpenStreetMap. Int J Digital Earth 1–16. doi:10.1080/17538947.2014.951081

Boyd D, Crawford K (2012) Critical questions for big data. Inf Commun Soc 15(5):662–679

Dodge M, Kitchin R (2013) Crowdsourced cartography: mapping experience and knowledge. Environ Plann A 45:19–36

Estima J, Painho M (2013) Exploratory analysis of OpenStreetMap for land use classification. In: Proceedings of the second ACM SIGSPATIAL international workshop on crowdsourced and volunteered geographic information. ACM, New York, NY, USA, pp 39–46

Fogue M, Garrido P, Martinez FJ, Cano J-C, Calafate CT, Manzoni P (2013) An adaptive system based on roadmap profiling to enhance warning message dissemination in VANETs. IEEE/ACM Trans Netw 21(3):883–895

Graham M, Shelton T (2013) Geography and the future of big data, big data and the future of geography. Dialogues Hum Geogr 3:255–261

Lin W (2014) Revealing the making of OpenStreetMap: a limited account. Can Geogr-Geogr Can pp 16–22

Mooney P, Corcoran P (2013) Analysis of interaction and co-editing patterns amongst OpenStreetMap contributors. Trans GIS 18(5):633–659

Poteete AR, Ostrom E (2004) Heterogeneity, group size and collective action: the role of institutions in forest management. Dev Change 35:435–461

Quercia D, Saez D (2014) Mining urban deprivation from foursquare: implicit crowdsourcing of city land use. IEEE Pervasive Comput 13:30–36

Schellekens J, Brolsma RJ, Dahm RJ, Donchyts GV, Winsemius HC (2014) Rapid setup of hydrological and hydraulic models using OpenStreetMap and the SRTM derived digital elevation model. Environ Model Softw 61:98–105

Stephens M (2013) Gender and the GeoWeb: divisions in the production of user-generated cartographic information. GeoJournal 78:981–996

Printed by Printforce, the Netherlands